Speech Quality of VoIP

Speech Quality of VoIP

Assessment and Prediction

Alexander Raake
Deutsche Telekom Laboratories, Germany

John Wiley & Sons, Ltd

Other Wiley Editorial Offices

John Wiley & Sons Inc., 111 River Street, Hoboken, NJ 07030, USA

Jossey-Bass, 989 Market Street, San Francisco, CA 94103-1741, USA

Wiley-VCH Verlag GmbH, Boschstr. 12, D-69469 Weinheim, Germany

John Wiley & Sons Australia Ltd, 42 McDougall Street, Milton, Queensland 4064, Australia

John Wiley & Sons (Asia) Pte Ltd, 2 Clementi Loop #02-01, Jin Xing Distripark, Singapore 129809

John Wiley & Sons Canada Ltd, 6045 Freemont Blvd, Mississauga, ONT, L5R 4J3, Canada

Wiley also publishes its books in a variety of electronic formats. Some content that appears
in print may not be available in electronic books.

Library of Congress Cataloging-in-Publication Data

Raake, Alexander.
  Speech quality of VoIP : assessment and prediction / Alexander
Raake.
    p. cm.
  Includes bibliographical references and index.
  ISBN-13: 978-0-470-03060-8 (cloth : alk. paper)
  ISBN-10: 0-470-03060-7 (cloth : alk. paper)
  1. Internet telephony. 2. Computer network protocols.   I. Title.
TK5105.8865.R33 2006
621.3850285 – dc22
                                        2006011847

British Library Cataloguing in Publication Data

A catalogue record for this book is available from the British Library

ISBN-13: 978-0-470-03060-8 (H/B)
ISBN-10: 0-470-03060-7 (H/B)

Typeset in 10/12pt Times by Laserwords Private Limited, Chennai, India
Printed and bound in Great Britain by Antony Rowe Ltd, Chippenham, England
This book is printed on acid-free paper responsibly manufactured from sustainable forestry
in which at least two trees are planted for each one used for paper production.

To Ilona.

Contents

Preface

To a large extent, I have carried out the work summarized in this book during my activity at the Institute of Communication Acoustics (IKA), Ruhr-University Bochum, Germany. I would like to acknowledge the support of all my colleagues at IKA. In particular, I wish to thank my friend and colleague PD Dr.-Ing. Sebastian Möller for numerous fruitful discussions and his ongoing support. I would like to thank Prof. Dr.-Ing. Dr. techn. h.c. Jens Blauert for supporting my work, and for providing the scientific and organizational framework necessary for carrying out interesting research work. I would also like to thank Prof. Dr.-Ing. Ulrich Heute for his interest in my work all along, and his constructive review of an earlier version of this book. I wish to thank Prof. Dr. phil. Ute Jekosch, not the least for the interdisciplinary flavor she has brought into my work. For their support in system implementation and verification, as well as auditory testing, I would like to thank Marcel Wältermann, Jan Krebber, Sebastian Rehmann, Jörn Opretzka, Joachim Riedel, Marius Hilckmann, Issiaka Yacoubo, and Nicolas Côté. For the great research atmosphere with a lot of humor, which I encountered in my early days at IKA, I would like to thank my "old" colleagues, in particular Prof. Dr.-Ing. Christoph Pörschmann, Dr. phil. Thomas Hempel, and Dr.-Ing. Renato Pellegrini. For the technical and organizational office support I am grateful to all the technicians and members of the office, in particular Peter Salzsieder and Edith Klaus.

I would like to thank Dr.-Ing. Jens Berger and Florian Hammer as well as other colleagues involved in ITU-T Study Group 12's standardization activities for many fruitful scientific and technical discussions. Furthermore, I would like to thank Frank Kettler and his colleagues from HEAD acoustics, Herzogenrath, Germany, for their support in user interface characterization. I would like to thank Dr. Brian FG Katz who was so kind to show a lot of patience while I was preparing a first draft of this book during my work at LIMSI-CNRS in Orsay, France.

Last but not least, I want to thank all the members of my family and friends who have followed this project with a lot of patience and support. In particular, I am indebted to my wife Ilona for her endurance, patience, time, love and encouragement throughout these years; to my father for all kinds of support; to Marc Hanisch as a colleague but more importantly for his great help as our friend; and to my friend Ramin Kaweh for the big-picture-discussions probably necessary for such a book...

List of Abbreviations

Definitions

A	Advantage Factor (if not stated otherwise)
$a_1 \ldots a_6$	curve-fitting parameters of parametric bandwidth impairment model
a_i	arrival time of a packet i
a_j	constant for preference mapping according to Elliptical point model with rotation
a_x	attenuation of level switching device, send path
a_y	attenuation of level switching device, receive path
α	scale parameter of Weibull-type delay distribution
α_j	orientation of the quality vector of a subject j in a multidimensional feature space
b	probability of losing a packet in the "bad" state of a Gilbert model type loss model
b_{jt}	weighting factor for linear dimension terms in case of preference mapping according to Elliptical point model with rotation
b_0	description of an auditory event
b_1, b_2	weighting coefficients in case of 2-dimensional space and vector model type preference mapping
b_i	jitter buffer delay
b_k	constant for preference mapping according to Elliptical point model with rotation
Bcl	clipping robustness factor
Bpl	packet loss robustness factor
$BurstR$	Burst Ratio (US Patent 6,931,017, 2005)
$C_{r,max}$	linear MOS-transformation with two anchor points: clean reference connection expected to yield highest quality
$C_{r,min}$	linear MOS-transformation with two anchor points: reference connection expected to yield lowest quality
$c_1 \ldots c_6$	constants for model of noise–packet loss interaction
C_r	linear MOS-transformation with one anchor point: clean reference connection
Cl	percentage of clipped speech (i.e. interrupted speech)
D	frequency–weighted version of $DELSM$
d_{jk}	weighted Eukliden distance between points j and k

d_i, D_i	(jitter buffer) playout delay
$\hat{d}_i$	mean packet delay
$DELSM$	frequency–dependent difference in sensitivity between the directed and diffuse sound [dB]
Δf	frequency difference [Hz]
Δf_G	critical bandwidth [Hz]
δ_{ij}	Kronecker symbol
Dr, Ds	sensitivity of the user interfaces (D) at send and receive side, respectively, towards the direct speech signal, vs. their sensitivity towards ambient noise
$e(k)$	residual echo signal after echo cancellation
e_{jk}	error term for preference mapping according to Elliptical point model with rotation
EL	(talker) echo loss [dB]
f	frequency [Hz]
f_c	center frequency [Hz]
f_{low}, f_l	lower cut-off frequency of bandpass-filter [Hz]
f_{up}, f_u	upper cut-off frequency of bandpass-filter [Hz]
F_s	sampling frequency [Hz]
$\mathbf{g}(k)$	impulse response of a Loudspeaker-Enclosure-Microphone-system
γ	shape parameter of Weibull-type delay distribution
G	signal–to–equivalent–continuous–circuit–noise ratio [dB]
$\mathbf{h}(k)$	impulse response of the adaptive filter employed in an echo-canceler
h_0	auditory event
I, $Itot$	impairment factor
I_{bw}	impairment factor for bandwidth restriction
I_{loss}	packet loss impairment factor
I_1	estimated instantaneous impairment level during change from burst to gap condition for 4-state Markov model type loss
I_2	estimated instantaneous impairment level during change from gap to burst condition for 4-state Markov model type loss
Icl	impairment factor for speech clipping (i.e. interruptions)
Id	delayed impairment factor, for impairments that are delayed relative to the speech signal
$Idle$	impairment due to listener echo
$Idte$	impairment due to talker echo
Ie	equipment impairment factor for the impairment due to speech coding (without errors or packet loss)
Ieb	impairment related to burst or "bad" period (used for 4-state Markov chain loss model)
Ieg	impairment related to gap or "good" period (used for 4-state Markov chain loss model)
Ie, eff	effective equipment impairment factor covering both the impairment due to speech coding and coding under packet loss
$iMOS$, $intMOS$	integral quality MOS
$Iolr$	impairment due to overly loud connection

Is	simultaneous impairment factor, for impairments that are simultaneous to the speech signal
Ist	impairment due to non-optimum sidetone
L_i	location of degradation period i within the entire stimulus or connection duration
Ls	vector containing the times identifying the mid-points of different loss periods
Lst	frequency-dependent sidetone path attenuation [dB]
$LSTR$	listener sidetone rating [dB]
m_0	overall number of perceived packets
m_i	number of times i packets got lost in consecution in a given packet trace
$\overline{MOS}(R)$	Segment-wise time-average (including exponential behavior) over R estimated with E-model for random loss, and subsequent transformation to MOS scale
μ	minimum (packet) network delay
μ_{01}	mean number of packets lost in a row (2-state Markov model)
μ_{10}	mean number of packets found in a row (2-state Markov model)
μ_{13}	steps of sojourn of a particular 4-state Markov chain in "bad" state
μ_{31}	steps of sojourn of a particular 4-state Markov chain in "good" state
$n(k)$	noise signal
Nc	circuit noise [dBm0p]
$Nfor$	noise floor [dBmp]
No	total equivalent circuit noise level [dBm0p]
Nor	equivalent circuit noise for room noise at receive side [dBm0p]
Nos	equivalent circuit noise for room noise at send side [dBm0p]
$\hat{v}_i$	delay variation expected from an observation
OLR	overall loudness rating between MRP and ERP [dB]
p	probability for transition from "found" to "lost" state in a two-state Markov chain loss model
p_{ij}	probability for transition from state i to state j
p_{FEC}	overall packet loss probability in case FEC is used
$\mathcal{P}$	transition probability matrix of an n-state Markov chain
P_i	state probability, i.e. probability to occupy state i
p_b	overall packet loss rate in the "bad" or burst state of a 4-state Markov chain loss model
p_g	overall packet loss rate in the "good" or gap state of a 4-state Markov chain loss model
p_n	network packet loss probability, in case that packet loss in the network and packet drop due to jitter buffering are distinguished
P_R	random packet loss percentage of a given section in case of adjacent sections of different random loss levels
Pb	packet loss percentage during the "bad" or burst periods of a 4-state Markov chain loss model
pc	conditional loss probability
ppl; p_{pl}	overall packet loss probability

Ppl	overall packet loss percentage
Pr	A–weighted sound pressure level of room noise at receive side [dB(A)]
Pre	room noise level at receive side after passing the listener sidetone paths
Ps	A–weighted sound pressure level of room noise at send side [dB(A)]
q	probability for transition from "lost" to "found" state in a two-state Markov chain loss model
Q	signal–to–quantizing–noise ratio [dB]
Q_{jk}	quality for a stimulus k as judged by a subject j
qdu	quantizing distortion unit
R	transmission rating (factor)
R_C	expected, combined transmission rating factor
RLR	receive loudness rating, measured between the 0 dBr point in the network and the ERP [dB]
$RLRset$	receive loudness rating of the telephone handset [dB]
Ro	basic signal–to–noise transmission rating factor
$Ro_{NB/WB}$	maximum transmission rating (factor) for wideband
$s(k)$	speech signal
$\hat{s}(k)$	transmitted speech signal
s_0	sound event
SLR	send loudness rating, measured between the MRP and the 0 dBr point in the network [dB]
$SLRset$	send loudness rating of the telephone handset [dB]
$STMR$	sidetone masking rating [dB]
T	mean one-way delay of talker echo path [ms]
t_b	average sojourn in the "bad" or burst state of 4-state Markov chain loss model
t_g	average sojourn in the "good" or gap state of 4-state Markov chain loss model
t_i; t_k	temporal borders of adjacent segments of given microscopic loss behavior within a packet trace or connection
T_j	orthogonal matrix yielding rotation of subject-specific multidimensional feature space
T_P	packet size [ms]
Ta	overall delay, measured between the talker's MRP and the listener's ERP
τ_1	"exponential decay time of instantaneous judgement, quality impoverishment"
τ_2	"exponential decay time of instantaneous judgement, quality improvement"
τ_j	"exponential decay time of instantaneous judgement in case of quality change"
$TCLw$	weighted terminal coupling loss, expressing the attenuation of an acoustic echo path across a given terminal [dB]
$TELR$	talker echo loudness rating [dB]
Tr	round–trip delay for listener echo [ms]

$\mathcal{W}$	matrix of identical row vectors representing the equilibrium distribution of an ergodic Markov chain
W_i	relative weight of a passage i within a connection for integral quality
WEPL	weighted echo path loss for listener echo [dB]
x	time instance [s]
$\hat{x}(k)$	echo signal estimate
$\tilde{x}, \tilde{x}(k)$	acoustic (talker) echo
$\mathbf{x_k}$	position of a stimulus k in a multidimensional feature space
$\mathbf{y_j}$	ideal point of a subject j in a multidimensional feature space
z_{bw}	bandwidth [Bark]

Abbreviations

3GPP	3rd Generation Partnership Project
A	Advantage Factor
ACELP	Algebraic Code–Excited Linear Prediction
ACR	Absolute Category Rating
ADPCM	Adaptice Differential Pulse Code Modulation
AMR	Adaptive Multi-Rate (Codec)
BRI	Basic Rate Interface
CCI	Call Clarity Index
CCITT	Comité Consultatif International Télégraphique et Téléphonique
CCR	Comparison Category Rating
CLID	CLuster IDentification Test
CNG	Comfort Noise Generation
CS-ACELP	Conjugate Structure Algebraic Code–Excited Linear Prediction
CT	Conversation Test
DAM	Diagnostic Acceptability Measure
DAT	Digital Audio Tape
dB(A), dBA	logarithmic unit of the A-weighted sound pressure level
dBm	unit for logarithmic magnitude of a signal
dBm0	unit for logarithmic magnitude of a signal, relative to the 0 dBr reference point of the network
dBm0p	unit for logarithmic magnitude of a signal, relative to the 0 dBr reference point of the network and weighted psophometrically according to ITU–T Rec. O.41 (1994)
dBmp	"unit for logarithmic magnitude of a signal, weighted psophometrically according to ITU–T Rec. O.41 (1994)"
DCME	Digital Circuit Multiplexing Equipment
DCR	Degradation Category Rating
DECT	Digital European Cordless Telecommunication
DRM	Diagnostic Rhyme Test
DSP	digital signal processor
DTX	Discontinued Transmission
EC	Echo canceller

EL	Echo Loss
ERB	equivalent rectangular bandwidth
ERP	Ear Reference Point (cf. ITU–T Rec. P.64, 1999)
ETSI	European Telecommunications Standards Institute
FEC	Forward Error Correction
FN	Nth Formant
GSM	Global System for Mobile communication
GSM-EFR	Enhanced Full Rate (GSM codec)
GSM-FR	Full Rate (GSM codec)
GSM-HR	Half Rate (GSM codec)
GPRS	General Packet Radio Services
GUI	Graphical User Interface
HATS	Head And Torso Simulator
HFT	Hands-Free Terminal
HR	Half Rate (GSM codec)
IEEE	Institute of Electrical and Electronics Engineers
IETF	Internet Engineering Task Force
iLBC	Internet Low Bit-Rate Codec
ILD	Interaural Level Difference
INDSCAL	Individual Difference Scaling
INMD	In-service Non-Intrusive Measurement Device
IP	Internet Protocol
IRS	Intermediate Reference System
iSCT	interactive Short Conversation Test
ISDN	Integrated Services Digital Network
ISO	International Organization for Standardization
ITU	International Telecommunication Union
ITU-T	"International Telecommunication Union – Telecommunication Standardization Sector"
LAN	Local Area Network
LBR	Low Bitrate Redundancy
LD-CELP	Low Delay Code-Escited Linear Prediction
LEM	Loudspeaker-Enclosure-Microphone system
LOT	Listening Only Test
LPC	Linear Predictive Coding
LTI	linear time-invariant
MDS	Multidimensional Scaling
MELP	Mixed Excitation Linear Prediction
MNRU	Modulated Noise Reference Unit
MOS	Mean Opinion Score
MRP	Mouth Reference Point (cf. ITU–T Rec. P.64, 1999)
MRT	Modified Rhyme Test
MSN	Multiple Subscriber Number
NB	narrowband
NIST	National Institute of Standards and Technology
OSI	Open Systems Interconnection

PAMS	Perceptual Analysis Measurement System
PCM	Pulse Code Modulation
PESQ	Perceptual Evaluation of Speech Quality
PLC	Packet Loss concealment
POTS	Plain Old Telephone Service
PSQM	Perceptual Speech Quality Measurement
PSTN	Public Switched Telephone Network
QoE	Quality of Experience
QoS	Quality of Service
rms	root mean square
RMSE	Root Mean Squared Error
ROHC	Robust Header Compression
RPE-LTP	Regular Pulse Excitation Long Term Prediction
RTCP	RTP Control Protocol
RTP	Real-Time Transport Protocol
SAR	Speaker Alternation Rate
SCT	Short Conversation Test
SD	Semantic Differential
SIP	Session Initiation Protocol
SMS	Short Message Service
SNR	Signal-to-Noise Ratio
SRAEN	Système de Référence pour la détermination de l'Affaiblissement Équivalent pour la Netteté
SUS	Semantically Unpredictable Sentences
TCP	Transmission Control Protocol
TIPHON	Telecommunication and Internet Protocol Harmonization Over Networks
TOSQA	Telekom Objective Speech Quality Assessment
UDP	User Datagram Protocol
UMTS	Universal Mobile Telecommunications Service
VAD	Voice Activity Detection
VoIP	Voice over Internet Protocol
VSELP	Vector Sum Excited Linear Prediction
WB	wideband
WLAN	Wireless LAN

Introduction

POTS is how the technical literature on Voice over Internet Protocol (VoIP) calls the speech communication service already enabled by the traditional wireline networks: Plain Old Telephone Service. This type of service is still possible with VoIP, but its implementation barely resembles its predecessor. What is so new about VoIP that makes appear so old, what was known as *the* telephone system for decades?

If people are asked today what VoIP is, not many of them will know the answer. In turn, many people have an established notion of what a "mobile phone" is. The features that may be associated with this term agree very well with what it refers to, namely a telephone that allows the user to be mobile, due to the physical independence of a wireline connection. In turn, the term 'VoIP' implicitly identifies the more abstract transport mechanism based on which speech data is packetized, addressed and transmitted to a recipient. No reference is made to the physical access to the network, since the physical layer is not concerned; VoIP is conceivable over mobile channels (wireless LAN, UMTS) as well as over wireline channels (Ethernet, etc.).

The employed packet-based speech transport mechanism, not the physical representation of the data on the network, is the novelty introduced by VoIP. It bears the potential of merging the vast IP-infrastructure that has been established over the last several years for local and wide area networks with speech transmission. Consequently, new as well as traditional network operators were the first to acknowledge the advantages of VoIP, such as:

- Easy extension to a larger number of users,

- re-usage of available resources,

- simple combination of voice-services with data- or other media-services,

- access to a speech communications market for companies that had specialized on computer networks.

The user, in turn, may appreciate VoIP for the reduced cost at which it is offered. VoIP can even be exploited free of charge as *internet telephony* using available programs and internet connections. However, VoIP is not only linked to a "perceivable" reduction of costs. More fundamentally, VoIP is bound to features of the transmitted speech signal that are perceived *auditorily*. This is of particular relevance, since the user may not be aware of the fact that VoIP is employed for a given connection. Because of the economical and technological advantages, VoIP is increasingly employed in the telephone core network, where it has begun to replace the circuit switched network technology.

Packet-based transmission may introduce *degradations* like packet loss or jitter, which are time-varying by nature. These degradations are typically associated with a considerable perceptual impairment of the transmitted speech (see e.g. Gros and Chateau, 2001; Holmes *et al.*, 2001). The impairment related to time-varying degradations is perceptually different from that due to stationary degradations (Mattila, 2001), since their occurrence cannot be foreseen by the user. Moreover, auditory perception gives higher relevance to sounds that change in time (see e.g. Greenberg, 2004; Wang *et al.*, 2003).

In turn, VoIP is also linked to *improvements* over circuit switched telephony. One important improvement associated with VoIP directly concerns the transmitted speech signal: Because of the flexibility of VoIP in terms of the employed bandwidth and codec, it enables wideband speech transmission.

Perceived degradations and improvements imply a reference, relative to which the degradation or improvement is perceived. This reference does not have to be explicit. Owing to the perceptual experience, human listeners have established a reference for the perceptual signal features they encounter in their everyday environment, such as the features of speech or of environmental or animal sounds (see e.g. Jekosch, 2000, 2005b). Based on this experience, they are able to detect changes to these features, without necessarily knowing the source signal, like the identity of a talker (cf. e.g. Brüggen, 2001). Similarly, users of technical applications that transmit sounds have established a reference for the features typical of these applications, such as the telephone, the radio, or the television (see e.g. Duncanson, 1969).

The result of the appraisal of the perceived speech with regard to a reference is referred to as *speech quality* (Jekosch, 2000, 2005b). Speech quality is of multidimensional nature; it is based on a certain set of perceived features that can be viewed to span a perceptual feature space (McDermott, 1969; McGee, 1964). The speech quality related to a telephone service is one of the factors that determines its acceptance (Möller, 2000). Hence, speech quality is of importance both from the user's, and from the network provider's point of view. Ultimately, it is the user who *decides* on the acceptance of a system or service. The perceptual basis for this decision is not directly accessible from the outside. Hence, information on the features or the quality of a transmission system as perceived by the user can only be obtained from his/her description. To this aim, auditory tests can be employed that ideally deliver a description that accurately reflects the perceptual reality of the user (Blauert, 1997). Ideally, the speech quality related to a telecommunication system should be assessed in a conversation test, since it reflects the actual usage of the system for speech communication (Möller, 2000).

A network planner needs to have an idea of the quality related to his/her network before it is actually set up. The planning process itself can be viewed as an engineering task seeking to optimize the trade-off between an over-engineered network and a network lacking acceptance. One source of information on the suitability of a particular network configuration is the set of expected values for certain instrumentally measurable parameters or network characteristics like transmission spectra, speech or noise levels, or the signal-to-noise ratio (SNR). However, planning decisions cannot be taken directly based on these parameters, without knowing their impact on the speech quality to be perceived by the users. Moreover, the behavior of certain transmission components like codecs or echo cancellers cannot easily be described on the basis of instrumentally measurable characteristics like an SNR.

More complex signal-based measures have been developed for quality estimation (see e.g. Berger, 1998; Hauenstein, 1997; ITU–T Rec. P.862, 2001). Instead of making classical physical measurements like that of an SNR, they are based on models of human audition; knowledge of the signal processing carried out by the auditory system is applied to evaluate the quality of a given speech sample that has been transmitted over the system under consideration. Initially, signal-based measures have been developed to estimate quality in comparison to an unprocessed reference signal, but more recent models can be employed when no reference signal is available (ITU–T Rec. P.563, 2004). In the network planning phase, such models may be used to estimate the quality related to some network components that are already accessible at that point. However, they are of little help, when the signal that results after transmission over the entire network is not yet available.

In the planning phase of a network, the speech quality related to each possible configuration cannot be determined in auditory tests. Such tests, and especially conversation tests are time-consuming and expensive, and they require the network to be set up. As a practical way forward, so-called network planning models have been developed (ETSI Technical Report ETR 250, 1996; Johannesson, 1997). They map instrumentally measurable characteristics of the network to an estimate of the speech quality to be perceived by an average user of the system. The relationship between the model input parameters or characteristics and the resulting speech quality estimates is determined prior to the model development in quality perception tests. Components that cannot be described by classical level or spectrum measurements are also partly considered by network planning models. On the basis of the quality ratings obtained in auditory tests, specific impairment values may be determined for the respective components (Möller and Berger, 2002). These impairment values and the other parameters or characteristics used as input to the model are combined to form the final quality estimate. An example of such network planning models is the so-called E-model, which is the model currently recommended by the International Telecommunication Union (ITU–T Rec. G.107, 2005).

The existing network planning models were initially developed for traditional, circuit-switched telephone networks operated with handset telephones. Hence, their predictions of the quality to be expected with a particular VoIP network configuration were of limited validity. In a similar fashion, current signal-based quality measures show deficiencies in their predictions of VoIP quality (see e.g. Möller, 2000; Pennock, 2002). This lack of appropriate modelling solutions is due to the more fundamental lack of a comprehensive understanding of the related speech quality; speech quality under time-varying distortions on one hand, and under the effects related to an extension or reduction of the transmission bandwidth on the other hand.

The speech quality of VoIP networks can parametrically be modelled in a valid and reliable way only when appropriate parameters for the description of the physical network properties have been defined. More importantly, the development of parametric planning models for VoIP requires that the quality perception process can quantitatively be related to these physical parameters. In order to determine the application range of such a model, the factors that may impact the expectation of the user, i.e. the reference against which the perceived features are evaluated, have to be taken into consideration.

Overview of the Book

VoIP means transmitting speech over computer networks. In contrast to classical telephony, where research into the relation between physical transmission parameters, the resulting speech signal and the related speech quality has a longer tradition, speech quality of VoIP has only recently become an issue.

The present book tries to merge knowledge of the technical characteristics of VoIP networks with the author's original research on the perception of speech transmitted across them. The book is separated into two main parts: The first three chapters provide an overview of the available knowledge on principal, relevant aspects of speech and speech quality perception (Chapter 1), of speech quality assessment (Chapter 2), and of transmission properties of telephone and VoIP networks on one hand, and of the related perceptual features and resulting speech quality on the other hand (Chapter 3). Owing to the review-type nature of this first part, all main aspects of speech quality in VoIP networks are addressed. However, particular emphasis is laid on those physical characteristics and perceptual features that distinguish VoIP from other speech communication networks.

The first part provides the motivation for the original research described in the second part of this book:

Packet loss is identified as the degradation most typical of VoIP. The other impairments specifically linked to packet-based transmission, such as jitter and delay, may ultimately translate into packet loss (or discard) in one way or another. Moreover, the perceptual features associated with packet or frame loss are shown to be different from the features of the traditional circuit-switched wireline networks. On the basis of the available literature, it is shown that integral quality, i.e. the impression users have at the end of a connection, can be related to the quality perceived at different moments of the connection by some type of weighted time-averaging.

Wideband speech is identified as one of the main differences from traditional telephony with regard to the user's expectation. The perceived features of a speech signal transmitted across a system can be identified using multidimensional assessment techniques (Chapter 2). The desired features of speech stimuli can be determined by comparing the perceived features with the quality assigned to the stimuli by the users. For wideband telephony, the users have not yet established a fixed set of desired features. Consequently, different external factors may more easily impact the desired features, so that the quality of speech sounds that are presented in an acoustically identical fashion may vary.

Starting from the analysis of VoIP quality elements and features, the following chapters are concerned with the original research conducted by the author. The aim is to quantify and predict speech quality under packet loss on one hand, and in case of wideband transmission on the other hand:

Chapter 4 investigates by means of auditory listening and conversation tests, how differences in the packet loss distribution impact the integral quality perceived by subjects. Therefore, a perception-based classification of packet loss into loss showing a particular short-term behavior (*microscopic* loss behavior), and loss showing a particular long-term behavior (*macroscopic* loss behavior) is proposed (see Chapter 3 for a more detailed definition). For a given speech codec, the impact of different types of microscopic and macroscopic loss behavior on speech quality is quantified starting from the fundamental properties of the loss distribution. The quantitative relationship is formulated as a parametric extension of the E-model, and is shown to provide valid predictions for a number of loss conditions. Since speech quality concerns conversational speech, and time-varying distortions may interfere with the communication flow, the impact of the interactivity of a conversation on the speech quality under packet loss is investigated. Finally, a link is established between the multidimensional analysis of speech quality on the one hand, and speech quality perception under combined degradations on the other hand. In two extensive series of conversation tests on speech quality under packet loss and additional degradations, the restriction of multidimensional quality tests to the listening situation is circumvented. It is proven that different physical characteristics of the network as well as certain perceptual features of the transmitted speech may interact in their impact on speech quality. The observations are applied to the task of quality prediction by developing a first modification of the E-model that better accounts for one of the interactions.

Wideband telephone speech is shown to be of particular interest for speech quality research (Chapter 3). On the basis of the considerations on perceived features and desired features, Chapter 5 investigates the impact of linear distortion and wideband transmission on speech quality by means of auditory tests. At first, the quality improvement over narrowband telephony (approximately 300 to 3400 Hz) that can be achieved with wideband transmission (50 to 7000 Hz or beyond) under optimal conditions is quantified. Starting from this improvement, the quality degradation of wideband speech due to bandwidth limitations is studied. Using the results of a listening test, a parametric quality prediction model is developed for bandlimited speech. The model reflects properties of the human auditory system related to spectral processing. Not only linear, but also non-linear distortion of wideband speech is addressed in this chapter. In analogy to the narrowband case, the impairment due to (wideband) speech coding is addressed by deriving corresponding wideband equipment impairment factors. After the discussion of the impact of linear and non-linear distortion on perceived quality, the impact of additional factors on the desired features is investigated by means of listening tests.

1

Speech Quality in Telephony

1.1 Speech

Language can be regarded as a communication system particular to humans. Its signs can be available in written or acoustic form[1]. Speech is a subsystem of language, that is, the '[...] communication by means of language in the acoustic channel' (Sebeok, 1996). Human interlocutors, for example, in a telephone conversation, communicate by exchanging speech signs. Thus, they are able to convey abstracted information in acoustic, that is, physical form (with little effort). The acoustic speech signal present at the recipient's ear, the *sound event*, causes an *auditory event* in the listener's brain (Blauert, 1997). On its way from sound event to auditory event, speech is processed at different levels in the brain, starting from the preprocessing provided by the auditory periphery. Ultimately, meaning is established on higher processing levels from the percept and additional contextual information, such as the emotional state of the listener.

The auditory event related to a particular sound event as well as the meaning the listener attaches to it are not accessible from the outside. Hence, knowledge on the perception related to specific sound events can only be gained through introspection (in case of own perception) or the description provided by another listener[2]. The science of formally studying auditory perception is referred to as *psychoacoustics*. Because of the absence of instrumental means for directly accessing perceptional events, human subjects are the only appropriate measurement instruments. A systems-analytic view of auditory experiments was proposed by Blauert (1997). A corresponding schematic representation of a listener in an auditory test scenario is depicted in Figure 1.1 (Blauert, 1997, pp. 5–12). Here, s_0 denotes a sound event and h_0 the related auditory event; b_0 refers to a description of the auditory event h_0.

As in all measurements, the description is typically given as an assignment of numbers on an appropriate measurement scale, for example, as a value b in certain scale units, to the measurement object (Stevens, 1951, p. 22). As in instrumental measurements, the measurement method has to be chosen in such a way that b (i.e. the rating or judgment) *validly* and

---

[1] Or in other more specific forms such as sign language.

[2] Alternatively, knowledge on perception can be indirectly gained from the reaction of subjects, or their performance on specific tasks.

Speech Quality of VoIP: Assessment and Prediction   Alexander Raake
© 2006 John Wiley & Sons, Ltd

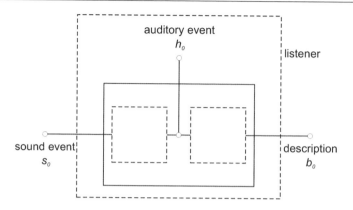

Figure 1.1: Schematic representation of a test subject in an auditory test, see Blauert (1997).

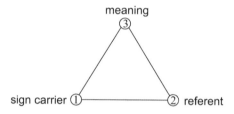

Figure 1.2: Semiotic triangle in general form according to Nöth (2000).

reliably quantifies the measurement object, the auditory event h_0. Here, validly means that the test method actually measures what it is intended to measure. Reliably means that the method is accurate and produces results without large scattering, and similar to the results obtained when the test is repeated (Guilford, 1954, pp. 349–357). The applied rating scale itself constitutes a sign system for conveying information on the perceptional event.

It has to be noted that signs are not signs *per se*: A sign becomes a sign when meaning can be associated with its carrier by an interpreting subject. Hence, '[...] a sign is a mental unit which is processed as standing for something other than itself, as pointing at something else' (Jekosch, 2005a). The science of signs is generally referred to as *semiotics*. Different diadic and triadic sign models have been proposed to capture the relation between the sign carrier and the associated meaning (Ogden and Richards, 1960; Peirce, 1986). Triadic sign models of semiotic theory differentiate three constituents (correlates) of a sign, which form the so-called semiotic triangle, shown here in a general representation (Figure 1.2, according to Nöth, 2000).

In case of language, the *sign carrier* or sign vehicle is the written word or the acoustic speech signal, thus the form, in which the sign is presented. The *referent* is the (possibly abstract) object the sign stands for. The *meaning* is the sense made of the sign by its interpreter. Hence, used as the correlate of a sign, *meaning* can be regarded as the role the sign plays, or the function it has for a sender or receiver. Here, the situation in which the sign 'happens' – for example, during a telephone conversation – is an important factor both for the reception and for the conveyance of the sign (Jekosch, 2000, 2005a,b).

Most of this book is concerned with the acoustic speech signal, i.e. the *form*, and the perception of its quality. However, many of the perceptual experiments were carried out as conversation tests, which include the actual application of speech: communication. Communication, between a person and other persons or objects in the outside world of the person, is the main subject of semiotics (Jekosch, 2005a). From the point of view of semiotics, it becomes clear that the three constituents of a speech sign cannot be separated, neither in terms of speech production nor in terms of speech perception. For example, only in case of small degradations of transmitted speech (affecting primarily the carrier), will it remain intelligible without additional listening effort, and the referent and sense may remain unaltered. The role of content (i.e. referent) and meaning (Figure 1.2) for speech quality perception is specifically addressed in Section 5.4.2 of Chapter 5.

1.1.1 Speech Acoustics

By speech, linguistic information is conveyed in the form of pressure waves traveling from the speaker's mouth to the listeners' ears. The longitudinal waves of atmospheric pressure variation are time-varying by nature. They reflect the variation pattern of the vocal tract articulators, such as the vocal folds (glottis), lips, tongue and jaw used for conveying the linguistic information. Hence, speech sounds can be viewed as a carrier signal filtered by the adjacent tubes the vocal tract is formed of, and amplitude-modulated by the articulatory movements modifying the physical properties of these tubes. The carrier signal produced by the glottis is a broadband signal with a strong tonal component, the pitch of which is due to the periodically opening and closing vocal folds. Hence, the glottis signal has a broadband line spectrum. It is characterized by the spacing of the lines, corresponding to its inverse pitch period, generally referred to as *fundamental frequency, F0*. For an overview of acoustic phonetics see, for example, Vary *et al.* (1998, pp. 5–28) or O'Shaughnessy (2000, pp. 35–107).

Speech can transmit information to one or more listeners as a result of the combination of both *temporal* as well as *spectral* properties, which are used by the recipients for decoding the information. The acoustic speech signal is characterized by short periods of a particular acoustic behavior, typically referred to as *phones*. They correspond to the acoustic realization of the smallest meaningful contrastive units in the phonology of a language, the *phonemes*. Phonemes can be coarsely divided into two classes, the vowel type and the consonant type. Vowel phonation is characterized by an unrestricted airflow through the vocal tract, while consonant phonation is due to a restriction of the airflow at some point in the vocal tract. During periods of 'stationary' acoustic behavior – related to a certain position of the articulators – characteristic acoustic properties of the different phones can be extracted from the spectrum of the speech sound. For example, the different vowel sounds of a particular language can be differentiated from the peaks they show in the amplitude spectrum. The first three peaks result from the three major resonances of the vocal tract, corresponding to the first, second and third *formants* F1, F2 and F3[3]. In the speech perception process, they are used by the auditory system as cues for vowel discrimination.

---

[3]The formants lie in different frequency ranges, depending on the type of vocalic sound and on the speaker: F1 is typically in the range between 0.2 and 1.2 kHz, F2 in the range 0.7 to 3.3 kHz, and F3 in the range 1.5 to 3.7 kHz.

Information is coded in the *temporal variation* of the smallest building units of speech. Different phones are concatenated into phone sequences by a speaker in order to form the words to be expressed. In practice, the resulting phones differ from the phonemes the speaker intended to articulate, both in articulation and in acoustics. This is due to the fact that subsequent articulatory gestures overlap and ultimately yield an articulation that depends strongly on the phonetic context of the phones (*coarticulation*). In phonology, words are traditionally divided into *syllables*, that is, phoneme sequences typically containing one vocalic sound. Linguistic information is not only coded in the particular composition of these sequences but is also carried by the segmental duration[4] of the phones in the syllables (Greenberg, 1999; Klatt, 1976). For example, certain words may be stressed more than others by the lengthening of particular phonemes. Phone durations vary broadly in the range between 20 and 240 ms, depending, for example, on the phone type, and on whether the phone is stressed (Umeda, 1975, 1977, as an example for read American English: 50 to 240 ms for vowels and 20 to 150 ms for consonants). Average syllable durations lie in the range from 100 to 300 ms (Arai and Greenberg, 1997) depending on the durations of the phones the syllable is composed of. The control of duration is typically combined with variations in fundamental frequency and speech level. This interplay is generally referred to as *prosody*. Although prosody may provide additional contextual information supporting word identification, its main functions are stressing information, indicating the purpose of a particular utterance (i.e. whether it is an exclamation or question), or providing information on the emotional state of the speaker (e.g. Murray and Arnott, 1993). Higher order formants (FN, $N > 3$), the fundamental frequency F0 and properties of the speaking style may also be used by listeners to identify the speaker (e.g. Mersdorf, 1996).

1.1.2   Speech Perception

From the above considerations it can be concluded that in order to process and understand speech signals, the auditory system has to (a) be able to resolve overlapping spectral cues and (b) decode the information ultimately associated with the dynamics of speech.

The stages involved in speech processing by the brain range from the auditory pre-processing over phonetic, phonological, lexical, and syntactic to semantic analysis, until meaning can finally be extracted (for an overview see O'Shaughnessy, 2000, pp. 141–172). In audition, these processing stages do not necessarily have to be sequential; some may act in parallel, and others may be skipped entirely (see e.g. McAdams, 1993, pp. 149–150). The following considerations depart from the incoming acoustic speech signal, and discuss aspects of lower- and higher-level speech perception considered relevant for this book.

1.1.2.1   Spectral Processing

To provide the functionality of spectral resolution, the ear is equipped with a mechanism for transforming spectral information into place information (see Zwicker and Fastl, 1999, pp. 23–60, for an overview on the information processing in the auditory system). After excitation of the eardrum by the incoming speech signal, the middle ear's ossicles supply the necessary impedance matching from the airborne sound outside the ear to the liquid-borne sound propagation in the inner ear (Figure 1.3). The cochlea in the human inner ear

[4]The durations of phones and syllables play a role in the impact of packet losses on the quality perception of packetized speech, as will be discussed in Chapters 3 and 4.

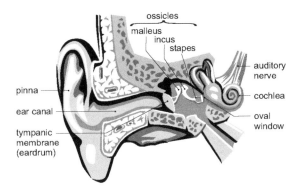

Figure 1.3: Schematic illustration of the outer, middle and inner ear. Copyright 2006 A. Raake, COREL® and their licensers. All rights reserved.

consists of three parallel tubes (or channels, referred to as *scalae*) filled with liquid, which are rolled up in the form of a snail. The basilar membrane separates two of the channels from the third (the scalae vestibuli and media from the scala tympani), and bears the organ of Corti. The organ of Corti hosts the sensory cells (*hair cells*) that convert the analog sound information into neural spike trains further processed at different levels in the brain.

The above-mentioned frequency–place transformation is achieved in the cochlea by exploiting the properties of traveling waves. The incoming sound waves are coupled via the eardrum and ossicles to the *oval window* connecting to the cochlea, where they excite the traveling waves. In case of a pure tone of a given frequency f, the traveling wave shows an excitation spanning over a certain area of the basilar membrane. The traveling wave yields a maximum excitation at a particular location, corresponding to the frequency of the tone transformed into place coordinates. A tone complex is thus decomposed into traveling waves showing main excitations at certain places on the basilar membrane, which correspond to the frequency components the tone complex is composed of. In case of tones of higher frequency, the maximum displacement of the basilar membrane occurs closer to the oval window, and tones of lower frequency lead to excitations closer to the end of the cochlea, the apex.

The relation between frequency and place is neither linear nor logarithmic. In pitch perception experiments, however, it was found that the pitch-ratio scale[5] shows a linear relationship to the place of main excitation on the (rolled-out) basilar membrane (Figure 1.4). For describing different phenomena of psychoacoustics, it has proven to be very useful to apply a transformation of frequency onto a scale showing a linear relation to the place of basilar membrane excitation. One of the most widely used examples of such a scale is the *Bark scale*, which relates frequency to a measure called *critical-band rate* (Zwicker, 1961; Zwicker and Fastl, 1999, pp. 149–164). It is measured in units of *Bark* and transforms the frequency range most relevant for human audition (0 to 16 kHz) to critical-band rates ranging from 0 to 24. Integer numbers of the critical-band rate correspond to the starting

---

[5]The scale on which the perceived pitch of tones can be displayed in such a way that ratios on the scale correspond to the ratios of tone perception, for example, answering the question how many times higher or lower a tone 'A' is perceived as a tone 'B'.

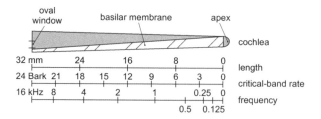

Figure 1.4: Schematic cross section of the rolled-out cochlea showing the basilar membrane (Zwicker and Fastl, 1999, pp. 149–164, mod. from Fig. 6.11). Below the figure, schematic scales are provided for location on the basilar membrane [mm] (top), critical-band rate [Bark] (middle) and frequency [kHz] (bottom).

points of segments that subdivide the basilar membrane into equally long *critical bands*, beginning at the apex (Figure 1.4).

The concept of critical bands is closely related to different psychoacoustic phenomena, such as spectral masking. The relation between critical bands and masking can be illustrated by the following experiment (Zwicker and Fastl, 1999, pp. 149–173): A test tone of a certain frequency f is presented centered between two band-pass noise signals, whose cutoff frequencies are separated by Δf. Then, for increasing width Δf of the noise gap, the masked threshold, i.e. the sound pressure level necessary for the test tone to be just audible, remains constant until a certain critical bandwidth is reached. For values of Δf above this *critical bandwidth* Δf_G, the masked threshold decreases. The critical bandwidth Δf_G determined this way is a constant associated with the center frequency f, the frequency of the test tone. Transformed from frequency to place, the different critical bandwidths correspond to constant distances on the basilar membrane, or in terms of critical-band rate, to constant increments of one Bark. Obviously, the processing performed by the cochlea can be viewed as a spectral decomposition of the incoming signal by an array of overlapping band-pass filters. Another psychoacoustic phenomenon related to the critical bands is associated with loudness perception: for equal sound pressure levels, the loudness of a band-limited noise signal is perceived as constant until the bandwidth is increased beyond the critical bandwidth associated with the center frequency of the noise. For larger bandwidths, the loudness increases: In loudness perception the excitation within the region of one critical band is grouped (Zwicker and Fastl, 1999, pp. 149–173).

A convenient approximation of the experimentally determined relation between critical-band rate and frequency is given by Equation (1.1) (Zwicker and Fastl, 1999, pp. 158–160):

$$z/\text{Bark} = 13\arctan(0.76\,f/\text{kHz}) + 3.5\arctan\left[(f/7.5\text{kHz})^2\right] \tag{1.1}$$

In some cases, an inverse relationship for deriving frequency from critical-band rate may be useful. A good fit of experimental data (Zwicker and Fastl, 1999, pp. 158–160) was found by the author of this book using Equation (1.2).

$$f/\text{Hz} = 1285.93\left(\frac{e^{(z/\text{Bark})^{2.64}}}{1836.93} - 1\right) + 93.3\frac{z}{\text{Bark}} \tag{1.2}$$

Both formulae play a role in the model of speech quality in case of bandwidth restrictions, which is presented in Chapter 5. The spectral processing provided by the hearing system is of particular relevance for the processing of speech sounds. For example, the formants revealed in a spectral representation of a vowel type phoneme are replicated in the excitation pattern on the basilar membrane. By the hair cells, the excitation associated with the formants is translated into firing rates processed further on higher levels in the brain (Young and Sachs, 1979). Here, this information can be used as (spectral) cues supporting the identification of lexical units.

1.1.2.2 Temporal Processing

The processing of spectral cues like harmonics used to detect certain phonemes are paralleled by the temporal processing performed by the auditory system. The temporal speech cues that can be exploited by the auditory system can directly be deduced from the speech acoustics summarized in Section 1.1.1. Rosen (1992) suggested a framework for describing the temporal cues of speech:

- The coarse structure, that is, the envelope of the speech signal, is associated with cues on the syllable level, corresponding to time frames greater than 20 ms, with syllable durations typically in the range from 100 to 300 ms (Greenberg, 2004).

- The periodicity provides pitch information ($2-20$ ms $\Rightarrow F0 = 50-500$ Hz).

- The fine structure contains information on the identity of the speaker (time frames below 2 ms).

After the preprocessing by the auditory periphery, the auditory nerve transmits the temporal information faithfully to the brain (Wang *et al.*, 2003). On higher processing levels, however, sparse acoustic events (like voice onsets) are marked with precise spike timing, while rapidly occurring acoustic events are transformed into firing rate–based representations (Lu *et al.*, 2001).

Obviously, a high relevance is given to the exact representation of envelope information and timely coding of sparse acoustic events like onsets. A syllable-centric view on speech perception underlining the importance of the envelope structure has been proposed by different authors (e.g. Greenberg, 2004; Mehler and Segui, 1987; Warren, 1999, pp. 73–76). A possible reason for less precise temporal coding of the periodicity and fine structure information, and instead representing it in a transformed way as firing rates, may be the necessity of integration with the information from other, slower senses, like the visual or tactile (for physiological details on temporal auditory processing, see e.g. Wang *et al.*, 2003; Young and Sachs, 1979).

1.1.2.3 Speech Perception and Auditory Memory

Different theories of speech perception have been reported in the literature, which are accompanied by different views on human auditory memory (for an overview see O'Shaughnessy, 2000). One theory assumes a dual perception process: bottom-up auditory processing and top-down phonetic processing (e.g. Mehler and Segui, 1987; Samuel, 1981).

According to this model, the auditory process analyzes acoustic features and stores them in auditory (echoic) memory (150–350 ms; see e.g. Baddeley, 1997; Massaro, 1975). The phonetic process is then assumed to yield syllable perception relying on features stored in echoic auditory memory. The phonetic process is driven by the expectation of the listener, mainly resulting from lexical contextual information.

This implies a storage-type memory, with different types of 'stores' at different stages of speech perception (see Crowder, 1993, for a more detailed description). These 'stores' can be distinguished as follows (e.g. Baddeley, 1997; Cowan, 1984):

Echoic memory: Peripheral storage of auditory features for durations of 150–350 ms. A classical experimental technique to determine the capacity of this type of storage was developed by Massaro. In his tests on backward recognition masking, he found that the second of two similar short sounds (e.g. pure tones of similar frequency, see Massaro, 1975) presented in fast succession prevented the identification of the first, when the delay between the second and the first sound was less than around 250 ms. Backward recognition masking was strongest for very short delays and decreased until the threshold of 250 ms was reached. The criticism of this paradigm is that the 250 ms may not represent the decay time or duration of the storage, but the duration needed to extract the information from the 'store' (Cowan, 1984).

Short-term auditory memory: Storage for longer durations (2–20 s), presumably applying some form of recoding of the auditory information. This 'store' is related, for example, to the *recency* and *suffix effects* (Crowder, 1993): In memory tests, it was shown that the last of acoustically presented verbal items like digits were remembered best (recency). The maximal number of items that can be stored was found to correspond to the 'magical' number seven reported as a limit for different types of human information processing tasks (Miller, 1956). The so-called suffix effect was observed when, after the presentation of the list of items, an additional, redundant speech item (suffix) was presented, which was considerably degrading the retention of the items presented before (Crowder and Morton, 1969). The suffix effect was found to be restricted to speech type suffixes similar to the list items in terms of pitch, voice quality and spatial location, without the meaning of the suffix being of any relevance for the effect to occur. Obviously, the suffix erases or inhibits the auditory storage of the last item(s). Neither suffix nor recency effects were found in serial recall experiments with visual verbal stimuli (modality effect). The modality and suffix effects are not fully explained: the suffix effect was found to also occur in case of lip-read interference of word lists presented acoustically, or of lists presented in a lip-read fashion (instead of both suffix and list being presented acoustically; for details see, for example, the literature reviews by Cowan, 1984; Crowder, 1993). It was thus hypothesized that the short-term auditory memory involved is related to mental mechanisms generally concerned with language perception and analysis (Crowder, 1993). However, the recency effect was also found in other tests on auditory perception, like tests on the loudness perception of nonstationary signals (Susini *et al.*, 2002) or tests on the perception of time-varying speech transmission quality (Gros, 2001; Gros and Chateau, 2001). In the latter tests, recent periods of bad quality were observed to have a larger influence on the final quality judgments than previous periods of bad quality (see Chapter 3). In Baddeley (1997), some examples of the recency effect are also summarized for other than the auditory modality.

Long-term auditory memory: This memory spans over periods of time up to several years or a lifetime, and allows recognition and identification of, for example, musical instruments, melodies and voices. It also refers to a memory of acoustic-lexical items, for example, used for comparison with the auditory features extracted from recently processed speech units during speech comprehension[6]. Similar to the recency effect described above, a small effect of primacy can be observed in serial recall tasks, showing better recall for the first in a particular list of items. While recency is generally ascribed to short-term auditory memory, the primacy effect is thought to be related to long-term auditory memory: from the recall tests reported by Glanzer and Cunitz (1966) and Postman and Philips (1965) it can be concluded that the primacy effect results from a verbal representation of the first item(s) in memory, while the 'classical' recency effect is related to the storage of the item's features in short-term auditory memory.

Instead of this storage-based view on human auditory memory, Crowder (1993) advocates a procedural paradigm: particular perceptual events and possible additional cognitive processes (e.g. mood, attention) yield activation of different regions in the brain. Depending on the type of signal, different, possibly parallel, stages are involved in the perception process. During perception, auditory periphery information is partly recoded into other forms of representation at different processing levels (e.g. into verbal representations). The related activation patterns in the different areas of the brain reflect memory: the information is retained in those areas of the brain that were active during the initial processing (learning).

1.1.2.4   Comprehension

The main function of speech is communication. The sender of a speech message wants to make herself/himself understood, and the recipient intends to understand what the speaker wanted to convey. In the literature, the result of the speech recognition process is referred to as *comprehension* (see e.g. Jekosch, 2000, 2005b, p. 103). Miller (1962) has described the speech comprehension process as a combination of 'decision units' (which are certainly not consecutive) carrying out the phonetic, lexical, syntactic, semantic and pragmatic decisions taken in the course of everyday speech communication. In the following, a list of terms related to speech comprehension is provided for the purpose of this book. It is synthesized from the terms used in the literature and corresponding considerations by Möller (2000, pp. 26–27) and Jekosch (2000, pp. 100–105). The different terms and a simplified illustration of the interrelations are shown in Figure 1.5.

Comprehensibility addresses how well the speech signal allows content to be related to it. It concerns only the form of the speech sign. It reflects the ability of the carrier to actually convey information. With regard to transmission systems, the term *articulation* is sometimes used to quantify the capability of the speech link to faithfully transmit information (e.g. French and Steinberg, 1947; Möller, 2000, pp. 26–27).

---

[6]Obviously, the long-term memory beyond auditory memory is composed of different levels of encoding, like lexical, semantic and pragmatic stages.

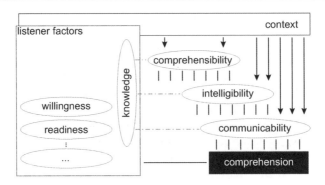

Figure 1.5: Simplified illustration of the terminology used to describe the factors involved in comprehension, and their interrelations. Note that the depicted process is not a linear one.

Comprehensibility relates to the identification of phonemes and phoneme clusters without a lexical activation. Often, the expressions *segmental intelligibility* or *syllable intelligibility* are used synonymously for comprehensibility as it is defined here. According to these definitions, articulation is a prerequisite for comprehensibility.

Intelligibility refers to how well the content of an utterance (i.e. the referent, see Figure 1.2) can be identified on the basis of the form (i.e. the carrier). Intelligibility, according to this definition, strongly depends on lexical, syntactic and semantic contexts. Considerations on the role of context for intelligibility are provided in Benoît (1990). A model for quantifying the role of context for intelligibility was proposed by Bronkhorst *et al.* (1993, 2002), relating intelligibility to the recognition of smaller speech units such as syllables or phonemes (i.e. to *articulation/comprehensibility*).

Communicability means that a speech message is such that it can serve to communicate, that is, can fully be understood by a recipient, ideally as it was intended by the sender (assuming that none of the interlocutors has the intention to deceive the other). It is related to functional aspects of speech and the entire communication process. Depending on the semantic and situational context (e.g. the topic of the conversation, the relationship between the interlocutors, etc.), communicability as it is used here requires a certain degree of intelligibility, which in turn requires a certain level of comprehensibility.

Communicability is closely related to perceived quality: speech quality can be considered as the perceived degree of communicability relative to the desired or expected degree of communicability (see also Section 1.2). From Figure 1.5 it becomes clear that speech communication links are conceivable where comprehensibility and intelligibility may be perfect, but communicability is not: For example, in case of a long transmission delay, the transmitted speech may be highly intelligible, but communicability may be affected severely. As timing plays an important role for the speech communication process, any impairment of the conversation time structure will modify comprehension (see e.g. Brady, 1968, for an analysis of timing aspects in telephone conversations).

Note that in the speech quality-related literature, the term communicability is sometimes used in a slightly different way, namely to express the extent to which a communication *link* (and not the exchanged messages) is able to provide an efficient medium of communication (see e.g. Wijngaarden *et al.*, 2001, 2002).

Comprehension is the result of the speech perception process. It presupposes that a perceived utterance is communicable, and that the recipient is ready and willing to comprehend.

The terms 'context' and 'knowledge' refer to different factors, depending on the level in the process of comprehension (see Figure 1.5): For *comprehensibility*, context refers to aspects such as coarticulation, and knowledge refers to phonological aspects like phoneme or syllable recognition, assisted by phonotactic knowledge, for example, on whether certain phone clusters are elements of the set of clusters characteristic of the respective language. For *intelligibility*, context refers to prosodic and syntactic, as well as to lexical and semantic aspects; knowledge, on the other hand, refers to the ability of the listener to extract the content from a speech message (i.e. lexical, syntactic and semantic knowledge). *Communicability* requires contextual information related also to the sender of the message. Here, context refers to the semantics of the utterance, and the entire situation the utterance was spoken and perceived in (i.e. related also to pragmatic aspects). Then, knowledge refers to the competence of the listener to fully understand what the speaker has said (including knowledge of the speaker, the situation, etc.).

When auditory tests are to be designed to measure either comprehensibility, intelligibility or communicability, context and knowledge associated with respective higher-level concepts should be excluded from the test. For example, in a comprehensibility test, the context information is reduced by using nonsense words, in order to prevent – as far as possible – any lexical access by the listener.

1.1.2.5 Restoration of Missing Sounds

Already on low levels, auditory perception is capable of employing context information to recover missing information, and to yield meaningful recognition results: in the perceptual illusion of *phonemic restoration*, listeners perceive speech samples with some phonemes replaced by noise or other sound segments as complete[7]. For example, in the classical study by Warren (1970), the first /s/ in 'legislatures' was replaced by sounds like a cough. The sound used for replacement is typically perceived as being an additional sound, which can only poorly be localized within the utterance. Dependencies of restoration on different factors were observed:

- Context: Better restoration was found for longer words than for shorter ones. Also, a small effect of word frequency was observed (Samuel, 1981). Moreover, restoration is better in meaningful sentence contexts than for independent speech units or words (Bashford and Warren, 1987; Bashford *et al.*, 1988). The context also determines

---

[7]This effect plays a role in mechanisms applied to recover lost packets in packet-based speech transmission networks like VoIP (see Chapter 3).

whether intelligibility is increased by phonemic restoration: if the underlying speech material consists of complete linguistic units such as sentences, the insertion of noise into the silence gaps of periodically interrupted speech leads to an increase of intelligibility (Powers and Wilcox, 1977). In turn, if no context is provided, intelligibility is not improved (see Miller and Licklider, 1950, who used monosyllabic word lists).

- Confirmation: The higher the spectral similarity between the missing speech sound and the sound used for replacement, the better (or the more certain) is the restoration (Bashford and Warren, 1987).

- Duration: In case of periodic interruptions of read discourse, it was found that filling the interruptions with white noise leads to illusionary continuity up to maximal durations of around 300 ms (Bashford and Warren, 1987; Bashford et al., 1988). With speech material presented at different speaking rates, it was confirmed that the maximally restorable duration for periodically interrupted speech approximately equals the average word duration (Bashford et al., 1988). Speech interrupted by periodic silence gaps − in turn − was perceived as unnatural already for durations around 10 ms. Only for interruption durations longer than 100 ms was the perceptional effect clearly identified as being due to interruptions (Bashford and Warren, 1987; Bashford et al., 1988).

Several authors have explained phonemic restoration with the dual perception process described in Section 1.1.2.3 (top-down phonetic and bottom-up acoustic processing). As the phonemic restoration by white noise was found to be more efficient for consonants than for vowels, a great importance of the bottom-up process, i.e. of the delivery of auditory cues, was hypothesized (Samuel, 1981).

The motivation for this explanation is the fact that phonemic restoration is related to a more fundamental perceptual illusion, which is typically referred to as *continuity illusion* (e.g. Carlyon et al., 2002) or more generally *temporal induction* (see Warren, 1999, pp. 134–154 for an overview). The simplest form of temporal induction occurs, for example, when two spectrally identical short noise signals (e.g. of 200 ms duration), one louder than the other, are presented alternating with each other. In this case, the softer noise seems to continue behind the louder one. Under certain conditions, the continuity illusion also occurs when the two sounds are perceptually different (*heterophonic continuity*, see Warren, 1999, pp. 137–141). The first study mentioning this type of continuity illusion was reported by Miller and Licklider (1950). In their tests, they presented listeners with speech signals periodically interrupted at different interruption rates (50% duty cycle). When the interruptions were filled with white noise at interruption rates between 10 and 15 times per second, the speech signal seemed to continue behind the noise. Miller and Licklider referred to this phenomenon as 'picket fence effect', in analogy to the visual modality: when watching a landscape through a picket fence, the fence interrupts the view periodically. However, the landscape is seen to continue behind the fence.

The continuity illusion is closely related to the principle of closure described by the Gestalt psychologists (for an introduction to the laws of 'Gestalt' (German: *Gestaltgesetze*[8]) see e.g. Katz, 1969, pp. 33–39): For example, in case of a basic geometric form like a square, one edge of which is covered by another form, the visual system 'closes' the interruption, and the square is perceived as a continuous unit.

---

[8]Founded among others by Koffka, Köhler, Wertheimer and Von Ehrenfels.

The auditory processing aspects of phonemic restoration, where missing *linguistic* information, can be summarized as follows (items (1)–(3) are the basis for the underlying continuity illusion; Bregman, 1990; Warren, 1999):

(1) The on- and off-transitions of the deleted sound are masked by the restoring sound.

(2) The restoring sound defines its own limits and not the edges of the sound it restores.

(3) By streaming processes, the neural activity caused by the restoring sound is partly associated to the interrupted sound, rendering the restoring sound softer than when it is presented in isolation (Bregman, 1990; Warren, 1999, pp. 134–154). The newly associated neural activity implies that the missing sound is still there.

(4) After recoding of the auditory feature information, pattern recognition processes associate a linguistic unit with the restored section, if sufficient context information is provided.

1.1.2.6 Human Adaptation to 'Noisy' Communication Channels

Apart from perceptual mechanisms, humans have additional means of enhancing the comprehensibility and communicability of speech, by adapting their communication behavior to the environment. If some information gets lost and cannot be restored by perception, interlocutors may recover the loss by question and reconfirmation, that is, by re-sending the respective message[9]. In noisy environments, interlocutors adopt a Lombard speaking style, that is, raising their voice and stressing syllables differently (e.g. Köster, 2003; Lane *et al.*, 1961, 1970). Related strategies apply when speaking to hearing-impaired or foreigners: for example, using clear speech, i.e. overarticulated and slowed-down louder speech, a talker can enhance the comprehensibility of his utterances (e.g. Payton *et al.*, 1994). During telephone conversations, impairments like long transmission delays, which have been recognized by the conversation partners, may be compensated for by adopting a walkie-talkie type conversation strategy: in order to avoid erroneous turn-taking, the listener may wait until the talker has finished, before he speaks himself; vice versa, the talker can try to code continuous information into one continuous message, instead of awaiting frequent back-channeling (i.e. confirmations) from his counterpart (Krauss and Bricker, 1966). To overcome severe loudness loss on a transmission system, a talker can raise his voice, and the listener can press the handset to his ear in order to improve the acoustic coupling (e.g. Krebber, 1995).

1.2 Speech Quality

Obviously, a multitude of information is contained in the speech signal typical of everyday communication, as it may happen, for example, during a telephone conversation. Although context and perceptual mechanisms as well as adaptation of the communication behavior yield reliable comprehension even in unfavorable conditions, deteriorations of spectral and/or temporal cues of the speech signal degrade the contained information: be it the linguistic information (the actual message), or paralinguistic information regarding the speaker

---

[9]Assuming they are aware of the loss!

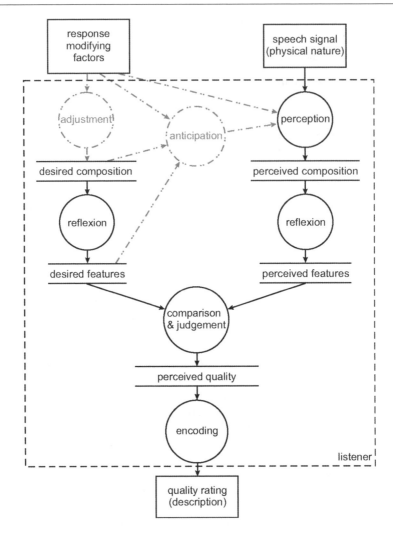

Figure 1.6: Quality: Perception and rating process as seen from the listener's perspective (based on ideas from Jekosch, 2000, 2004, 2005b)). In the figure, circles represent processes, two horizontal parallel lines storages and rectangles the inputs and outputs to the listener. Note that the comparison of features can be assumed to be carried out both on a feature-by-feature- and on an integrated 'Gestalt' level.

identity or his emotional state. Even if comprehension is not at stake, other factors such as annoyance, for example, due to time-varying unwanted sounds additional to the wanted signal, modify the user's appreciation of a particular connection. In this context, the term *speech quality* is typically used, extending the comprehension-oriented view on speech to additional factors governing speech perception. In this section, a working definition of the quality-related terms is presented, providing the scientific paradigm for the research described in this book.

1.2.1   Definition of Quality

Quality is the

'result of [the] judgment of the perceived composition of an entity with respect to its desired composition'.
(Jekosch, 2000, 2005b, pp. 15).

Here, the ***perceived composition*** is the

'[t]otality of features of an entity',
(Jekosch, 2005b, pp. 16).

and a ***feature*** is

'[a] recognizable and nameable characteristic of an entity.'
(Jekosch, 2005b, p. 14).

If the entity under consideration is a sound, the 'perceived composition' refers to the perceived *auditory* composition, and the 'desired composition' to the desired *auditory* composition of the sound (see Figure 1.6; instead of perceived or desired 'composition', the terms 'perceived nature' and 'desired nature' are sometimes used, e.g. Jekosch, 2004). Two sounds can be distinguished on the basis of their loudness, pitch, duration, timber and spaciousness (see Letowski, 1989). The totality of these attributes or *features*[10] describes the *perceived composition, perceived nature* or *character* of a sound (Jekosch, 2004, 2005b; Letowski, 1989, respectively). In this book, the term 'sound' refers to speech sounds.

The listener is a part of a particular communication situation, communication either with one or several other persons, or, in a more abstract sense, with (parts of) his/her environment (e.g. in case of nonspeech sounds). Because of the framework of communication, the listener anticipates the percept to some extent (see Figure 1.6). On the basis of *modifying factors*[11] such as the context and situation in which the sound occurs, and on personal factors such as her/his mood or motivation, the listener relies on mental, *desired characteristics* of the percept[12]. If the listener has prior experience with the particular nature of the sound, the *desired* characteristics correspond to a stored schema, which is accessed on the basis of the particular context, and potential preceding auditory percepts related to the same act of communication. In a reflection process[13], the listener decomposes the desired characteristics of the sound and identifies the desired (or expected) features. Correspondingly, the perceived characteristics are reflected and transformed into a set of perceived features. Finally, the judgment on the comparison of the desired and the perceived features constitutes the perceived quality.

In this context, the ***desired composition*** is defined as

'[the] totality of features of individual expectations and/or relevant demands and/or social requirements.'
(Jekosch, 2005b, p. 16).

---

[10]For simplification of the notation, this book uses the term *feature* sometimes to refer to the feature itself (i.e. its identification), but sometimes also to additionally refer to its markedness or magnitude (owing of course to the multidimensional 'nature' of many features). In case of possible ambiguities, a differentiation will be made.

[11]The modifying factors lead to an *adjustment* of the desired nature, see Figure 1.6.

[12]The *desired characteristics* (i.e. the expectation of the user) will be addressed in more detail in sections 3.7 and 5.4

[13]The reflection process may be more or less 'conscious', for example, depending on whether the communication situation is a directed one, as in the case of an auditory quality test, or an undirected one, as in the case of a natural telephone conversation.

1.2.2   Speech Quality Assessment

In speech quality tests, the test subjects are asked to describe their quality perception, or in other words, to *assess* the quality of a particular speech sample or system used for its transmission: they are sought to assign numbers to objects (Blauert and Jekosch, 2003). In the same way as it was mentioned for auditory listening tests in Section 1.1, the subjects make their judgment on a rating scale provided by the experimenter. This process of 'encoding' ultimately leads to a quality rating, the output of the speech quality assessment process depicted in Figure 1.6.

With reference to the above quality definition, 'assessment'[14] can be considered as a subordinate term to the more generic term 'appraisal': assessment assumes a measurement, where the test subjects' perception process is to some extent directed by certain test design factors like the rating procedure or the test scenario being used. Consequently, the directives given to the test subjects are part of the modifying factors depicted in Figure 1.6. They exert an influence on the desired nature of the perceived speech sounds as well as on the perception process.

In contrast, during an everyday conversation, undirected, individual perception takes place. This undirected perception is subject to the particular attention listeners or interlocutors pay to certain of the different signals they are faced with; only those the individual person considers relevant for the communication situation are actually perceived ($\equiv$ *stimuli*; Jekosch, 2005a). In the case of directed perception (as in a speech quality test), different stimuli or features may appear more relevant to the subjects than others. For example, in an everyday conversation among friends the content, that is, what is said, may be more important than in a laboratory listening test focusing primarily on the form of certain speech signals. Hence, one of the main goals of speech quality tests – that is, of auditory tests in general, see Section 1.1, – is to appropriately choose the test design and directives, in order to achieve valid and reliable measurement results, as far as possible reflecting the quality perception of users during undirected communication.

There are further correspondences to the more general auditory test situation discussed earlier: Figure 1.7(a) shows a listener in an auditory quality test situation, in analogy with the auditory test on particular auditory features of a sound depicted in Figure 1.1. The bold lines in Figure 1.7(a) represent the configuration as it is displayed in Figure 1.1. In this case, the test subject delivers a description of all or of certain features of the perceived sound. Such a test not focusing on quality judgments but on the description of perceived (quality) features, is referred to as *analytical* type of speech quality test. If an actual description of the perceived quality is sought, the process of comparison and judgment comes into play (Figure 1.6). The corresponding type of speech quality tests is referred to as *utilitarian* type (Figure 1.7(b); Hecker and Guttman, 1967; Quackenbush *et al.*, 1988, pp. 15–16). In this type of tests, quality is typically rated on a unidimensional quality or impairment scale. The outcome of such tests is particularly interesting for applications like network planning or efficient preference testing to determine the best of different implementations of a network component or system, or of other types of systems, such as hearing aids.

At this point, a second dichotomy for auditory tests in addition to the *analytical–utilitarian* one can be mentioned: test methods can be distinguished according to

---

[14] '*Assessment*: Measurement of system performance with respect to one or more criteria. Typically used to compare like with like, whether two alternative implementations of a technology, or successive generations of the same implementation...' (Jekosch, 2000, 2005b, p. 109).

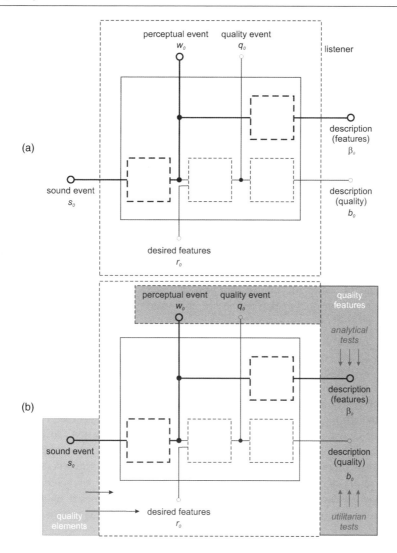

Figure 1.7: (a) Schematic representation of a test subject in a listening quality test (combining the concepts by Blauert, 1997; Jekosch, 2000, 2004, 2005b). (b) Additional illustration of the terms *quality elements* and *quality features* (see Jekosch, 2000, 2004, 2005b), and of the differentiation between *analytical* and *utilitarian* type speech quality tests (Hecker and Guttman, 1967; Quackenbush *et al.*, 1988, pp. 15–16).

whether they are *subject-oriented*, that is, are carried out to gather information on human perception, or whether they are *object-oriented*, that is, investigate how the sound produced by or transmitted across certain systems is perceived (Letowski, 1989). In summary, the two dichotomies lead to four types of quality tests, as depicted in Table 1.1 (for an overview of some relevant analytical and utilitarian test methods see Chapter 2, Section 2.1).

Table 1.1: Quality tests: The four different applications resulting from the two dichotomies *analytical–utilitarian* and *subject-
oriented–object-oriented* (combination of Quackenbush *et al.*
(1988, pp. 15–16), and Letowski (1989). In the latter, the terms
heuristic and *diagnostic* are used instead of *utilitarian* and *analytical*, with different definitions).

	Subject-Oriented	Object-Oriented
Utilitarian	Quality perception	Assessment of system quality
Analytical	Quality features and their perception	Quality features and acoustic or system correlates

The main emphasis of this book is on object-oriented tests, seeking to relate the quality
perceived by users of certain speech communication networks to instrumentally measurable
network parameters. Some of the research presented here can also be considered as subject-
oriented, as aspects of quality perception are addressed[15].

1.2.3 Quality Elements

Up to this point, speech quality has been discussed mainly from a user's perspective. The
remainder of this book discusses the implications of the design and planning of networks
involving Voice over Internet Protocol (VoIP) transmission on speech quality as perceived
by users. Although looking at quality from the perspective of the system, a network planner
has to have the envisaged users in view if the network is to reach broad acceptance. If speech
quality is regarded from the point of view of network planning or monitoring, the effect of
certain network components or of certain parts of the network infrastructure on quality is
of interest. For all the factors that have an impact on the quality perceived by the user and
are in one way or another related to aspects of the design, technical realization or usage
of the particular telecommunication system or service, Jekosch has proposed the following
definition (see also Figure 1.7):

A *quality element* is the

'Contribution to the quality

- of an immaterial or a material product as the result of an action/activity or a process in one
  of the planning, execution or usage phases.
- of an action or of a process as the result of an element in the course of this action or
  process.'

(according to Jekosch, 2005b, p. 22, modified from DIN 55350, Part 11).

From the point of view of perception, a complementary definition can be given for the
quality-relevant perceptual features: a *quality feature* is

---

[15]In general, it has to be noted that tests that are *a priori* subject-oriented may also provide object-oriented
information and vice versa, depending on the interpretation of the results.

'[a] recognized and designated characteristic of an entity that is relevant to the entity's quality.'
(Jekosch, 2000, 2004, 2005b, p. 17).

The dichotomy of *quality features* and *quality elements* is used as an important tool to structure the content of this book: In Chapter 3, the quality elements of speech communication networks involving VoIP are summarized, and an overview of the related quality and quality features is provided. The author's research on the impact of the most important of these quality elements on quality and quality features is described in Chapters 4 and 5. Therefore, we employ simulation tools to generate the quality elements in a laboratory situation, as described in Appendix B On the basis of the description of the quality elements and the knowledge of the related quality, modeling approaches will be discussed that quantify the relationship between the quality elements and the quality features.

1.2.4   Speech Quality and Quality of Service

Different terms are typically used in the literature to refer to the (speech) quality of speech communication systems (see Möller, 2000, p. 11): 'Mouth-to-ear' or 'end-to-end' quality refers to the quality of the entire system from the mouth of the talker to the ear(s) of the listener. The term 'integral quality' is used when the quality due to the totality of quality dimensions (or features) is considered. Another term frequently used in the literature is 'overall quality', which is used synonymously for mouth-to-ear by some authors, and for 'integral quality' by others. To avoid this ambiguity[16], the term *integral quality* will be used in this book in cases where the quality resulting from the totality of quality features is referred to. When referring to the (integral) quality instantaneously judged by subjects – for example, during listening to a speech signal degraded by time-varying distortions – the term *instantaneous quality*[17] will be employed (see Gros and Chateau, 2001; ITU–T Rec. P.880, 2004)[18]; the (mathematically obtained) time average of the instantaneous quality will be referred to as *average instantaneous quality*. Consequently, in this book, *integral quality* is the quality subjects relate to an entire conversation or speech passage, taking the history and evolution of the conversation or passage into consideration; it corresponds to the final quality judgment obtained at the end of the conversation or listening sample. More generally, in this book the term *speech quality* refers to the quality perceived in a conversational situation. If a listening-only situation is referred to, the term *speech transmission quality* is used.

Speech quality is only one of the different factors ultimately determining the acceptability of a telecommunication service. To refer to all aspects related to the acceptability of a service, the term *quality of service* (QoS) is typically employed following the definition provided in ITU–T Rec. E.800 (1994):

Quality of Service (ITU–T Rec. E.800, 1994):

'The collective effect of service performance which determines the degree of satisfaction of a user of the service.'

QoS, according to this definition, is composed of four aspects, namely, *service support*, *service operability*, *serveability* and *service security* (ITU–T Rec. E.800, 1994). *Service*

---

[16]Adopting the terminology and argumentation used by Möller (2000).

[17]Instead of 'instantaneous integral quality'

[18]Here, the terms *instantaneous judgment*, *time-varying speech quality* and *continuous speech quality evaluation* are used.

support is related to services like directory assistance or technical assistance; *service oper-ability* refers to how easily a service can be operated by a user; *serveability* comprises aspects like the accessibility and retainability of a service (i.e. how faithfully a service can be obtained and be provided for a given period of time), as well as the level of speech quality it supplies; finally, *service security* is concerned with issues such as the protection against unwanted access to, or monitoring of, the transmitted data by third parties (e.g. 'spoofing' or 'sniffing' of data in packet-based systems[19]).

QoS can be looked at from two different perspectives: that of the service provider, and that of the user of the service. In order to reflect the user's perspective of QoS, the QoS framework has recently been complemented by the framework of *quality of experience* (QoE) (ITU–T Delayed Contribution D.197, 2004). In this context, an implicit distinction is made between the quality element side and the quality feature side. According to the proposed redefinitions, QoS now refers to the network, that is, the quality element side, and QoE refers to what the user actually perceives of QoS:

Quality of Service (revision proposed in ITU–T Delayed Contribution D.197, 2004):

> 'The collective effect of *objective* service performance which *ultimately* determines the degree of satisfaction of a user of the service.'

Quality of Experience (new definition proposed in ITU–T Delayed Contribution D.197, 2004):

> 'A measure of the overall acceptability of an application or service, as perceived subjectively by the end user.'

In order to reduce complexity, the term QoS will be used throughout this book, pointing out whether the user's or the service provider's perspective is considered by employing the concept of *quality elements* (service provider) and *quality features* (user).

A taxonomy integrating speech quality into the framework of QoS for telephone ser-vices was developed by Möller (2000). It was recently modified in order to better match a corresponding taxonomy of QoS developed for telephone-based services involving spoken dialogue systems (Möller, 2005a,b). According to this taxonomy (see Figure 1.8), QoS is composed of three factors, which constitute the quality elements of the service: *speech communication factors* concern the actual communication between the (two or more) inter-locutors over the speech transmission system. According to the above QoS definition from ITU–T Rec. E.800 (1994), these factors contribute to the serveability of the system. *Service factors* cover service support and security, and parts of service operability and serveability. They summarize the impact of the service characteristics, however, excluding the speech communication factors. *Contextual factors* do not directly form a part of the QoS definition provided above. They relate to nonphysical aspects of the service, like the costs for the user (investment, monthly and per-call charges, etc.), and the contract conditions (e.g. the contract period or the period of notice).

Similar to the modifying factors depicted in Figure 1.6, the contextual factors have an impact on the user (i.e. on his attitude, emotions or motivation), and hence on the desired nature of the transmitted speech. Because of the user's role for all the intermediate levels of appraisal shown in Figure 1.8, and for the ultimate service acceptability, the user is

---

[19]Sniffing: Network-traffic monitoring, possibly with access to the actual data content. Spoofing: The other end-point is deceived by his interlocutor who pretends an incorrect identity. Examples are email- or caller-ID-spoofing. On a lower level, network traffic could, for example, be rerouted to a malicious recipient.

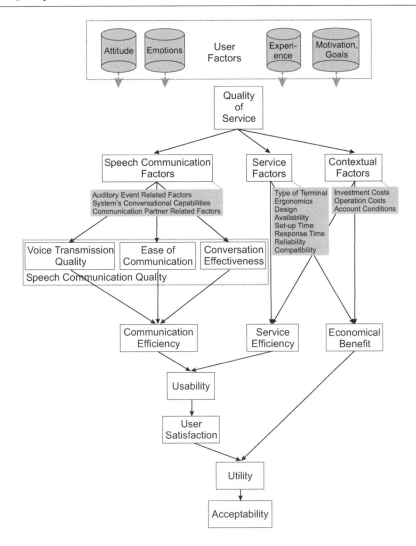

Figure 1.8: Quality of Service (QoS) taxonomy developed by Möller (2005a,b). The gray frame in the middle of the picture highlights the three constituents of speech communication quality (voice transmission quality, ease of communication and conversation effectiveness).

depicted on top of the figure. Different user factors are shown that contribute to the quality perception, like his attitude or goals.

The speech communication factors contribute to the quality features and to the corresponding speech quality[20] perceived by the user. They result from the quality elements, which in turn depend on the decisions taken during the network planning phase. The speech communication factors can be further subdivided into three constituents:

---

[20]With reference to Möller (2000), the term *speech communication quality* is used in Figure 1.8 instead of *speech quality* as it is used in this book.

Voice transmission quality considers all those quality elements that ultimately lead to an impact on quality effective already in a pure listening situation, that is, to factors like noise on the line, a loss of signal level introduced by the transmission, or transmission errors like packet loss. In this book, voice transmission quality is referred to as *speech transmission quality*.

Conversation effectiveness refers to the quality elements affecting the conversational capabilities of the system, that is, factors only relevant in a conversational situation. Examples are talker echo and transmission delay.

Ease of communication concerns conversation partner related factors, which also include a potential adaptation to an adverse acoustic environment (i.e. Lombard speech, etc., see Section 1.1.2.6).

As already stated in Section 1.1.2, all components of speech (communication) quality together determine the communication efficiency[21], that is, the resources a user has expended in relation to the accuracy and completeness with which he/she has performed a particular communication task (ETSI Guide EG 201 013, 1997), as well as his/her level of awareness of it. The service efficiency describes the resources expanded during the usage of the service, disregarding all issues related to the actual communication. Together, communication and service efficiency determine the usability of the service, that is, the aptitude of the system to be used for completing a specific task. Utility, the next step in the taxonomy, can be considered as the trade-off between the system's usability and the related costs. Ultimately, the acceptability measures the ability of the service to find acceptance. It is typically determined as the relation between the number of actual users and the number of potential users. It has to be emphasized again that it is the user who dictates every step in the chain, from the system- and service-related factors to the service's acceptance.

An example from the domain of early VoIP networks may serve for illustration: the first means to establish an internet telephone connection for a user not technically inclined were based on software tools quite complicated to handle. Consequently, the service[22] factors were counterproductive for a high usability. Moreover, communication efficiency was very low, as time-varying degradations like packet loss and delay jitter on the one hand, and more importantly severe amounts of mean transmission delay on the other, made effective communication difficult. The potential economical benefit of making long-distance calls free of charge was counterbalanced by the small amount of potential conversation partners, who had to be equipped with the same software client, and, in particular, be online at the same time: the serveability was very low. As a consequence, during the early tests carried out in the framework of this book, the number of subjects with prior experience with Internet telephony was rather low (around 4–8%, in 2001), slowly increasing to around 15–20% for later tests (in 2004), owing to both the improvement of the serveability and service operability of Internet telephony 'services', and their increasing recognition by potential users (see Chapter 4).

---

[21]In Section 1.1, the term *communicability* was used to describe the capability of a particular conversation to lead to a certain level of communication efficiency. Correspondingly, the *communicability of the system* summarizes both the voice transmission quality and the conversation effectiveness provided by the system.

[22]Early days internet telephony cannot be regarded as actual service, since the existing network structure and Internet service was 'reused' on the basis of software clients not stemming from dedicated service providers assuring aspects of QoS.

2

Speech Quality Measurement Methods

Different methods exist for measuring the speech quality of a network or its components. In order to address the measurement of speech quality, the terms 'assessment' and 'evaluation' are typically used in the literature. These terms will be briefly summarized here, prior to a description of typical measurement techniques:

Evaluation is the 'determination of the fitness of a system for a purpose – will it do what is required, how well, at what cost and so on. Typically for [a] prospective user, may be comparative or not, may require considerable work to identify user's needs' (Jekosch, 2005b), and thus is the evaluation of the system for a given purpose (adequacy). (Möller, 2005a).

Assessment is the 'measurement of system performance with respect to one or more criteria. Typically used to compare like with like, whether two alternative implementations of a technology, or successive generations of the same implementation' (Jekosch, 2005b). Hence, assessment relates to the performance of the system (Möller, 2005a).

Although the underlying measurement motivations related to assessment and evaluation are distinct (Jekosch, 2005b), the two concepts are not entirely orthogonal[1].

Speech quality measurement techniques can be further classified according to how well diagnostic information can be obtained:

- Glass-box approaches give access to diagnostic information. With glass-box methods the source of a certain quality level can be identified, and the relation between system configuration and quality observed. For example, parameter-based models are typically glass-box models, since they build on a 'diagnostic' description of the system (see Section 2.2.2).

---

[1]For example, in a system evaluation, the user's experience with other versions of the system leads performance aspects to play an indirect role in the tests. Conversely, a speech-quality-related performance assessment should necessarily take the actual application into consideration, and will then indirectly address how well the system suits a given purpose.

Speech Quality of VoIP: Assessment and Prediction   Alexander Raake
© 2006 John Wiley & Sons, Ltd

- Black-box approaches do not give direct indication of the source of low quality. Certain signal-based models are examples (see Section 2.2.1). For black-box methods, the system under test is described only by its input and output.

Current instrumental speech quality measurement techniques cannot necessarily be classified in a strict manner according to this black-box/glass-box dichotomy. Instead, different methods enable more or less diagnostic information to be obtained. Sometimes the terms 'white-box' and 'grey-box' are used in the literature in these cases (Möller, 2005a, pp. 86–89).

Methods of speech quality measurement can also be distinguished according to the applied type of measurement apparatus into *auditory* and *instrumental* methods. Since perceived quality is internal to the user, auditory quality tests are ultimately the only means of validly and reliably assessing quality (see Section 1.2). As auditory speech quality tests are time-consuming and costly, instrumental quality measurement methods have been developed. Within the range of network configurations these models have been designed for, and proven to provide valid and reliable quality predictions, they can be used as a replacement or complement of auditory tests. The present chapter gives an overview of different auditory and instrumental quality measurement methods relevant for this book.

Note that in the literature, instrumental measurement techniques are often referred to as 'objective', and auditory methods as 'subjective'. However, as pointed out earlier, the perceiving and judging subject can be considered as a measurement system (see Blauert, 1997); the subject is involved in the process of instrumental model development, and the model attempts to mimic the perceiving subject. Moreover, it is the experimenter who ultimately ('subjectively') interprets the measurement results. Thus, the 'subjective'/'objective' differentiation appears to be inappropriate. Consequently, we will not use these terms here, but instead employ the differentiation according to the measurement apparatus into 'auditory' and 'instrumental' methods.

2.1 Auditory Methods

According to Jekosch (2000, 2005b), a speech quality test is

> '[a] routine procedure for examining one or more empirically restrictive quality features of perceived speech with the aim of making a quantitative statement on these features'. (Jekosch, 2005b, p. 91)

Using this definition, auditory quality test methods can be broadly categorized as utilitarian and analytic test methods on the basis of whether the totality or a certain subset of perceived features of speech is to be investigated (analytical), or whether single individual features or integral quality are to be measured (utilitarian, see also Section 1.2.2).

Utilitarian test methods employ a unidimensional quality rating scale, so that a direct comparison between the quality of different speech communication systems can be made. The usage of one single scale and a direct assessment of quality or preference has also been referred to as *isometric approach* (Quackenbush *et al.*, 1988, p. 72–73). In turn, analytical methods are based on the multidimensional analysis of test stimuli: they aim at identifying and quantifying the perceptual features underlying speech quality. If possible, analytical methods also intend to uncover the acoustic characteristics, that is, the quality elements, which are correlated with the different quality features (Hecker and Guttman, 1967; Quackenbush *et al.*, 1988, p. 15–16).

In order to capture the entire range of underlying auditory features, and to ensure a common understanding of these features, the reliability and validity of analytical methods can be significantly improved, when trained listeners are used (e.g. Mattila, 2001; Möller, 2000, pp. 105–120)[2]. In turn, tests of the utilitarian type are typically carried out with naive subjects. These should ideally represent the user group the system or service under investigation is aimed at (e.g. age range, social and cultural background, etc.).

It should be noted that this *utilitarian–analytical* categorization of auditory test methods can be considered as being of *utilitarian* type itself, as it employs a unidimensional view on test methods. A multidimensional, that is, analytical view on speech quality assessment methods was provided by Jekosch (2005b, pp. 105–111).

2.1.1   Utilitarian Methods

Utilitarian methods have three main goals: 'To be reasonably efficient in test administration and data analysis, to measure speech quality on a uni-dimensional scale, and to have good reliability in the test method'. (Quackenbush *et al.*, 1988, pp. 15–16). Utilitarian methods can be classified according to the categorization generally applicable to psychometric methods (Möller, 2000, pp. 48–49):

- The applied scaling method and scale level[3].

- The presentation method used[4].

- The test 'modality' (listening-only, conversation, talking and listening).

2.1.1.1   Comprehension Tests

Depending on how much context information is provided, different test methods studying the comprehension enabled by a transmission system can be distinguished (see Section 1.1.2.4).

Comprehensibility tests (often referred to as *articulation tests*) seek to quantify the system's impact on the identification of small speech units such as phonemes or syllables. For example, the impact of context can be inhibited by using nonsense, however, phonotactically correct monosyllabic words (e.g. Cluster identification test (CLID) for speech synthesis systems, see Jekosch, 1992; for more examples see Jekosch, 2005b, pp. 93–103, pp. 113–141). Comprehensibility tests are typically carried out as open response tests[5].

Slightly more context information is provided when meaningful words are used. Then, the identification of individual units like phonemes is indirectly studied on the basis of the *intelligibility*[6] of the words. An example of this type of method is the family of rhyme tests, where the answers are not collected as open responses, but as choices from different

---

[2]These subjects should, however, not be *expert* listeners, that is, listeners with particular knowledge of the underlying signal processing, and thus potentially biased in their view.

[3]The scale levels are: 1) *nominal* (determination of equality), 2) *ordinal* (determination of greater or less), 3) *interval* (determination of equality of intervals or differences), and 4) *ratio* (determination of equality of ratios) (e.g. Stevens, 1951, p. 25).

[4]E.g. adjustment, single stimuli, paired comparison (e.g. Stevens, 1951, p. 43).

[5]Open response: the subjects are asked to write down what they have heard, as opposed, for example, to choosing one of several possible predefined answers (multiple choice).

[6]Linking identification with content, see Section 1.1.2.4.

similarly sounding words. Examples are the Modified Rhyme Test (MRT; House *et al.*, 1965; German version: Sotscheck, 1982), and the Diagnostic Rhyme Test (DRM); Voiers *et al.*, 1975).

Another type of intelligibility tests is the (open response) SUS test, which employs syntactically correct but Semantically Unpredictable Sentences (Benoît, 1990). Hence, individual words are meaningful, but entire sentences are not. SUS contradict the typical listening habits of subjects, who anticipate words completing the sentences to be meaningful (Jekosch, 2005b, p. 117). Consequently, the sensitivity of the SUS test procedure is relatively high. In turn, the search for semantic congruency may mislead the subjects, who replace words they missed depending on the semantic context information.

Other methods have been proposed to increase the limited sensitivity of intelligibility tests, for example, in order to distinguish between different high-quality transmission systems. A method relying on semantically unpredictable sentences was proposed by Nakatani and Dukes (1973). Their speech interference test is intended to provide an indirect measure of the speech quality associated with a particular transmission system. According to this approach, the intelligibility related to the system under test is compared to that related with a reference system. The particularity of the method is that both the reference and the system under test are presented together with interfering speech of different *interfering speech levels*. For both the reference system and the system under test, that level of interfering speech is determined, which leads to 50% sentence intelligibility. Then, the relative quality can be expressed as the difference between the two interfering speech levels.

Other comprehensibility and intelligibility test methods have been reported in the literature. For more information, the interested reader is referred e.g. to Quackenbush *et al.* (1988), Jekosch (2000, 2005b) or Möller (2000).

2.1.1.2 Listening Quality Tests

Many utilitarian quality tests are carried out as listening tests (LOTs; listening only tests). They can be distinguished according to whether each stimulus is presented in isolation (absolute rating), or whether stimuli are presented in pairs (paired comparison). For paired comparison tests, two subcategories can be distinguished on the basis of the presentation method: either both stimuli of one pair are constant and the relative quality or degradation is assessed, or the degradation of a reference stimulus has to be adjusted to the degradation of the stimulus under test (isopreference test). The different methods are described in more detail in the following.

Absolute Category Rating Tests
In telecommunications, absolute category rating (ACR) tests are most commonly used to assess integral quality (ITU–T Rec. P.800, 1996). Different scales are recommended, depending on the focus of the particular test (e.g. listening quality, listening-effort, or loudness preference, see ITU–T Rec. P.800, 1996).

The scale most frequently used is the 5-point ACR quality scale (Figure 2.1). It is typically referred to as *MOS scale*, owing to the fact that the mean over the ratings obtained on this scale is called the *Mean Opinion Score*. A similar scale, however, without numerical annotation, is recommended for conversation tests (CTs) (ITU–T Rec. P.800, 1996).

Quality of the speech:

excellent	good	fair	poor	bad
5	4	3	2	1

Figure 2.1: 5-point ACR-scale (MOS scale, ITU–T Rec. P.800, 1996).

The scale is often interpreted as an interval scale (i.e. mean and standard deviation are calculated), but strictly speaking its characteristics only satisfy the requirements of an ordinal scale (i.e. calculating the frequency of votes per category, the median and the percentile values).

For absolute quality listening tests, the usage of different sequences of two to five independent, short, meaningful and simple sentences is recommended (ITU–T Rec. P.800, 1996). With sentence durations of 2–3 s, overall sequence durations ($\equiv$ stimuli durations) of below 10 s are commonly achieved. Both the choice of the underlying text material and the user interface chosen for stimuli presentation may have significant effects on the realism of the telephony context simulated by listening tests in a laboratory environment (see Chapter 5, Section 5.4).

The number of different talkers used for source material recording on one hand, and the type of degradations to be assessed on the other determine the generalizations that can be drawn from the tests: for example, the quality related to certain circuit configurations can depend significantly on the talker. Hence, the number of talkers should be chosen according to the systems under test (Möller, 2000, p. 54).

Other variants of category rating tests have been proposed for special applications: for example, in ITU–T Rec. P.835 (2003), a threefold rating procedure is recommended for the assessment of noise-suppression algorithms. It is intended to provide some additional information on the source of low speech quality. Hence, judgments of integral quality collected on the 5-point ACR scale are complemented by two additional separate judgments on the distortion of the speech signal, and of the intrusiveness associated with the noise signal. To reduce the impact of the order in which the scales are presented, the recommended procedure involves a randomized presentation of the three scales after each stimulus.

Degradation and Comparison Category Rating Tests

Listening tests involving degradation category rating (DCR) or comparison category rating (CCR) typically enable a more fine-grained resolution of small quality differences than the ACR method. This is achieved primarily by the underlying paired comparison technique.

In DCR, each stimulus is preceded by a clean reference stimulus representing top-line quality. The subjects are asked to rate the degradation of the test stimulus relative to the clean reference.

CCR employs pairs of *test* stimuli: the quality of the second stimulus is rated relative to the first. Both stimuli are randomly selected from the set of all test stimuli. CCR and DCR use category rating scales similar to the 5-point ACR scale. For more details on the test procedures and scales see ITU–T Rec. P.800 (1996).

Isopreference Tests

Similar to DCR and CCR tests, isopreference tests are conducted as paired comparison listening tests. However, they do not involve ratings, but an adjustment procedure: for each test stimulus, subjects adjust the parametric degradation of a reference stimulus to yield the same quality level as that of the test stimulus. If the quantitative relation between the reference degradation and speech quality is known, the speech quality of the test stimulus can be determined from the 'equally preferred' degradation setting of the reference. The original version of the isopreference method was based on 'isopreference contours', determined for different combinations of speech and noise levels (Munson and Karlin, 1962). Each isopreference contour was numerically related to a particular 'Transmission Preference' level. The method was simplified by Rothauser *et al.* (1968), who have also suggested to employ multiplicative Gaussian white noise as reference degradation, instead of the additive noise originally used. The so-called 'threshold method' described in ITU–T Rec. P.800 (1996) is a variant of the isopreference method.

A comparison of isopreference tests conducted with two different types of reference degradations, and test stimuli degraded by different low-bit-rate codecs and narrowband filters can be found in (Möller, 2000, pp. 121–129).

Continuous Evaluation

In case of time-varying degradations like packet loss, somewhat more diagnostic information on the relation between instantaneous quality events and integral quality may be sought. Therefore, Gros and Chateau (2001) proposed a particular listening test method extending the method for instantaneous rating of time-varying video quality (see ITU–R BT.500-8, 1998) to the evaluation of time-varying speech quality (ITU–T Rec. P.880, 2004). In principle, this continuous rating procedure pertains to the absolute category rating type, but instead of asking for single integral quality ratings, subjects are asked to assess speech quality continuously with the help of a slider. The rating scale is a continuous version of the 5-point ACR scale (Figure 2.1), and is depicted in Figure 2.2. After the continuous judgment of each stimulus, subjects are asked to rate the *integral* quality on the 5-point ACR scale typically used in listening tests, in order to provide information on the relationship between the instantaneous ratings and integral quality (Figure 2.1: in this case preannotated with numerical labels).

During the test, the slider position is to be sampled at least every half second, in order to capture even relatively fast rating changes. As the method is to be applied to evaluate time-varying speech quality, the usage of long speech samples of 45–180 s duration is recommended (compare also Jekosch, 2000, pp. 166–186).

The instantaneous ratings obtained for each stimulus are averaged over the subjects[7], to obtain a mean instantaneous rating profile. Similarly, the integral quality ratings are averaged over subjects to yield integral MOS-values. For more details on the test procedure see ITU–T Rec. P.880 (2004). A description of tests from the literature which made use of this method can be found in Section 3.3.5.7.

---

[7]Assuming that each subject has heard the same stimulus for each condition, that is the same speech file degraded by the same impairment profile. Otherwise, averaging may lead to arbitrary results.

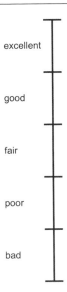

Figure 2.2: Continuous category rating scale for continuous quality evaluation (ITU–T Rec. P.880, 2004).

Third Party Listening Tests

A test method which tries to mimic a conversational situation, but actually uses a listening-only paradigm, is the third party listening procedure (ITU–T Rec. P.831, 1998; ITU–T Rec. P.832, 2000). It was developed as a compromise between collecting data on conversational capabilities of networks and network components such as echo cancelers, and reducing the costs and efforts associated with actual conversation tests. Instead of listening to individual samples uttered by one talker, the listener follows a conversation between two talkers, and is put in the position of one of the two. Hence, the listener does not talk himself/herself, but is passively listening.

Two versions of the test method can be used, one more realistic than the other. The more realistic procedure (A) involves recording of the conversations with a Head and Torso Simulator[8] (HATS). The listener uses headphones, and speech is presented monaurally, for example, to the preferred listening ear. Masking effects that occur, for example, during phases where both talkers speak (double talk), are implicitly considered by this method: during the recordings of the test stimuli, the artificial ears of the HATS pickup the utterances of the distant talker via the user interface. The utterances by the talker whose position the listening subject has taken are directly picked up by the HATS via the air paths. This way, the acoustic conditions accurately reflect a conversational situation. In the second version (B), no HATS is used, so that the acoustic conditions are much less realistic. However, the effort necessary to perform the complex recordings of (A) is greatly reduced, while relative quality information can still be obtained.

---

[8]A dummy head with artificial ears (including microphones) and an artificial mouth. The head is mounted on a torso to approximate the sound shadowing introduced by a real person.

The lack of realism is the main criticism of third party listening tests: the listener does not hear her/his own voice, and does not involve the same cognitive processes normally activated during real conversations. However, since the timing of speech activity can be controlled well with this method, the procedure allows more diagnostic information to be collected than CTs do. Applications range from the assessment of echo cancelers under single and double talk conditions to the assessment of entire communication links. The type of ratings collected during the test depends on the application. For example, the evaluation of echo cancelers typically involves separate category ratings for disturbances due to echo (during single and double talk), due to clipping or switching (during double talk), and for the quality of background noise transmission (ITU–T Rec. P.831, 1998).

2.1.1.3 Talking and Listening Tests

Talking and listening tests are closer to a real conversational situation than third party listening tests, as the test subjects are listening and talking. Since it involves talking, this test method has extensively been used for echo canceler evaluation (ITU–T Rec. P.831, 1998).

In a test of this type, a user produces utterances corresponding to certain parts of real telephone conversations, such as the greetings exchanged in the beginning of a call. The conversation partner at the other end is simulated by a HATS, which produces utterances to mimic a telephone conversation. In this procedure, the lack of realism is not caused by the artificial passive listening as in the third party listening test, but by the artificial conversation partner and conversation task. Moreover, diagnostic data on the performance of devices like echo cancelers during double talk are difficult to obtain, because the exact timing of the utterances necessary in this case cannot easily be implemented.

2.1.1.4 Conversation Tests

As the term implies, 'speech communication systems' should enable two-way communication between users. This dedicated usage is best reflected when CTs are chosen for speech quality assessment. However, this is only one of the advantages of CTs, which can be summarized as follows (Möller, 2000, pp. 59–60):

- The test situation reflects the actual usage of telecommunication systems (see above).

- A wider range of quality elements can be assessed, including those relevant only in a conversational situation (such as delay or echo).

- The subjects focus on the content, i.e. what is said, instead of judging quality mainly based on the form.

- The longer exposure to each circuit condition reduces the effect of the relative reference observed in listening tests: in LOTs, the frequent change between stimuli may lead to a stronger impact of just preceding stimuli.

- Both the longer duration per condition and the general naturalness of the situation evoke an internal long-term telephone reference rather than a strongly context-dependent one.

- The variation of talkers is larger than in listening tests (where typically source speech material from 4 to 8 talkers is used), so that signal processing devices such as codecs, which may be sensitive to particular voices, can reliably be assessed.

- The longer duration of presentation enables the assessment of strongly time-varying distortions.

Many of these advantages are counterbalanced by the time-consuming and expensive test procedure. In turn, with the choice of appropriate conversation scenarios, the time necessary for each conversation can be reduced, and thus the number of conditions per test session be increased (conversation test scenarios are addressed in more depth below).

Some drawbacks of CTs, however, are related to the method as such, not to the applied test conversation scenario:

- Analytical information cannot readily be obtained, as conditions cannot easily be listened to again, or even listened to in pairs.

- Because of the mixed attention between conversation and rating task, conversation test results are more realistic, but also less critical (see e.g. Gros, 2001; Gros and Chateau, 2002).

- The overall number of test conditions that can be assessed without listener fatigue is smaller than in listening tests.

The general procedure for CTs is described in ITU–T Rec. P.800 (1996). The rating scale most widely used is the 5-point ACR-scale (MOS). For CTs, ITU–T Rec. P.800 (1996) recommends a version of the ACR-scale which dispenses with the number labels shown in Figure 2.1.

Conversation Scenarios
For CTs, it is necessary to involve the conversation partners in an appropriate conversation task using predefined *conversation test scenarios*. Apart from free conversations, different types of conversation scenarios have been described in the literature (e.g. Kitawaki and Itoh, 1991; Möller, 2000; Wijngaarden *et al.*, 2002). These tasks range from interactive games (e.g. TNO test tasks, described in Wijngaarden *et al.*, 2002), jointly solving certain military tasks (e.g. ACE-95 test, described in Wijngaarden *et al.*, 2002), to finding locations on city maps, identifying differences between two versions of pictures (e.g. DERA and Kandinsky tests, described in Wijngaarden *et al.*, 2002, and Möller, 2000, pp. 75–77, respectively), proofreading of texts (e.g. CRC test, described in Wijngaarden *et al.*, 2002), and the rapid exchange of random numbers (e.g. Kitawaki and Itoh, 1991).

The main shortcoming of many of the above test scenarios is the lack of naturalness. Since the conversation situation as such provides the advantage of high naturalness, lack of naturalness due to an unrealistic scenario may be counterproductive to the initial advantage of CTs. Especially in the light of telephone conversations, reading random numbers or playing a game represents very specific types of conversation tasks, which do not reflect an everyday telephone usage. To overcome the lack of naturalness in telephone-oriented tests, the so-called Short Conversation Test (SCT) has been developed by Wiegelmann (1997) and Möller (2000). The corresponding test scenarios represent real-life telephone scenarios like ordering a pizza or reserving a plane ticket. They lead to natural but semistructured, comparable and balanced conversations of a short duration of approximately 2–3 minutes (as opposed to 8–10 minutes necessary for several of the other scenarios mentioned above).

In the framework of this book, the SCT scenarios were complemented by a new type of conversation test scenarios, in order to provoke more interactive conversations between the test subjects. An additional requirement was to maintain the advantage of the SCT-scenarios, namely to lead to *realistic* conversations: most conversation tests described in the literature employ scenarios that are rather unrealistic for actual telephone conversations; as mentioned above, the test tasks range from playing games over the phone to reading random numbers as fast as possible (see Section 2.1.1.4; e.g. Kitawaki and Itoh, 1991; Wijngaarden *et al.*, 2001). In order to cope with the cognitive load related to several of these test scenarios, and also to teach subjects to detect certain degradations, a training of the subjects is sometimes carried out (Wijngaarden *et al.*, 2002). Such training may further reduce the naturalness of the telephone situation.

Attempting to both provide more interactive conversations, and maintain a certain level of realism, interactive Short Conversation Test scenarios (iSCT scenarios) were developed for the research described in this book. The increase in interactivity was approached on the basis of the following factors (Hammer *et al.*, 2004a; ITU–T Delayed Contribution D.221, 2004):

- The scenarios were developed iteratively using informal CTs to evaluate the sensitivity towards delay.

- The conversation discipline was lowered: the subjects knew each other and called each other by their real first names.

- The subjects were instructed to perform the tasks as quickly as possible.

- A combination of natural, 'telephone typical' situations with simple interactive tasks was chosen. The naturalness of the scenarios was verified by interviewing the subjects after the entire CT ('Conversations perceived as natural?, Yes/No'). In case of the iSCTs, a similar percentage of 'Yes' answers as for the SCTs was obtained (app. 78% vs. 83%).

Examples of these tasks consist in the rapid exchange of numerical or lexical data motivated by the scenario, such as room numbers and email-addresses for new employees of a large company (see Appendix F). Both partners have a table of written text items in front of them, with one column common to both subjects used as an identifier to locate the required information. The second column contained the complementary, known items to be communicated to the conversation partner, and the empty third column was to be filled in with the information obtained from the partner. In order to prevent the subjects from too quickly applying a 'walkie-talkie' like strategy (only one subject speaks at a time), one element in the list of each talker was not found in the other one's list, with the position of that item in the list unknown to both interlocutors. This way, a short additional information exchange was provoked[9].

---

[9]For examples of tests carried out with the SCT scenarios, see Chapter 4, and for an example of a test carried out with the iSCT-scenarios, see Section 4.3.

2.1.1.5 Note on Absolute Quality Rating Scales

Several drawbacks of the ACR scale shown in Figure 2.1, the MOS scale, have been pointed out in the literature (Möller, 2000, pp. 68–72, for an overview)[10]:

(a) The scale may be used differently by different subjects. The discrepancies are due to varying interpretations of the scale labels by individual subjects (both comparing results between countries and comparing results within one country).

(b) On average, subjects do not perceive the categories of the scale as equally spaced. The perceived spacing varies between languages (Jones and McManus, 1986).

(c) The sensitivity of the scale is restricted:

   o The outer categories (i.e. 'excellent' and 'poor') may be avoided by subjects, in order to account for eventual better or worse conditions still to come (compression).

   o If one stimulus is rated using one of the outer categories, subsequent more extreme stimuli fall beyond these outer limits. Nevertheless, they cannot be rated differently than by using the same category (saturation).

   o Only five categories are available, and no ratings in between categories are permissible. If the outer categories are avoided, even less categories are actually used.

To explicitly avoid the drawbacks described under (c), other scales have been proposed, which are based on the same categories as the MOS scale. The 11-point scale depicted in Figure 2.3 is a continuous scale with additional numerical labels. The values 0 and 10 typically serve as anchors. Since the verbal scale labels range only from 1–9, it can be expected that saturation effects are reduced with this scale. In Figure 2.3, the English version is compared to the German and the French ones, since the latter two have been applied in the tests described in Section 5.4.2.

Another continuous scale with the labels of the MOS scale was proposed by Bodden and Jekosch (1996), see Figure 2.4. Similar to the 11-point scale, this 7-point scale is intended to avoid saturation effects. In turn, in this original version no numerical annotation is provided to the subjects. The scale has been extensively used as the basis for the tests on the quality of spoken dialog systems carried out by Möller (2005a).

Instead of rating absolute quality, Möller (2000) has proposed to use the so-called CR-10 degradation scale in integral quality tests. The CR-10 scale was developed by Borg (1982) for applications such as perceived exertion or pain judgment (see Appendix I of ITU–T Rec. P.833, 2001). It is a category scale with ratio properties. To achieve the ratio property, the verbal labels are not positioned in an equidistant manner. Instead, they are distributed such that the exponential growth function normally obtained from magnitude estimation tasks can be replicated.

Möller (2000) has shown that – in principle – the CR-10 ratings can linearly be transformed onto the quality scale of the E-model, the network planning model described in Section 2.2.2.2.

---

[10]Note that the criticisms (a) and (b) have been observed for the ACR-scale *without* additional numerical annotation (in contrast to the version depicted in Figure 2.1).

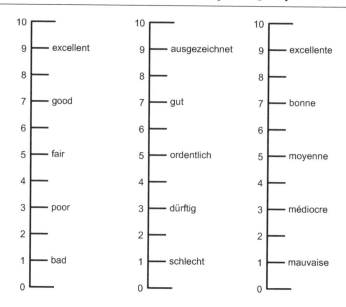

Figure 2.3: Continuous 11-point rating scale (IEC Publ. 268-13, date unknown); English, German and French versions.

Figure 2.4: 7-point quality scale (in German, see Bodden and Jekosch, 1996).

Note that in this book, the MOS scale was employed in spite of the known disadvantages it is associated with. However, one of the main goals of this book was to determine quantitative relations between a parametric description of quality elements on one hand, and speech quality ratings on the other. Since most test results reported in the literature have been collected on the MOS scale, its application at least as an additional scale is necessary to enable across test comparisons. However, in order to backup some of the findings described in this book, the 11-point scale (Figure 2.3) or the CR-10 scale (Borg, 1982; ITU–T Rec. P.833, 2001; Möller, 2000, Appendix 1) was used in addition to the MOS scale.

2.1.2 Analytical Methods

The main goal of analytical methods is to analyze the features that underlie the perception of certain stimuli. The validity of analytical methods is bound to two assumptions, which were formulated by Quackenbush *et al.* (1988, pp.72–73):

- Both the selected listeners and the user group of interest can equally well discriminate a set of features in the test stimuli, independent of their preferences for the features.

- The features perceived by the test subjects are somehow correlated with the quality perceived by the members of the envisaged user group.

As mentioned earlier in this chapter, analytical methods aim at identifying and quantifying perceptual dimensions, their relation to quality, and ideally also their relation to underlying signal properties.

2.1.2.1 Attribute Scaling

The first type of multidimensional analysis method is that of attribute scaling. It employs a set of rating scales on which subjects are to describe what they have heard. Ideally, the set of scales enables all features to be properly captured by the subjects. Owing to the usage of several, descriptive scales instead of one scale to directly assess preference (isometric approach), attribute scaling has been described as 'parametric approach' (Quackenbush *et al.*, 1988, pp. 72–73). The number of dimensions necessary to describe the entire set of features is typically reduced using factor analysis. The reduced set of descriptors enables the identification of the underlying feature-space.

The validity of attribute scaling depends on an important factor[11]: the attributes of the rating scales have to be understandable to the subjects, and the individual scales generally be agreed upon with regard to the measurand they are supposed to measure (Mattila, 2001, p. 48).

Two drawbacks of the usage of verbal attributes can be observed:

(a) The attributes may direct the listeners' attention into a particular direction, potentially degrading the validity of the method.

(b) The differing meaning of the attributes for different subjects may lead to less precision than, for example, achieved in multidimensional scaling (MDS) tests (see Section 3.6).

In turn, the advantage of attribute scaling over nonverbal multidimensional analysis methods such as multidimensional scaling (see Section 2.1.2.3) is that it provides a direct description of the resulting feature-space.

The problem of ambiguous attributes can be reduced when each (continuous) scale is labeled with antonym pairs at its extreme points. This scaling technique is generally referred to as *semantic differential* technique (SD; see Osgood *et al.*, 1957). A way to interpret semantic differential scaling data is to calculate the mean over the subjects for each scale. When the (horizontal) scales are displayed one below the other, the connection of the mean scale values provides a 'polarity profile' for the respective test condition.

MDS and attribute scaling are often used in combination, MDS to accurately determine the feature space, and attribute scaling to find an appropriate description of it.

2.1.2.2 Diagnostic Acceptability Measure (DAM)

A formalized attribute scaling procedure called the *Diagnostic Acceptability Measure (DAM)* has been developed by Voiers (1977). It makes use of an isometric scale for quality ratings, two 'metametric' scales for pleasantness and 'intelligibility' ratings[12], as well as 17 parametric scales for the description of the underlying perceived features (an overview of the DAM is given in Quackenbush *et al.*, 1988, pp. 67–82). The parametric scales distinguish

---

[11] In addition to the assumptions generally applicable to analytical methods as stated above.

[12] If intelligibility is rated and not assessed in a performance test, the obtained results reflect 'listening effort' rather than intelligibility.

between the perception of the speech signal and the perception of the background. The total of 20 scales are the *listener response scales* of the DAM. The ratings obtained on the listener response scales are transformed onto an intermediate set of 10 scales, the analytical *reporting scales* of the DAM. Another transformation of the 20 response scales yields ratings on the four total quality scales of the DAM ('intelligibility', 'pleasantness', 'acceptability' and 'composite acceptability', the latter being the main output of the method).

Another particularity of the DAM is the selection procedure for the highly trained test subjects to be used in the test. They are chosen on the basis of the correlation of their ratings for a large set of training stimuli with the mean ratings obtained in the past (see Quackenbush *et al.*, 1988, for details).

At the time of its introduction, the DAM was evaluated for all available types of digital voice systems. Its validity for assessing speech quality under today's range of degradations has not been investigated to date.

2.1.2.3 Multidimensional Scaling

Another form of multidimensional analysis is MDS. In theory, it makes the least assumptions on the features of the stimuli to be assessed. In its most basic form, it focuses only on the perceptual *differences* between stimuli. As such, MDS intends neither to deliver *descriptions* of the features found to be dissimilar nor to determine *preference* for any of these features (for an overview of MDS see e.g. Kruskal and Wish, 1978; Shepard *et al.*, 1972).

The rationale behind MDS is as follows: if a set of dissimilarities between all objects under consideration is available, MDS determines the associated space, in which the objects can be positioned so that their distances in this space correspond to the initial set of dissimilarities. Dissimilarity data are typically processed in a matrix form, where each entry stands for the dissimilarity between the objects denoted by the line and column indices. MDS techniques can be distinguished according to two criteria:

1. In *classical* MDS, only one dissimilarity matrix is used. Nonclassical procedures rely on multiple dissimilarity matrices. An example is *weighted* MDS, where each subject is represented by an individual dissimilarity matrix. This method is also referred to as *Individual Difference Scaling* (INDSCAL; Carroll and Chang, 1970).

2. *Metric* MDS involves a dissimilarity representation that is linearly related to the distances in the space representation of the output objects. The term 'metric' refers to the Euclidean distance metric underlying the space representation. In *nonmetric* MDS, the dissimilarities of objects are typically expressed using rank orders, which are not linearly related to distances in the object space (i.e. high numbers for very dissimilar objects, and low numbers for very similar ones).

A 'classical' and metric example of MDS is the derivation of a two-dimensional, physical map of objects such as cities, of which only the relative distances are known as input data (here representing a metric dissimilarity matrix). Using MDS, the map can easily be determined. It may differ from reality in two ways: as only one dissimilarity matrix forms its input, a classical MDS solution can (i) be rotated, and (ii) be inverted (i.e. mirrored).

For all types of MDS, the optimal space configuration of the objects is determined by minimizing some type of difference criterion between the input dissimilarity data and the distances obtained from the chosen output space configuration. This is normally done iteratively by trying out different numbers of dimensions, and determining the one with which the difference criterion between input and output data can be minimized best. A measure typically used to express the deviation between input and output configurations is the so-called 'Stress' (Kruskal and Wish, 1978). It employs the normalized sum-of-squared distances between the *metric* object distances and a (in this case nonmetric) representation of the dissimilarities.

In weighted MDS, it is assumed that different subjects weight the dimensions differently in their importance for dissimilarity. Consequently, two different spaces form the output of the INDSCAL procedure: the stimulus or object space, and a subject space. The latter has the same dimensionality as the stimulus space, and each subject is represented in it by a vector. Its orientation and length are measures of the weight different subjects give to different dimensions in their contribution to dissimilarity.

In auditory research, MDS typically involves paired comparison listening tests (e.g. Bappert and Blauert, 1994; Mattila, 2002a; McDermott, 1969). The subjects are asked to assess the difference between the paired stimuli, ideally presented in all possible combinations. The assessment can be carried out using similarity scales, on which subjects rate the degree of similarity (e.g. McDermott, 1969). Alternatively, triadic comparison techniques may be applied (nonmetric). In this case, three stimuli are presented in triads (e.g. Hall, 2001). Subjects simply judge which two of the three stimuli sound most similar, and which two sound most dissimilar. From the sum of scores given to the different objects in each pair (e.g. 2 for most dissimilar and 0 for most similar), rank orders can be determined (see above: non-metric MDS).

2.1.2.4   Preference Mapping

The multidimensional analysis of speech quality is typically carried out in order to understand the relation between fundamental features of the presented stimuli and the integral quality or preference they are associated with (Hall, 2001; Mattila, 2001, 2002a; McDermott, 1969). The identification of the relation between a multidimensional representation of stimuli and their preference is referred to as *preference mapping* (Carroll, 1972)[13].

Different types of dimensions can be distinguished, according to how they are related to integral quality or preference. Carroll describes two such categories (Carroll, 1972, pp. 208–48):

1. Dimensions that show a vector type of behavior with regard to quality (i.e. 'the more of that dimension, the better').

2. Dimensions for which an 'ideal point' exists associated with highest quality.

Examples of dimensions with a vector-type behavior are the 'naturalness' of a speech signal (the more, the better), or the 'noisiness' (here in the sense 'the less, the better')[14].

---

[13]Note that an *external* and an *internal* mode of preference mapping can be distinguished (Carroll, 1972). In external preference mapping, the stimulus space is determined independent of the preference data. Internal preference mapping can be used to relate preference data to a multidimensional representation *obtained* from this preference data, for example, using principal component analysis or MDS. In this section, only external preference mapping is addressed.

[14]Some small amount of noise may be desired, for example, in order to mask other artifacts in the signal (dither) or to assure the listener that the line is still active (comfort noise).

Examples of dimensions with an ideal point are the 'loudness' of the transmitted speech, or the dimensions related to the limitation of the frequency band, described by attributes like 'dark–bright' or 'low–high' (see also Gabrielsson and Sjogren, 1979).

It should be noted that the preferred multidimensional configuration can be considered as an indication for what has been defined as *desired nature* by Jekosch (2005b; see also Sections 1.2 and 3.7). Hence, for dimensions of vector-type preference relation, the desired magnitude of the particular feature is 'as much as possible' of it, and for ideal-point type of dimensions, the desired feature corresponds to the ideal point.

In the case of the vector model, preference mapping can be illustrated as follows: the quality scale of a particular subject is considered as a vector in the multidimensional feature space, pointing to the direction of increasing preference (see Figure 2.5, (A) for an example in two dimensions; Carroll, 1972). In this model, the quality related to a particular point in the feature space is proportional to the projection of the point onto the subject's quality vector. If $\mathbf{x_k} = (x_{k1}, x_{k2})^T$ is the multidimensional representation of a stimulus k, and the preference or quality vector of subject j has an angular orientation α_j, the related quality Q_{jk} is proportional to the projection q:

$$Q_{jk} \sim q = x_{k1} \cos(\alpha_j) + x_{k2} \sin(\alpha_j) = b_1 x_{k1} + b_2 x_{k2} \tag{2.1}$$

Consequently, the dimensions are *linearly* related to the quality a certain subject perceives: the overall quality can be estimated from the linear combination of the different dimensions (Carroll, 1972; Mattila, 2001). The coefficients of the dimensions b_1 and b_2 reflect their importance for quality, and depend on the orientation of the quality vector. Since the orientation of the quality vector may be subject-dependent, the weighting of the dimensions may be subject-dependent as well.

If the dimensions of the feature-space of a given subject j have 'ideal points' (Figure 2.5, (B)), the quality Q_{jk} of stimulus k, associated with a certain point $\mathbf{x_k}$ in the feature space, is related to the squared Euclidean distance between the point $\mathbf{x_k}$ and the subject's ideal point $\mathbf{y_j}$ ('unfolding model', see e.g. Carroll, 1972):

$$Q_{jk} \sim d_{jk}^2 = \left(\mathbf{v_j}\left(\mathbf{y_j} - \mathbf{x_k}\right)\right)^2. \tag{2.2}$$

Here, $\mathbf{v_j} = (v_{j1}, v_{j2})^T$ are the weighting coefficients of subject j for the two dimensions, which are accounted for in the Euclidean distance calculation.

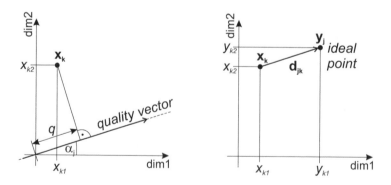

Figure 2.5: Illustration of preference models (Carroll, 1972). A: Vector model. B: Ideal point model.

Now, Carroll's *generalized* theory assumes that each subject considers one position in his multidimensional feature space to be ideal (Carroll, 1972), similar to what is described above. However, in this so-called 'generalized unfolding model' or 'Elliptical point model with rotation', a rotation of each individual's feature space is applied, to optimize the agreement with the preference data obtained from each subject. From an application point of view, this 'generalized' ideal point can be viewed as the subjects' trade-off between the vector-type and ideal-point-type dimensions. The mapping coefficients can be determined using a multivariate, quadratic regression (Carroll, 1972; Mattila, 2001, 2002a, pp. 204–208). The necessary rotation is performed by a matrix transformation of each individual's feature-space.

The model starts from the subjects' individual points of view: the preference scale values Q_{jk} obtained from the jth subject for the kth stimulus are predicted from the squared Euclidean distance between the stimulus representation $\mathbf{x_k}$ and the subject's ideal point $\mathbf{y_j}$, according to Equation (2.3):

$$Q_{jk} = a_j d_{jk}^2 + b_j + e_{jk}. \tag{2.3}$$

Here, a_j and b_j are constants, and the term e_{jk} accounts for errors. d_{jk} is the (weighted) Euclidean distance, as in Equation (2.2).

The rotation is performed with an orthogonal matrix T_j, yielding modified positions of the stimulus representation and the ideal point. Owing to the applied rotation, two-way interactions between the different dimensions, and also linear terms of the dimensions are produced (for more details see Carroll, 1972; Mattila, 2001, pp. 204–206):

$$Q_{jk} \approx \sum_t \sum_{t'} r_{tt'}^j x_{kt} x_{kt'} + \sum_t b_{jt} x_{kt} + c_j. \tag{2.4}$$

Here, t and t' are indices for the dimensions; the factors $r_{tt'}$ represent the weighting coefficients for the interaction between dimensions, including the squared terms; b_{jt} are the weighting factors for linear dimension terms, similar to the vector model.

Obviously, the mapping is subject-dependent, reflecting the fact that different subjects may have different points of optimal quality, and therefore weight different dimensions differently. These considerations also underline the fact that preference (or quality) depends on the respective context, which determines the weighting of the underlying dimensions (see also Section 1.2).

2.2   Instrumental Methods

In order to reduce the necessity for time-consuming and costly perception tests to measure the quality of networks and systems, much effort has been spent on the development of alternative, instrumental methods. It has to be noted, however, that the speech quality of today's telecommunication systems cannot readily be determined solely on the basis of basic signal measures like the signal-to-noise ratio. Although quality may be *correlated* with different instrumentally measurable signal characteristics, it is typically not possible to establish a simple relationship between these instrumentally measurable magnitudes and the quality perceived by a user of the system.

To account for the more complex interdependencies, different types of quality estimation and prediction models have been developed (for an overview see Möller and Raake, 2002). Each model has its proper domain of application, and range of network conditions it has been designed for. Consequently, there is no universal quality model that can be applied in all circumstances. In order to better determine what model to chose for a given application, the models can be categorized according to different criteria (Möller and Raake, 2002):

- The application aimed at (network planning, codec optimization, network operation, etc.).

- The predicted quality features (integral quality, intelligibility, conversational aspects, etc.).

- The network components and configuration under consideration (entire connection mouth-to-ear, codecs, etc.).

- The respective model input parameters (measured parameters like noise levels, entire speech signals, etc.).

- The extent to which psychoacoustic knowledge and empirical data have been incorporated.

On the basis of the application domain, three main types of models can be distinguished:

1. Signal-based measures.

2. Network planning models.

3. Monitoring models.

2.2.1 Signal-based Models

Signal-based measures were initially developed to assess the quality of single network components like codecs. Currently available models can also be applied to larger network segments, and include degradations such as background noise transmission, or transmission errors such as packet loss (e.g. PESQ; ITU–T Rec. P.862, 2001). If such models are employed within the range of applications they have been designed for, they deliver valid and reliable quality estimates. This is achieved with a – compared to other model types – relatively accurate modeling of the signal processing performed by the human hearing system. Since the output of signal-based models such as PESQ is a unidimensional quality index, and the input is the speech signal, diagnostic information on the sources of low quality cannot directly be obtained. Most signal-based models can therefore be considered as black-box models. Model extensions enabling more diagnostic information to be obtained have recently been proposed and are now being developed further in ITU–T Study Group 12 (ITU–T Contribution COM 12–04, 2004).

'Classical' variants of signal-based measures simulate the paired comparison situation described in Section 2.1.1.2 (although they aim at predicting ACR test results). They base their quality estimates on the comparison of a clean source signal as reference with the source signal processed by the system under test. After loudness equalization and

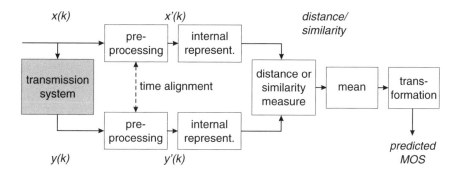

Figure 2.6: Principle of signal-based measures, according to Hauenstein (1997), Type I.

time-alignment, the two signals are transformed into an internal, psychoacoustics-based representation. In the following steps, difference or similarity vectors are determined from the two representations, which are time-averaged and mapped onto a final quality estimate (Figure 2.6).

Since this type of signal-based measures operates on the differences between unprocessed and processed samples, they can be viewed as impairment or degradation models. While the internal representation typically involves knowledge on auditory processing, the final mapping function is determined purely empirically, on the basis of the results of a large number of listening tests.

Examples of signal-based measures employing a reference signal are:

- PESQ (Perceptual Evaluation of Speech Quality; ITU–T Rec. P.862, 2001),

- PSQM (Perceptual Speech Quality Measure; ITU–T Rec. P.861, 1996),

- TOSQA (Telekom Objective Speech Quality Measure; Berger, 1998), and

- the models proposed by Hauenstein (1997) and Hansen and Kollmeier (2000).

The model currently recommended by the ITU–T is PESQ (ITU–T Rec. P.862, 2001), which is an optimized combination of the two algorithms PSQM (Beerends *et al.*, 2002; ITU–T Rec. P.861, 1996, PESQ's psychoacoustic model) and Perceptual Analysis Measurement System (PAMS) (Rix and Hollier, 2000; Rix *et al.*, 2002, PESQ's time-alignment). A wideband version of PESQ has recently been defined in ITU–T Rec. P.862.2 (2005). See Section 3.9 for more details on applying such signal-based comparative measures to VoIP.

The model type illustrated in Figure 2.6 assumes electric interfaces to the transmission system under test. Alternative methods that also include the electro-acoustic user interfaces, and can consequently be applied to the entire transmission chain mouth-to-ear, are currently under discussion in standardization bodies like the ITU–T.

Another type of signal-based measure operates without a reference signal. Here, a reference signal is generated from the transmitted speech signal using linear predictive coding (LPC) techniques. The artificial reference is used in the same manner as a real reference signal (see Figure 2.7). Consequently, these measures simulate an auditory ACR test, where a reference internal to the user is evoked (Section 3.7). Several 'problems' are

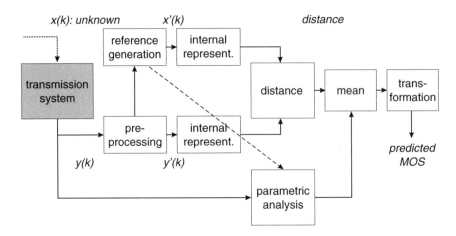

Figure 2.7: Principle of single-ended signal-based measure (according to ITU–T Rec. P.853, 2004).

known to be related to this approach, for example in case of high additional noises, time-varying characteristics, unnatural voices or a codec working similarly to the LPC technique used for reference generation. These problems are addressed separately by these *single-ended measures* in a parametric analysis, and are considered in the final averaging and transformation processes (ITU–T Rec. P.563, 2004, see Figure 2.7).

An alternative approach to estimate speech quality from the signal consists in linking signal characteristics to perceived quality *dimensions*. If the relation between the quality dimensions and integral quality is known, for example from a corresponding preference mapping model determined beforehand from analytical tests, speech quality can be derived from the signal in an analytical way (Halka, 1993; Quackenbush *et al.*, 1988). In an ongoing research activity, this approach is currently investigated for its application to networks involving VoIP (Heute *et al.*, 2005). Here, the aim is to develop different instrumental measures each sensitive only to one individual dimension, which can then be extracted quantitatively from the signal. By combining the output of each individual measure using the previously established preference relation, speech quality is estimated. There is an additional advantage of such dimension-based models over less analytical ones: assuming that such a model is to be extended to new network elements or types of degradation, the effort of model development is considerably reduced. With such a method, additional training can be avoided, if new types of network degradations or elements are introduced. A brief description of the underlying multidimensional analysis can be found in Section 2.1.2.

2.2.2 Parameter-based Models

For network planning, it is desirable to have a quality estimate at hand before the network has been set up. Such quality predictions can significantly support the selection of certain components or configurations. Different models have been developed for this purpose. They estimate quality from instrumentally measurable characteristics of the system. Since parametric models allow diagnostic information on the system impact on quality to be obtained, they can be considered as glass-box models.

2.2.2.1 Submod

The SUBMOD model, developed by British Telecom, was first described by Richards (1974). In a later, enhanced version it was renamed CATNAP (ITU–T Suppl. 3 to P–Series Rec., 1993). For parameters measured using in-service, nonintrusive measurement devices, a specific model version has been defined, which is called *Call Clarity Index* (CCI; ITU–T Rec. P.562, 2000; monitoring: See Section 2.2.3). The SUBMOD model is based on spectral input parameters and an algorithm similar to that used in telephony for calculating loudness (see Section 3.3.10). The model enhancements to include conversational effects such as talker echo have not been published, and enhancements to include the effect due to time-varying distortions are based on the respective enhancements of the E-model described later in this book. Consequently, the SUBMOD model will not be addressed further in this book. The interested reader is referred to Möller (2000) for more detailed information on the model and the underlying loudness rating principle.

2.2.2.2 E-model

The model currently recommended by the ITU–T for network planning is the so-called E-model (ITU–T Rec. G.107, 2005). It is mainly empirical by nature, and was developed on the basis of large amounts of auditory test data (both LOTs and CTs). The E-model takes the impairments due to talker echo and transmission delay into consideration and can hence be applied to predicting quality in a conversational situation. It was compiled as a combination of different quality prediction models in the framework of the European Telecommunications Standards Institute (ETSI Technical Report ETR 250, 1996; Johannesson, 1997). The E-model is widely used to estimate quality during the planning of networks, and also during their operation (monitoring, see below). The model has been subject to extensive validation, and has been found to deliver realistic quality estimates for individual and within limits also for combined degradations (Möller, 2000; Möller and Raake, 2002).

Input Parameters
The E-model is based on a parametric description of telephone networks, see Figure 2.8 (for details see ETSI Technical Report ETR 250, 1996, and ITU–T Rec. G.107, 2005). Note that some of the parameters like loudness ratings are measured relative to a virtual reference point in the middle of the connection, the 0 dBr point.

The parameters can be related to the quality elements and quality features discussed in Chapter 3. In Section 3.3, additional parameters will be presented, parts of which are currently discussed in bodies such as the ITU–T as candidates to parametrically capture the effect of certain more recent quality elements.

SLR, RLR, OLR [dB]: Loudness ratings for the sending and receiving parts of the transmission path, and overall loudness rating, respectively. Loudness ratings express the frequency-weighted attenuation of the transmitted speech signal in comparison to a reference channel (ITU–T Rec. P.79, 1999; for more details on the loudness rating principle typically employed in telephony see also Section 3.3.10). Send loudness rating (SLR): attenuation between the speaking user's mouth and an electric interface in the network. Receive loudness rating (RLR): attenuation between an

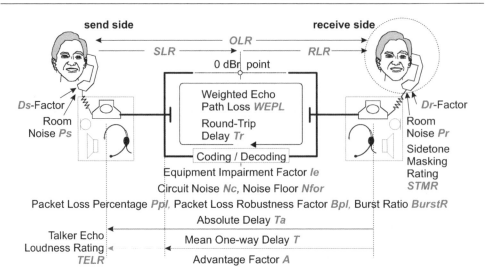

Figure 2.8: Summary of agreed-upon instrumentally measurable parameters acting as quality elements in telephony (modified from ITU–T Rec. G.107, 2005, for more details on the parameters see also ETSI Technical Report ETR 250, 1996).

electric interface in the network and the listening user's ear. Overall loudness rating: $OLR = SLR + RLR$.

Ta [ms]: Mean absolute delay of the transmission path (ITU–T Rec. G.114, 2000).

T [ms], $TELR$ [dB]: The mean one-way delay T is the delay responsible for *talker* echo in case of signal reflections at the far-end. The talker echo loudness rating (TELR) describes the weighted attenuation the echo signal undergoes on its way from the talker's mouth to the talker's ear. It is calculated as $TELR = SLR + RLR + EL$, where EL is the so-called echo loss, assuming a purely electric echo due to reflection on the line[15]. EL can be calculated as the weighted average of the frequency-dependent echo return loss, according to a procedure described in ITU–T Rec. G.122 (1993). In case of acoustic echoes due to coupling of the loudspeaker signal into the microphone of a user interface, EL is replaced by the weighted terminal coupling loss $TCLw$. It is calculated in a manner similar to EL, however, from the return loss measured on the electric connectors of the user interface.

Tr [ms], $WEPL$ [dB]: The round-trip delay Tr causes *listener* echo that is attenuated by the Weighted Echo Path Loss ($WEPL$). $WEPL$ is a parametric descriptor of the weighted average of the frequency-dependent loss the listener echo undergoes (ITU–T Rec. G.126, 1993).

Nc [dBm0p], $Nfor$ [dBmp]: Circuit noise (Nc) and noise floor ($Nfor$), that is subscriber line noise, stand for the narrow- and wideband Gaussian noise sources present on the

---

[15]Then, the reflected signal is attenuated by the employed user interface at send side (SLR), further attenuated by the actual reflection (EL), and once again attenuated on the receive path in the user interface at send-side (RLR).

transmission line. Nc is measured as the weighted absolute noise level relative to the 0 dBr point of the network (psophometrically weighted, see ITU–T Rec. O.41, 1994), and expresses the noise contributed by the different circuits the line is composed of (in tandem operation). $Nfor$ is measured as the psophometrically weighted absolute noise level at the receive side, and is the noise caused by the subscriber line.

Ps, Pr [dBA]: A-weighted room noise levels at send- and receive-side, respectively.

Dr, Ds: Sensitivity of the user interfaces at send and receive side towards the direct speech signal, versus their sensitivity towards ambient noises (for details see e.g. ETSI Technical Report ETR 250, 1996).

$STMR$ [dB]: The Sidetone Masking Rating ($STMR$) quantifies the attenuation of the talker's own voice signal coupled back to his handset listening ear. The coupling of ambient noise across this path, the Listener Sidetone Rating $LSTR$, can be determined as $LSTR = STMR + Dr$.

With the approach developed in this book for the parametric modeling of the impairment due to packet loss, three new parameters have been added to the model (ITU–T Delayed Contribution D.027, 2005; ITU–T Delayed Contribution D.044, 2001; ITU–T Rec. G.107, 2005):

Ppl [%]: (Average) Packet Loss Percentage.

Bpl [%]: Packet Loss Robustness Factor – robustness of the applied codec and packet loss concealment (PLC) algorithm against packet loss.

$BurstR$ Burst Ratio (see Section 3.3.5.1; ITU–T Delayed Contribution, D.020, 2001) – parameter expressing the relation of the average number of packets lost in a row for a certain dependent, i.e. bursty packet loss, over the average number of consecutively lost packets in case of random loss of the same overall loss rate. The parameter is explained further in Section 3.3.5.1.

We address the new parameters Ppl, Bpl and $BurstR$ as well as the related impairment model in more depth in Section 4.1.1.

Impairment Factor Principle

The underlying assumption of the E-model is that different individual degradations can be transformed onto a particular scale and are – on this scale – additive:

> 'The models basic principle is the fact that evaluation of psychological factors (not physical factors) on a psychological scale is additive' (ITU–T Suppl. 3 to P–Series Rec., 1993, p.36).

The idea of impairment additivity was not invented with the E-model or the models it is compiled from, but observed in perception tests performed by Allnat and co-workers on the degradation of video-pictures (Allnatt, 1975; Lewis and Allnatt, 1965). They showed that additivity of the studied impairments held, if an appropriate, perceptual scale was used. In their case, independent impairments were additive when expressed as a perceptual magnitude on a ratio scale.

In the E-model, the Transmission Rating Scale or R-scale is used to express quality. The scalar input parameters listed in the previous section are grouped into different classes of impairments, and are transformed onto the R-scale as additive *Impairment Factors*:

$$R = Ro - Is - Id - Ie, eff + A \qquad (2.5)$$

The R-scale for the Transmission Rating Factor R ranges from 0 to 100 (100 corresponding to optimum and 0 to worst quality). The classes of degradations are represented as follows (the complete set of formulae is presented in Appendix E):

- *Ro* describes the speech quality due to the basic signal-to-noise ratio, as it results from the speech level and the levels of the different noise sources on the line, be it line noise or room noise at send or receive side.

- The Simultaneous Impairment Factor Is accounts for the effect of degradations simultaneous to the transmitted speech signal, such as signal-correlated noise or excessive speech levels.

- The Delayed Impairment Factor Id stands for the impairments that are delayed with respect to the transmitted speech signal, such as transmission delay or echo.

- The Effective Equipment Impairment Factor Ie, eff comprises the Equipment Impairment Factor Ie, which quantifies the speech impairment due to low-bit-rate coding. To include the effect of packet loss, the loss-free Ie has recently been extended to the loss-dependent Ie, eff, on the basis of the model presented in this book, Section 4.1.1 (ITU–T Delayed Contribution D.027, 2005; ITU–T Delayed Contribution D.044, 2001).

- The Advantage Factor A quantifies the advantage of access related to a particular system: for example, a user of a mobile service may be more tolerant towards a degraded channel than a user of a wire-line network, owing to the advantage of mobility (see Section 3.7).

For its default settings (ITU–T Rec. G.107, 2005, Table 2), corresponding to a clean ISDN-channel operated with handset telephones, the current version of the E-model predicts a transmission rating of $R = 93.2$.

As mentioned earlier in this chapter, in most auditory speech quality tests performed in telecommunications, the 5-point ACR scale is applied ('MOS scale', see Figure 2.1). The transmission rating factor R predicted by the E-model can be converted to a MOS-rating using Equation (E.33) of Appendix E (ITU–T Rec. G.107, 2005, Annex B). The relation between R and MOS is depicted in Figure 2.9.

2.2.3 Monitoring Models

In order to control the performance of an existing network, and thus to identify and solve possible problems before a large number of users are dissatisfied, the so-called monitoring models can be used. Monitoring can be carried out offline or online:

Intrusive measurement (offline): specific test calls are set up and measurement signals such as noise or speech are transmitted across the network. From the comparison of the output and input signals, a direct quality estimate or quality-relevant network parameters can be obtained.

Nonintrusive Measurement (online): at a specific point of the network, a measurement signal is acquired during normal network operation. From this signal, network or conversation parameters relevant to quality can be derived.

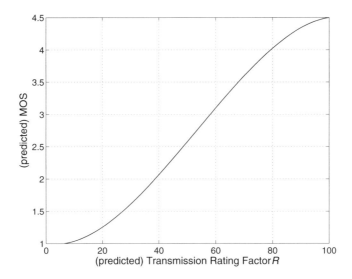

Figure 2.9: Relation between MOS and R-scale, according to ITU–T Rec. G.107 (2005). Modified from ITU–T Rec. G.107 2005, for more details on the parameters see also ETSI Technical Report ETR 250, 1996.

Monitoring devices based on nonintrusive measurements are referred to as *INMDs* (*In-service Non-Intrusive Measurement Devices*, ITU–T Rec. P.562, 2000). A more detailed description, for example of the accessible parameters in both measurement cases, can be found in Möller and Raake (2002). For both intrusive and nonintrusive measurements, modified versions of the parametric network planning models were developed. For example, the Call Clarity Index is the INMDs version of the SUBMOD model, and similarly the nonintrusive E-model is the monitoring version of the E-model (ITU–T Rec. G.107, 2005; ITU–T Suppl. 3 to P–Series Rec., 1993). A comprehensive overview of measurement techniques related to INMDs can be found in Ludwig (2003).

Monitoring approaches that derive network parameters can also be complemented by signal-based measures that are employed to estimate certain properties of the transmission, such as the equipment impairment factor introduced by the codec. Alternatively, the signal-based, single-ended measure recommended as ITU–T Rec. P.563 (2004) can be employed to yield a quality estimate based on a signal measured in the network. Some of the different options to carry out quality monitoring are illustrated in Figure 2.10. As shown in the figure, intrusive and nonintrusive speech quality monitoring techniques allow different parts of the transmission chain mouth-to-ear to be assessed. For example, with nonintrusive measurements some aspects related to the user interface employed at the send side, or the two conversation partners (speech activity, double talk, etc.) can be captured. In turn, the availability of an input (reference) signal makes intrusive measurements often more precise in terms of the determined characteristics of the channel. Figure 2.10 indicates the variations of the parameter-based models that can be employed in the two cases, and the signal-based approaches that can help to provide additional information. The figure also lists some of the parameters that can be captured in intrusive and nonintrusive measurements.

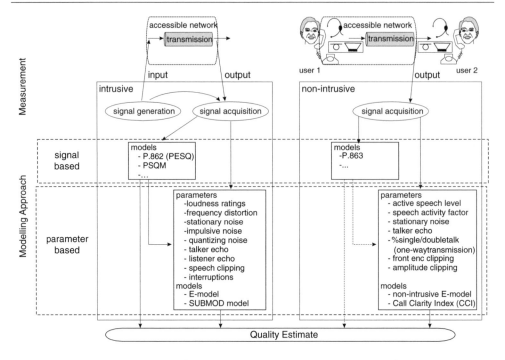

Figure 2.10: Overview of monitoring approaches typically employed in telecommunications (see text for details).

Table 2.1: Classification of Monitoring Approaches.

		Quality model	
		Glass-box (Parameter-based)	Black-box (Signal-based)
Measurement	Intrusive	e.g.: G.107 (E-model), SUBMOD	e.g.: P.862 (PESQ), TOSQA
	Non-intrusive	e.g.: Non-intr. E-model, CCI (P.562)	e.g.: P.563

In the light of the concept of glass-box and black-box measurement approaches mentioned in the beginning of this chapter, the different monitoring models can be classified as described in Table 2.1.

Since several years, efforts have been made to monitor speech quality with an INMD starting from trace data collected in VoIP networks. Similarly to the signal-based approaches discussed above, the information read from IP-headers can be used as input to parametric quality prediction models such as the E-model (e.g. Clark, 2001). Examples of such approaches are discussed further in Section 3.9.2 and Chapter 4.

2.3 Speech Quality Measurement Methods: Summary

In this chapter, both different instrumental and auditory methods for measuring speech quality were summarized. Auditory speech quality tests are the only means of making the quality perception internal to users accessible from the outside. As a matter of their importance for speech quality measurement, but also of the relevance for the research described in this book, several different assessment methods were described. They were categorized with regard to whether they aim at a utilitarian quality measure, such as a unidimensional quality index, or whether they aim at an analytical and/or diagnostic understanding of the observed speech quality.

The most important scales for speech quality tests have been summarized in this chapter, spanning from the traditional 5-point ACR scale, the so-called MOS scale, to continuous versions of this scale to particular degradation scales borrowed from pain and exertion assessment. In order to link analytical measurement results to the results obtained with utilitarian methods, means of preference mapping have been described as well. A quality measurement approach combining the analytical derivation of quality features underlying quality perception and the measurement of integral quality provides the best insight into the quality formation process: quality is considered as the result of a comparison of perceived features to desired features.

The chapter describes a number of dedicated test methods that reflect the requirements of modern telephone networks: the features of VoIP speech may be time-varying by nature, and may even lead to a time-varying quality perception. Thus, a new approach for instantaneous quality testing has been developed in the recent past. This method is complemented by specific test methods like the so-called third party listening test, which aims at the controlled speech quality assessment of signal processing devices such as echo cancelers, in order to detect the degradations such devices may introduce in phases of double talk. Another dedicated test method was developed for the evaluation of noise-suppression algorithms: noise suppression not only tries to reduce background noise, but may also impact the transmitted speech signal, which is considered by specific judgements on speech and background noise transmission. The actual usage of a telephone network is best reflected by CTs. In this chapter, different CT scenarios were discussed, which lead to different degrees of conversation interactivity and of the realism of the conversation. In this context, a new type of CT scenarios has been introduced, the interactive SCT scenarios, which are intended to yield realistic but interactive conversations.

Since auditory tests are time-consuming and costly, and furthermore require an existing system to be available, they are increasingly replaced by instrumental quality models. Such models deliver a quality estimate from an instrumental measurement on the system or a system component. On the basis of a categorization of these models as signal-based, network planning and monitoring models, the most relevant approaches have been described. Emphasis is laid on the so-called E-model, which is the network planning model currently recommended by the ITU–T (ITU–T Rec. G.107, 2005). This model is of particular relevance for VoIP, since it (a) is parameter-based and thus allows diagnostic information on causes of low quality to be obtained, (b) is widespread for the planning of telephone networks and (c) is increasingly employed as the basis for approaches of quality monitoring in operating VoIP networks.

All quality measurement approaches – regardless of whether they are auditory or instrumental – can be classified according to the transparency of the system under test: if only input and output of the system are known, measurements are typically referred to as *black-box approaches*. If the internal system architecture and its relation to the obtained quality measure are entirely transparent, the measurement can be referred to as a *glass-box approach*. Finally, this dichotomy and another one classifying an instrumental measurement as being either intrusive (i.e. off line) or nonintrusive (i.e. online, that is during network operation) were used for a categorization of existing quality monitoring approaches.

3

Quality Elements and Quality Features of VoIP

The present chapter gives an overview of the quality elements of Voice over Internet Protocol (VoIP) networks. In addition, the chapter provides a summary of the existing knowledge on how these elements impact speech quality and the underlying quality features. The elements are not the only factors relevant for quality (see Section 1.2). Hence, modifying factors such as the context of using a particular network, and the corresponding expectations of the user are discussed as well, based on findings reported in the literature. Starting from these considerations, the final section develops the choices made regarding the original research presented in the remainder of this book (Chapters 4 to 5, and Appendix B).

A comprehensive overview of the quality elements and related features typical of traditional wireline, circuit-switched telephony is provided in Richards (1973). The analysis of the factors determining speech quality in telecommunications is further elaborated in the overview and studies by Möller (2000), where additional, more recent developments such as mobile telephony are taken into consideration. Hence, the following discussion focuses primarily on VoIP-related quality elements and features, providing complementary information to that found in Richards (1973) and Möller (2000). As is explained further in this chapter, many of the quality elements and related features known from the PSTN (Public Switched Telephone Network) retain their importance also in the context of VoIP networks. Consequently, these quality elements and the related quality features are included in the discussions of this chapter, however, with less detail than is provided for typical VoIP quality elements such as packet loss.

3.1 Speech Transmission Using Internet Protocol

Today's speech transmission networks can coarsely be divided into circuit-switched and packet-switched networks. In circuit-switched networks like the traditional PSTN or the

Speech Quality of VoIP: Assessment and Prediction   Alexander Raake
© 2006 John Wiley & Sons, Ltd

more recent digital ISDN (Integrated Services Digital Network), a physical connection is established and maintained for the entire duration of a call. The same holds for mobile communication systems of the first and second generation (like the GSM, the Global System for Mobile communications[1]). In all of these systems, speech data is transmitted synchronously across the link, i.e. using a clock common to both sender and receiver. Consequently, in wireline circuit-switched systems, most degradations are stationary by nature; examples of exceptions are time-varying ambient noise, fading radio channels, or degradations due to the adaptation behavior of components like echo cancellers (ECs).

In contrast, no permanent physical connection is established in packet-based networks. For VoIP, however, the communicating devices at the end-points build up a connection using corresponding protocols (such as H.323 or SIP: IETF RFC 3261, 2002; ITU–T Rec. H.323, 1998). The prior buildup of the link basically settles the agreement between the two end-points that speech data will be exchanged between them. Only at this stage of the communication setup, the connection-oriented Transmission Control Protocol (TCP) is typically applied. After the connection is established, coded speech data are packetized into packets that are sent from source to destination. At different levels of packetization, header information is added to the speech data payload, successively increasing the packet size. The subsequent packetization steps and the protocols involved are illustrated in Table 3.1. The resulting packet structure and header information are shown in simplified form in Figure 3.1.

Table 3.1: Protocols and media access technologies involved in VoIP packetization. OSI model: Open Systems Interconnection reference Model (ITU–T Rec. X.200, 1994); RTP: Real-time Transport Protocol (IETF RFC 3550, 2003); UDP: User Datagram Protocol (IETF RFC 768, 1980); IP: Internet Protocol; WLAN: Wireless Local Area Network (IEEE Std 802.11, 2005).

Protocol/ Technology	Main Header Information and Task	OSI Model Layer
RTP	Timestamp: For example, interpacket delay can be calculated as the difference between the respective timestamps. Sequence number: Indicates whether packets are missing or arrived out of order. Payload type: Signalizes the type of payload, for example, speech coded according to G.729A (see text).	'[...] Transport protocol implemented in the application layer' (Tanenbaum, 2003)
UDP	Destination and source ports: Indication of the sending and receiving applications.	Transport layer
IP	Source and destination addresses: Packet routing.	Network layer
Ethernet, WLAN, and so on	Media access.	Sublayer of data link layer

---

[1]The difference lies only in the physical access to the network, that is, the technology applied for the bottom, physical layer of the Open Systems Interconnection (OSI) reference model (e.g. described in ITU–T Rec. X.200, 1994)

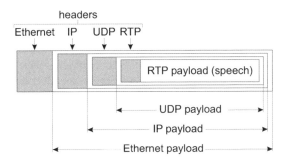

Figure 3.1: VoIP: Speech payload nested in a packet, with the headers added by different protocols (example for VoIP in an Ethernet-based Local Area Network (LAN); Tanenbaum, 2003, Figure 6-25 (b)). Note that the headers can be compressed to reduce the amount of data to be sent across the network (e.g. IETF RFC 3095, 2001).

Note that while the TCP used for connection setup is connection-oriented, requiring acknowledgments between endpoints, the User Datagram Protocol (UDP) typically used for the transport of the speech data is connectionless, and hence yields fewer and smaller packets (see Table 3.2).

On sending, the speech packets 'search' their way through the network, where they are routed from one node to the next based on the destination address they carry. Consequently, subsequent packets may take different paths on their way to the destination. In case of congestion at some point of the network, they may arrive out of order or simply with considerable, and/or varying delay (delay jitter). An efficient speech communication cannot be carried out, if the transmission delay becomes too large (see Section 1.1.2, and Chapter 4). Hence, packets arriving too late for timely playback may be discarded by the receiver (packet loss). Similarly, if a router in the network is faced with too many packets during a traffic-burst period, it may have to drop packets.

From these general considerations it becomes apparent that there are fundamental differences between wireline circuit-switched and packet-switched networks, which yield – from a user's perspective – different perceptual features. Some of these features are similar to those linked to time-varying distortions encountered in circuit-switched mobile communication channels (see Section 3.4). However, the latter are due to the intensity variations of the radio signal caused by the users' mobility, and hence are related to properties of the physical link, which are translated into time-varying distortions by the applied channel- and source (de-)coders. In wireline packet-based networks, time-varying distortions are due to the main characteristic of the applied network protocol and how it deals with the network

Table 3.2: Header sizes of different protocols involved in VoIP.

Protocol	RTP	UDP	TCP	IP
Packet header (minimum, [Bytes])	12	8	20	20

characteristics and traffic (corresponding to a higher level than the physical layer of the OSI reference model, see ITU–T Rec. X.200, 1994).

Note that the wireless access networks increasingly evolve towards IP. In recent years, several different wireless or cellular radio access technologies have been developed. One branch of these technologies consists of the 2nd generation (2G) circuit-switched cellular telephone networks such as GSM or D-AMPS (Digital Advanced Mobile Phone System). After intermediate steps such as the packet-based (but over circuit-switched GSM operated) GPRS (General Packet Radio Services), the 2G networks have been complemented by 3G (3rd generation) technologies that enforce the direct connection to packet-switched data-networks. An example of such technologies is UMTS (Universal Mobile Telecommunications Service; administered by 3GPP, the 3rd Generation Partnership Project), which is based on W-CDMA (Wideband Code Domain Multiple Access). Other examples are mobile access services realized using CDMA2000 (the successor of the American IS-95 (TIA/EIA-95-B, 1999), and a similar technology to W-CDMA, which both comply with ITU–R Rec. M.1457-5 (2006)). Another packet-based branch of wireless access technologies have evolved from the computer networking side. For example, in-house and other short-range applications may employ Wireless Local Area Network (WLAN) (e.g. IEEE Std 802.11, 2005). If not recovered, bit errors due to, for example, radio-signal variations may cause packet or codec frame loss as well as corrupted frames (e.g. Hammer *et al.*, 2004c, 2003), which are perceptually very similar to the time-varying packet loss distortion that may occur in wireline VoIP. The mechanisms involved between the bit-representation on the radio link and the IP packet world largely depend on the radio access technology and the subsequent protocols (see Section 3.3.6 for more details).

3.1.1 VoIP Applications

There are different applications of VoIP, such as:

- Companies apply VoIP in their intranet. Because VoIP has the advantage, among others, of reusing the existing network infrastructure for speech transmission (network convergence), additional costs related to installing and maintaining a parallel telephone infrastructure can be saved. This advantage of VoIP is supported by the relatively simple configuration of the network, for example, allowing new users to be added easily.

- On a larger scale, VoIP is applied by public service providers. Exploiting the advantage of network convergence, different types of networks can easily be combined using appropriate network gateways. Hence, telephone network providers may apply a mixture of VoIP and PSTN in their core network (backbone).

- Users connected to the internet, for example, by digital subscriber lines use software tools for cheap and easy speech communication with relatives, friends or other users who are currently online.

Obviously, several different network configurations involving VoIP are conceivable, as shown by the examples depicted in Figure 3.2. The users of such interconnected networks may be faced with different types of degradations that are characteristic of the technologies applied for individual network segments. Moreover, the integral quality of a speech link may result from the combined effects of stationary and time-varying degradations occurring simultaneously. Owing to the interconnection, the users may not necessarily know, what

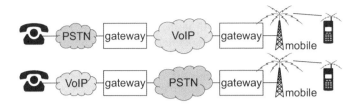

Figure 3.2: Examples of network configurations with either a PSTN or VoIP core network. The connection between the user interface and the adjacent network can be a wireline or wireless one (e.g. a wireless LAN when connected to the VoIP network, or a GSM network in case of connecting to the PSTN). The term 'gateway' is here used in a generic sense for a device connecting one network type to another.

type of networks a particular call is routed across. Consequently, they may be using a traditional wireline handset telephone connected to a PSTN end-office, although the network provider applies VoIP in his core network. As a result, the quality features the users perceive may differ considerably from the expected or desired ones.

3.1.2   VoIP Quality of Service

VoIP requires a certain level of quality of service (QoS) in order to yield acceptable performance. A large number of methods have been proposed to implement QoS in packet-based networks. These methods may be employed to realize different service levels offered to the users (Service Level Agreements, SLAs). A straightforward approach consists in some kind of packet prioritization. Here, two prominent examples are flow-based approaches such as RSVP (Resource ReSerVation Protocol, IETF RFC 2205, 1997), and class-based approaches such as differentiated services (DifServ, DS; IETF RFC 2474, 1998; IETF RFC 2475, 1998). For example, DifServ uses the type of service field of the IP packet header to differentiate four priority classes. DifServ-capable routers along the way of a packet handle it based on the type of service entry, and, e.g., let it gain priority over lower priority packets. Here, each class is provided with its own resources. An application of DifServ is assured forwarding, where 6 of the 8 bits of the IP header type of service field are used to differentiate four classes, and per class three discard probabilities for cases of congestion (IETF RFC 2597, 1999).

Another approach is MPLS (Multi Protocol Label Switching IETF RFC 3031, 2001), where an additional packet header enables a table-based look-up of the forwarding rules for a given packet. The combination of different table-entries constitutes a predefined path similar to the virtual circuits known from, for example, ATM (Asynchronous Transfer Mode). It makes this approach a connection-oriented one, somehow related to circuit-switched connections.

More VoIP-centric methods have been defined to adapt the network to QoS requirements (e.g. Bolot and Vega-García, 1996). A method often used aims at adapting the amount of data sent across the network to the current network traffic. Here, speech codecs such as, the adaptive multirate can be employed (AMR, ITU–T Rec. G.722.2, 2002). Other, complementary methods try to optimize the jitter buffer behavior in terms of packet loss (i.e. packet discard) and residual delay (e.g. Liang *et al.*, 2003; Ramjee *et al.*, 1994).

All of the examples mentioned earlier attempt to optimize QoS parameters of the network (i.e. packet loss, delay and jitter). In turn, additional measures attempt to make VoIP less affected by these parameters. Corresponding techniques consist in sending additional, redundant data (IETF RFC 2198, 1997; IETF RFC 2733, 1999), or in reducing the perceptual impact of losses during the decoding process (i.e. Packet Loss Concealment [PLC]). These measures will be addressed in more detail in Section 3.3.5.

A totally different method is the shaping of sender or gateway behaviors directly based on estimates of perceived speech quality. To this aim, a number of studies employ the E-model or E-model-based approaches (e.g. Jiang and Schulzrinne, 2002; Matta *et al.*, 2003; Narbutt and Davis, 2005; Sun and Ifeachor, 2004). For example, starting from parameters such as delay and packet loss, a particular jitter-buffer behavior may be derived that optimizes a speech quality estimate.

In spite of these and many more measures taken to enable VoIP over networks of high QoS, it will still take years before these measures can be fully employed across different network-trunks (e.g. across different international Internet Service Providers; ISPs). In the meantime, the speech quality of VoIP will still depend considerably on the characteristics of the individual underlying networks.

3.2 Overview of Quality Elements

In Figure 3.3, the network components applied in a user interface and corresponding network gateways are depicted. In the example, the user interface is connected via a network gateway to a VoIP network. At send side, background noise n and acoustic talker echo $\tilde{x}$

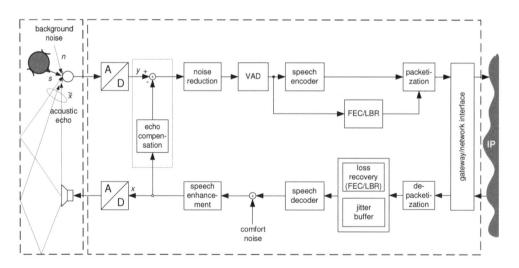

Figure 3.3: Schematic representation of the speech processing steps involved in a VoIP user interface and subsequent transmission (e.g. IP-phone with hands-free terminal functionality, or hands-free terminal connected to a VoIP-gateway). Modified from Vary *et al.* (1998), Figures 1.1 and 13.1.

are shown as possible quality elements, signals that are added to the speech signal. Following the analog to digital conversion (A/D, in narrowband (NB) telephony typically at a sampling-rate of 8 kHz), the microphone signal $y(k) = s(k) + n(k) + \tilde{x}(k)$ may undergo several processing steps before the actual packetization takes place. After cancellation of the potential talker echo $\tilde{x}$, the background noise $n(k)$ on the speech signal is reduced. Subsequently, it is detected whether the current segment (i.e. frame[2]) of the signal contains speech (Voice Activity Detection, VAD). In case no speech is present, a discontinuous transmission algorithm (DTX) decides that the parameters of an estimated background signal are transmitted instead of the actual signal. The smaller amounts of data that have to be send in this case help reduce the necessary transmission bandwidth, and the transmission energy consumed, for example, in mobile user interfaces. If speech is present, it is source coded and thus compressed according to a certain coding algorithm. If it is chosen to employ a sender-based error concealment strategy, either a redundant version of the encoded speech signal (Forward Error Correction, FEC) or an additional low bit-rate coded version of the speech signal Low Bit-rate Redundancy, LBR) may be added to the coded speech frames. The resulting data – including a few frames for background noise description – are packetized and transmitted over the IP-network. Here, due to the routing across different network paths, the packets may be subject to constant and time-varying delay (jitter), as well as to network packet loss.

At receive side, the unpacked and processed signal may either be decoded and converted back to analog form, or possibly transcoded into a coding format different from the one applied on the IP-path (codec-tandeming). This is the case, for example, if the transmission to the destination is routed across a circuit-switched cellular radio network like GSM. Then, the signal may undergo additional degradations due to the transmission across the radio-path (e.g. introducing bit errors).

The intermediate processing steps the signal undergoes before playback are illustrated in the loudspeaker-path of the send-side of Figure 3.3: For example, from the RTP packet header, delay variations are identified, which are compensated for by (jitter-)buffering. Loss during network transmission, as well as long packet delays and subsequent omission of some packets by the jitter buffer add up to a certain amount of intermediate packet loss. Some of the lost or omitted packets may be compensated for by the subsequent sender-based loss recovery measures (LBR or FEC, using the additional, redundant data). The resulting sequence of packets, without the 'effectively lost packets' (lost in the network or by the jitter buffer, and partly recovered by the loss correction), is unpacked and passed to the decoder.

An additional PLC mechanism may be applied directly at the decoder-level. For example, it may replace lost frames by estimated codec frames, e.g. using information from previously received speech data. In case of silence periods, a comfort noise signal is generated based on the transmitted background noise description. After decoding, the signal may be 'enhanced' by a subsequent speech enhancement algorithm (e.g. extending the lower transmission band limits). Finally, the signal is converted to analog form and played back to the receiver (here the sender) via the user interface (e.g. a handset telephone or a hands-free terminal, as it is depicted in Figure 3.3).

---

[2]Corresponding to a certain number of samples, depending on the applied coding algorithm.

3.3   Quality Elements and Related Features

After a first introduction to the quality elements and features related to VoIP, a more detailed overview of the elements[3], and an analysis of the relation to the quality features relevant to VoIP is provided in this section. In the course of the descriptions, instrumentally measurable parameters are presented, which can be used to capture the impact of certain elements on quality and quality features (see Section 2.2.2.2 for E-model-related parameters). As complementary reading, we refer to ITU–T Rec. Y.1541 (2006).

3.3.1   Voice Activity Detection (VAD)

To reduce the amount of data sent through the network, most IP-systems dispose of an optional mechanism to detect whether the current frame contains speech. Only if it does, is it encoded and sent across the network. In case no speech is to be sent, a comfort noise is commonly inserted in the periods of silence at receive side so that the channel is not entirely muted, and the user is assured that his call is still active.

Voice activity detection (VAD) and comfort-noise generation may both lead to a perceivable speech quality degradation. For example, VAD may cause clipping and hence loss of information (Gruber and Strawczynski, 1985). In addition, VAD interacts with the effect of packet loss: On one hand, with VAD the loss distribution and rate will be different because less data is sent across the network. On the other hand, packets lost in periods of no speech in a connection without VAD may lead to artifacts on the background noise transmitted from the send side, which could be avoided with VAD.

3.3.2   Codecs

To reduce the bandwidth costs of voice transmission, a variety of different codecs are in use. The reduction of speech data is typically achieved exploiting properties of speech production (e.g. estimating vocal tract parameters), and to a lesser extent of auditory perception (e.g. spectral masking). Coding algorithms can be classified as waveform coders (quantization of the actual waveform), parametric coders (based on a speech production model) and hybrid coders (as a combination thereof; see Vary *et al.*, 1998, p. 233 ff.).

Coding often results in a perceivable degradation of speech quality. A speech signal may even be coded multiple times (tandeming), for example, when it is transmitted across different types of networks using different voice compression techniques.

3.3.2.1   Narrowband Speech Codecs

In digital PSTNs, the G.711 is typically used (logarithmic pulse code modulation, PCM; 64 kbit/s). To split the available transmission bandwidth between users in phases of high network traffic, Digital Circuit Multiplication Equipment (DCME) can be applied. It may be

---

[3]In this context, the term quality element can refer to both individual network components like a particular codec, and certain physical characteristics of larger segments of the network, like the overall transmission delay. Note that the 'overall elements' can typically be described by instrumentally measurable parameters, such as the ones listed in Section 2.2.2.2 for the E-model. Individual components may contribute to different degrees to these overall characteristics, or may be bound to particular effects (such as switching in case of ECs), which cannot always be directly related to instrumental measurements.

operated, for example, with the G.726 codec, with a bit-rate chosen between 32 and 24 kbit/s, depending on the available bandwidth per connection (Adaptive Differential Pulse Code Modulation, ADPCM; possible bit-rates are 40, 32, 24 and 16 kbit/s, see ITU–T Rec. G.726, 1990). In mobile networks like the GSM, special codecs are used, such as, for example, the Enhanced Full Rate (EFR; ETSI GSM 06.60, 1996) and the Full Rate Codec (FR; ETSI GSM 06.10, 1988). In the alternative earlier 2G networks according to IS-54 (EIA/TIA/IS-54-B, 1990), a 8 kbit/s codec is employed (VSELP: Vector Sum Excited Linear Prediction).

In packet-based systems, codecs like the G.711 (s.a.), G.723.1 (dual rate codec operating at 5.3 or 6.3 kbit/s, respectively) or G.729 (CS-ACELP, Conjugate-Structure Algebraic-Code-Excited Linear-Prediction, 8 kbit/s) are typically used (ITU–T Rec. G.711, 1988; ITU–T Rec. G.723.1, 1996; ITU–T Rec. G.729, 1996). Codecs specifically designed for packet-based transmission are the Internet Low Bit-rate Codec (iLBC, 15.2 and 13.33 kbit/s; Andersen et al., 2002), and the AMR-NB (Adaptive Multirate Codec, Narrow-Band, working at eight source rates from 4.75 to 12.2 kbit/s; ETSI TS 126 071, 2002). These codecs will be applied at larger scale in future packet networks. The Adaptive Multi-Rate (AMR) codec is already used in UMTS networks. It enables an adaptation of the transmission bit-rate to the current radio-signal strength: The weaker the signal, the larger the bandwidth that can be assigned to channel-coding bits used for error correction by the receiver. Speech codecs operate on different frame sizes, which typically are in the range from 10 ms or less to 40 ms.

3.3.2.2 Wideband Speech Codecs

In the eighties, a 64 kbit/s wideband (WB) codec was standardized for the usage with ISDN (see also Section 3.3.13.5; Sub-band Adaptive Pulse Code Modulation (SB-ADPCM), operates at 64, 56 and 48 kbit/s; ITU–T Rec. G.722, 1988). For a more efficient resource usage, a lower bit-rate wideband codec was sought, applicable also to mobile networks. The result is the WB AMR codec (AMR-WB, Algebraic Code-Excited Linear Prediction (ACELP), ITU–T Rec. G.722.2, 2002), operating at bitrates of 6.60, 8.85, 12.65, 14.25, 15.85, 18.25, 19.85, 23.05 or 23.85 kbit/s. For the highest bit-rate, the bandwidth is approximately 50–6600 Hz, and the five bit-rates between 12.65 and 23.05 kbit/s provide a bandwidth of approximately 50–6000 Hz, decreasing for descending bit-rates (3GPP TR 26.976, 2002). Other WB codecs have been developed, such as the lower bit-rate G.722.1 (Modulated Lapped Transform [MLT], operating at 32 and 24 kbit/s, ITU-T Rec. G.722.1), and the bandwidth-scalable WB extension of the G.729 (bitstream interoperable with the G.729; ITU–T Rec. G.729.1, 2006).

3.3.2.3 Codecs: Quality Features

According to approaches by Schüssler (1987), Halka (1993) and Berger (1998), a speech transmission component like a speech codec can be viewed as a weak nonlinear system, depicted in Figure 3.4. Following this simplification, the codec introduces two types of distortions into the speech transmission path: Linear (frequency) distortions and nonlinear distortions, which are considered as uncorrelated with the linearly distorted signal. Based on this decomposition, Halka (1993) proposed a method to relate the respective signal components to perceived quality features, whose impact on speech quality is combined to an integral quality estimate.

The impact of a codec on perceived quality features has been addressed using different measurement approaches: In multidimensional scaling experiments, the underlying

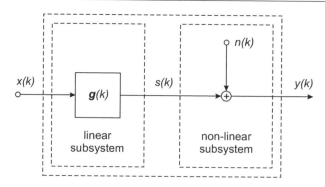

Figure 3.4: Simplified view of a transmission component like a speech codec as a weak nonlinear time-invariant system (weak non-LTI system), according to Schüssler (1987), Halka (1993) and Berger (1998, pp. 58–62).

perceptual feature space was found to be spanned by a set of dimensions, such as clearness (or intelligibility) and 'color of sound' (comprising attributes related to brightness, sharpness and fullness, see Bappert and Blauert, 1994), or naturalness, noisiness and the amount of low frequency-components (i.e. fullness; Hall, 2001). More details on quality dimensions are given in Section 3.4.

In telephony, the impact of coding algorithms on integral quality is typically captured using Equipment Impairment Factors (Ie; values for different codecs can be found in Table I.1 of ITU–T Rec. G.113 Appendix I, 2002). They quantify the impairment introduced by the codec on the quality-scale employed by the E-model (ITU–T Rec. G.107, 2005, see also Chapter 2). Most of the equipment impairment factors provided in ITU–T Rec. G.113 Appendix I (2002) were found in extensive quality tests carried out in the framework of the standardization work of ITU–T Study Group 12 (for an overview of related tests see e.g. ITU–T Contribution COM 12–69, 1998).

Until recently, speech samples degraded by artificial signal-correlated noise of different signal-to-quantizing-noise levels were used as reference in such tests (Modulated Noise Reference Unit, MNRU; ITU–T Rec. P.810, 1996), both for quality evaluation of low bit-rate codecs and codecs under transmission errors (see Appendix D, Section D.1). Because of the perceptual differences between the degradations due to signal-correlated noise and low bit-rate coding or coding under packet loss (e.g. Gros and Chateau, 2001; Hall, 2001), a different listening test procedure for the derivation of Ie-values is now recommended (ITU–T Rec. P.833, 2001; see also Appendix D). A similar procedure for deriving Ie-values using signal-based quality measures (see Section 2) has also been defined (ITU–T Rec. P.834, 2002; see also Möller and Berger, 2002).

The concatenation of different coding algorithms plays a crucial role for speech quality, potentially involving transcoding instead of reconversion to an intermediate representation such as logarithmic Pulse Code Modulation (log-PCM, ITU–T Rec. G.711, 1988). A concatenation, i.e. *tandeming* of codecs, has been shown to yield different levels of quality, depending on the order in which the coding algorithms are applied (e.g. ITU–T Contribution COM 12–69, 1998).

3.3.2.4 Quality Features of Wideband Codecs

As mentioned earlier, WB telephony over IP is achieved using WB lowbit-rate codecs. Both the G.722 and the AMR-WB at its highest bit-rate lead to an improved speech quality as compared with the standard log-PCM coded NB speech (ITU–T Rec. G.711, 1988, NB speech), mainly due to the increased naturalness (see Sections 3.3.13 and 3.4). Similar to NB codecs in case of NB speech, however, WB codecs may introduce a perceivable degradation of the transmitted *wideband* speech. The view of the codec as a weakly nonlinear time-invariant system can easily be extended to the WB case: The approach is independent of a bandwidth-notion.

A formal method for quantifying the impairment due to WB coding has not yet been defined (e.g. similar to the NB method defined in ITU–T Rec. P.833 (2001)). Different approaches are currently under discussion under SG 12 of the ITU–T, as a part of the development of a parametric WB network planning model (a wideband E-model). The quality advantage of WB over NB, and the impairments due to linear and/or nonlinear distortion (i.e. due to bandwidth restrictions and/or coding) are discussed further in Chapter 5.

3.3.3 Jitter

Owing to the routing of packets over different network paths and the asynchronous characteristics of the network, packets within one talkspurt[4] may arrive at their destination with varying delay. Jitter degrades speech quality significantly and has to be compensated for (e.g. Holmes *et al.*, 2001). This is usually done in the network at the receiver side by applying jitter buffers, which store packets for a static or dynamically managed amount of time before playback (e.g. Ramjee *et al.*, 1994; Rosenberg *et al.*, 2000). This measure introduces additional delay and potentially packet loss, for example, if packets arrive too late for playback. These drawbacks are, however, more acceptable than the degradation due to jitter. For more details on the relation between jitter, jitter buffering and packet loss, see Appendix A, Section A.2.2. Additional information on quantifying jitter may be obtained from ITU–T Rec. Y.1541 (2006).

In today's networks, jitter of high frequency, i.e. jitter within one talkspurt, can generally be disregarded as it is normally compensated for in the network, for example, (at each router) by slowing down packets that are ahead of schedule and processing those behind schedule at first, or at the receiver side by using jitter-buffers that add to packet loss and delay (for details see Appendix A.2, Section A.2.2). Low-frequency jitter, that is variations of delay that do not affect the timing of synchronous playback within one talkspurt – for example, resulting from timing drift of network routers, or adaptive jitter buffer algorithms working on a talkspurt-by-talkspurt level – lead to an absolute delay $T a$ that is not necessarily constant for the whole duration of a call. Consequently, 'stationary' degradations like echo may show a time-varying behavior as well, be it due to a time-varying delay or time-varying behavior of network elements like ECs.

3.3.4 Delay

Delay, or the 'overall transmission delay' results from the algorithmic delay of the applied encoder and error correction scheme, and from the delays introduced by the network, the

---

[4]'A talkspurt is a time period that is judged by a listener to contain a sequence of speech sounds unbroken by silence' (Brady, 1965).

Table 3.3: Contributions of VoIP processing steps to overall transmission delay (adopted from Thomsen and Jani, 2000).

Delay Source	Typical Range (ms)
Recording	10–40
Coder	10–20
Internet delivery	70–120
Jitter buffer	50–200
Decoder	10–20
Total	150–400

jitter-buffers and the decoder, or additional signal processing components like ECs. The delay related to different processing steps is summarized in Table 3.3.

3.3.4.1 Overall Transmission Delay: Quality Features

Large amounts of delay reduce the communicability of the system considerably and hence degrade its speech quality (Kitawaki and Itoh, 1991, e.g.). However, users may attribute the effect of delay to delayed reactions from the conversation partner instead of attributing it to the network, for example, interpreted as the other's lack of attention (ITU–T Contribution COM 12–62, 1990; Krauss and Bricker, 1966). Owing to this effect and the fact that delay is not always easily detectable, different studies reported a far less important impact of delay on *perceived* quality than described in Kitawaki and Itoh (1991). Examples of such studies are Möller and Jekosch (1998) and Karis (1991).

Obviously, the impact of delay depends considerably on the conversation situation and scenario (e.g. Kitawaki and Itoh, 1991; different conversation tasks were used in their study, ranging from free conversation to fast exchange of random numbers): If the topic involves a high interactivity[5] it becomes increasingly difficult to *purposefully* interrupt the other, when delay increases. Also, the amount of unwanted interruptions increases with delay, which – in addition to the decrease in interruptability – increases the potential of misunderstanding.

In a telephone conversation, users have a certain time window during which they expect a reaction from their interlocutor. According to Kitawaki and Itoh (1991), delay is detected when the actual reaction exceeds this time window. The size of this time window depends on both the utterance speed and the duration of the preceding talkspurt: The window widens both with increasing talkspurt duration and decreasing utterance speed. As higher interactivity implies both shorter talkspurt durations and faster utterance speed, delay detectability increases with interactivity. Interestingly, in Krauss and Bricker (1966) it was found that increasing delay in telephone conversations with a free conversation topic leads to increased utterance durations due to adaptation by the interlocutors. This adaptation may make delay more tolerable due to the associated widening of the time window during which users expect their partner's reaction. In turn, in case of more structured conversations as in the

---

[5] 'Interactivity' in this context is not used as communication *responsiveness* – '[...]making reference to the content, nature, form or just the presence of earlier reference', and thus reaching beyond *re*-activity (as described by Rafaeli, 1988). Instead, it is used here as a characteristic describing the dynamics of a telephone conversation, which can ideally be quantified by instrumental measurements (see also Hammer *et al.*, 2004a).

interactive exchange of random numbers, such an adaptation cannot be performed, which underlines why more interactive conversation tasks are more sensitive to delay.

The conversation flow parameters characterizing telephone conversations were extensively studied by Brady (1965, 1968). Kitawaki and Itoh (1991) applied the related metrics to the analysis of conversations under the effect of delay. Their analysis confirmed the assumption that more interactive conversation tasks – showing higher sensitivity to delay in the resulting quality ratings – show shorter talkspurt durations. More recent results on the interdependency between delay and double talk can be found in ITU–T Delayed Contribution D.214 (2004), showing a curvilinear increase of double talk percentage from 6% for 0 ms to 9% for 1000 ms delay.

Apart from the conversation topic or the degree of interactivity it creates, the impact of delay on quality also depends on the experience of the users with delay-type degradations. In Kitawaki and Itoh (1990), the threshold of delay detectability was reported to vary between 100 and 700 ms for trained or expert subjects, and between 350 and 1100 ms for untrained subjects. The large variation in turn is due to the impact of the conversation topic.

3.3.5 Packet Loss

Amongst the different quality elements, packet loss is the degradation, which makes VoIP perceptually most different from wireline telephone networks[6].

Packet loss can occur in the network or at the receiver site, for example, due to excessive network delay in case of network congestion. In wireless packet networks, packet loss often results from bit errors introduced on the radio link. In this case, not entire packets may be lost, but only parts of them. Depending on the importance of the lost information, the received parts of the affected packets may still be used during decoding (e.g. Hammer *et al.*, 2003).

The impact of packet loss at the network level depends on several factors, which will be summarized in the following subsections. The main elements influencing the impairment under packet loss are the *packet loss distribution*, the *packet size*, the packet *loss recovery* at packet level, using mechanisms like FEC (see subsequent text), the *speech decoding*, and respective measures for PLC applied in the decoder.

3.3.5.1 Packet Loss Distribution

Owing to the dynamic, time-varying behavior of packet-networks, packet loss can show a variety of distributions. The loss distribution most often studied in speech quality tests is random, or Bernoulli-like packet loss, defined by the overall loss probability p_{pl}. The packet loss percentage (Section 2.2.2.2) and the overall loss probability correspond to the same measure, with

$$\frac{P_{pl}}{\%} = p_{pl} \cdot 100 \qquad (3.1)$$

'Random loss' here means *independent* loss, implying that the loss of a particular packet is independent of whether or not previous packets were lost. Of course, randomness is a property inherent to all events described by a stochastic model. Reflecting the wide-spread usage of the term in the relevant literature, 'random loss' will nevertheless be used in this book to refer to Bernoulli or independent loss.

---

[6]Since jitter is generally avoided using jitter buffers, see Section 3.3.3.

'Random loss' does not represent the loss distributions typically encountered in real networks. For example, losses are often related to periods of network congestion. Hence, losses may extend over several packets, showing a dependency between individual loss events.

Different approaches are presented here that describe packet loss distributions showing dependencies between loss events. In the literature, dependent packet loss is often referred to as 'bursty'. The term 'bursty loss' is even more ambiguous than 'random loss' is, as it does not carry any information on how bursty 'bursty' is. In analogy to the argumentation provided for using 'random loss' instead of Bernoulli or independent loss, the term 'bursty loss' will be employed throughout this book, however, only when dependent loss is *generally* referred to. In more concrete cases, the burstiness of 'bursty' will be explicitly specified.

In the subsequent sections, both a model-based description of common packet loss statistics, and a perception-oriented description of loss as a sequence of periods of particular loss behavior are presented[7].

3.3.5.2 Loss Models

Packet traces collected for current VoIP networks[8] are commonly modeled by time-discrete state models, that is, Markov chains (see e.g. Bolot, 1993; Sanneck and Carle, 2000). The simplest model of this type yielding burst loss – packet loss for which the loss of a particular packet depends on whether the previous packet or packets were lost or received – is a two-state Markov chain, see Figure 3.5. Sparsely distributed packet losses – distributions of loss with long-term dependencies between loss-states – can only be modeled if a higher number of states is used (e.g. Yajnik *et al.*, 1999).

Two-state Markovian loss can be described with two parameters, for example, the two transition probabilities p (transition found–lost) and q (lost–found), as depicted in

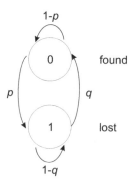

Figure 3.5: 2-state Markov loss model.

---

[7]Note that the loss metric proposed by the Internet Engineering Task Force (IETF) to parametrically describe packet loss will not be addressed in this book, since its suitability to quality modeling is still unknown (IETF RFC 3357, 2002).

[8]The sequences of packets lost and received on the way from one particular end-point to another.

Figure 3.5. The overall loss probability p_{pl} can be calculated according to Equation (3.2):

$$p_{pl} = \frac{p}{p+q} \qquad (3.2)$$

Here, p_{pl} corresponds to the probability of occupying state '1', that is, the lost state.

The loss distribution related to a 2-state Markov model can also be characterized by the number of packets, which are lost or found in a row. For example, the probability that the transition from '1' (lost) to '0' (found) occurs after k steps can be expressed as

$$P(k, q) = q\,(1 - q)^{k-1}. \qquad (3.3)$$

According to Equation (3.3), the number of steps k necessary to transit from 'lost' to 'found', that is, the number of consecutively lost packets, is a *geometrically* distributed random variable. This geometric distribution of consecutive loss events makes the 2-state Markov model (and higher order Markov models, see subsequent text) applicable to describing loss events observed in the Internet (e.g. Bolot and Vega-García, 1996; Bolot *et al.*, 1999).

The *average* number of packets lost in a row can be calculated as the expectation of k:

$$\mu_{10} = E\,\{k\} = \sum_{k=1}^{\infty} k \cdot q\,(1 - q)^{k-1} \qquad (3.4)$$

This expression can be rewritten as a geometric row, so that the expectation can be calculated as:

$$\mu_{10} = \frac{1}{q}. \qquad (3.5)$$

Similarly, the average number of consecutively found packets is

$$\mu_{01} = \frac{1}{p}. \qquad (3.6)$$

It follows intuitively from (3.2), that

$$p_{pl} = \frac{\mu_{10}}{\mu_{10} + \mu_{01}}. \qquad (3.7)$$

Now, if a given network packet trace is to be modeled using a 2-state Markov model, the transition probabilities p and q can be estimated from m_i, the number of times i packets got lost consecutively, and the overall number of received packets m_0 (with the sum of all m_i being the number of consecutive loss events, and the sum of all $m_i \cdot i$ being the overall number of lost packets):

$$p_{pl} = \frac{\sum_{i=1}^{\max(i)} i \cdot m_i}{m_0 + \sum_{i=1}^{\max(i)} i \cdot m_i} \tag{3.8}$$

$$q = \frac{1}{\mu_{10}} = \frac{\sum_{i=1}^{\max(i)} m_i}{\sum_{i=1}^{\max(i)} i \cdot m_i} \tag{3.9}$$

$$p = \frac{q p_{pl}}{1 - p_{pl}} = \frac{\sum_{i=1}^{\max(i)} m_i}{m_0} \tag{3.10}$$

The considerations regarding the 2-state Markov model are of interest also for random, i.e. *independent* packet loss: It represents a special case of 2-state Markovian loss, namely, setting

$$p = 1 - q = p_{pl}, \tag{3.11}$$

as, for independent loss, it is equally probable to loose a packet when departing from the lost or found state. For random loss, the numbers of packets lost (μ_{10}) and received in a row (μ_{01}) are

$$\mu_{10} = \frac{1}{1 - p_{pl}} \tag{3.12}$$

$$\mu_{01} = \frac{1}{p_{pl}}. \tag{3.13}$$

At this point, another useful measure needs to be mentioned, which expresses the tendency of a particular loss distribution for consecutive packet loss *independently* of the overall loss rate, the *Burst Ratio* (*BurstR*), see ITU–T Delayed Contribution, D.020 (2001); US Patent 6,931,017 (2005). It can serve to quantify the (short-term) burstiness of packet loss. The Burst Ratio is defined as

$$BurstR = \frac{\text{average no. of consecutively lost packets}}{\text{average no. of consecutively lost packets for random loss}}. \tag{3.14}$$

We can identify different types of loss behavior from this definition: If *BurstR* = 1, the loss type is 'random loss'. In case that *BurstR* < 1, the loss type is 'sparse loss', showing a *lower* probability of losing a packet, when the previous packet was lost. Finally, *BurstR* > 1 indicates bursty loss, where the probability of losing a packet is higher, when the previous packet was lost[9].

---

[9]Note that the additional differentiation between sparse and bursty loss distinguishes two types of dependent loss, we have generally referred to as bursty so far. All considerations in this book will be focused on random loss or bursty loss with *BurstR* > 1.

With Equations (3.2), (3.5), (3.12) and (3.14), the Burst Ratio for a 2-state Markov model is calculated as

$$BurstR = \frac{\mu_{10}\big|_{2-state}}{\mu_{10}\big|_{random}} = \frac{1}{p+q}. \qquad (3.15)$$

A slightly extended version of the 2-state Markov model is the *Gilbert model* (see Figure 3.6(a); Gilbert, 1960). While in the 2-state model it is assumed that the '1' state implies certain loss, the Gilbert model associates a loss probability $b \leq 1$ with this state, here referred to as the 'bad' state[10]. The Gilbert model can be illustrated as follows: For each step ending at the 'bad' state, a dice is tossed leading to a loss with probability b. In this case, μ_{10} as described by Equation (3.12) corresponds to the average number of consecutive packets in the 'bad' state, of which $\mu_{10} \cdot b$ are lost on average. Hence, the overall loss rate for the Gilbert model can be deduced from Equation (3.2) as:

$$p_{pl} = b \cdot \frac{p}{p+q} \qquad (3.16)$$

A further generalization of the Gilbert model is the Gilbert-Elliott model, which is depicted in Figure 3.6(b) (Elliott, 1963). Here, the 'good' state, too, is associated with an independent loss probability (g). In analogy to the Gilbert model, the overall loss for the Gilbert-Elliott model is given by:

$$p_{pl} = b \cdot \frac{p}{p+q} + g \cdot \frac{q}{p+q}. \qquad (3.17)$$

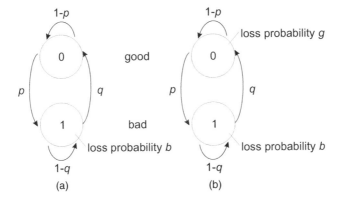

Figure 3.6: Gilbert model (a): 2-state Markov model with independent loss probability b for the 'bad' state (the loss probability for the 'good' state is 0; Gilbert, 1960). (b): Gilbert-Elliott model; 2-state Markov model with both independent loss probabilities for the 'bad' (b) and the 'good' state (g; Elliott, 1963).

---

[10]Note that the term *Gilbert model* is frequently misused in the literature to denote a 2-state Markov model, which can be considered only as a special case of the Gilbert model, using $b = 1$.

Obviously, both the Gilbert model and the 2-state Markov model shown in Figure 3.5 can be regarded as a special case of the Gilbert-Elliott model, with $g = 0$ for the Gilbert model, and additionally $b = 1$ for the 2-state Markov model.

Packet traces collected in real IP-networks can be more accurately modeled with a higher number of states (Sanneck and Carle, 2000; Yajnik *et al.*, 1999). In Sanneck and Carle (2000), an extended 2-state model, namely, an n-state model, is proposed that allows accurate modeling of the distribution of the number of consecutively lost packets, for up to $(n - 1)$ packets lost in consecution (corresponding to a *run-length*[11] of $\{n - 1\}$). Owing to both the disadvantage of a high number of independent parameters $(n - 1)$ and the inaccurate modeling of the number of consecutively *received* packets (Jiang and Schulzrinne, 2000), this n-state model will not be dealt with further in this book.

A simplification of an n-state model, still capable of capturing some longer-term loss dependencies, is the 4-state Markov model proposed by Clark (2001). It is depicted in Figure 3.7[12]. In this model, a 'good' and a 'bad' state are distinguished, which typically represent phases of higher and lower packet loss. Both for the 'bad' and the 'good' state, an individual 2-state Markov model represents the dependency between consecutively lost or found packets, as described in the previous text. The two 2-state models make for four (two each) independent transition probabilities. Two further probabilities characterize the transitions between the two 2-state models, leading to a total of six[13] independent parameters

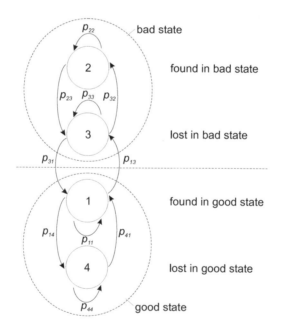

Figure 3.7: 4-state Markov model (used e.g. by Clark, 2001).

---

[11]A 'run' is the consecutive repetition of a particular event, for example, of persisting in a certain state of a Markov chain, or of loosing a packet. Then, the run-length is the number of times the corresponding event occurs in consecution (e.g. Gilbert, 1960).

[12]Note that only certain sparse, individual loss events may not be covered by this model, making it applicable to a wide range of network conditions.

Table 3.4: Average numbers of consecutively lost and found packets for the 'good' and 'bad' states of a 4-state Markov model as depicted in Figure 3.7.

State	No. of Consecutively Lost Packets	No. of Consecutively Found Packets
'good'	$\dfrac{1}{1 - p_{44}}$	$\dfrac{1}{1 - p_{11}}$
'bad'	$\dfrac{1}{1 - p_{33}}$	$\dfrac{1}{1 - p_{22}}$

for this particular 4-state Markov model. In analogy to the considerations on the 2-state model, i.e. to Equations (3.4), (3.5) and (3.6), the average numbers of consecutively lost and found packets – for the good and the bad state – can be calculated as summarized in Table 3.4.

The transition probabilities between the 'good' state and the 'bad' state are p_{31} and p_{13}. The average number of steps for which the chain stays in the 'bad' or the 'good' state (steps of *sojourn* in these states) are expressed by the parameters μ_{31} and μ_{13}, respectively. According to the calculation given in Appendix A, Section A.1, they can be determined to:

$$\mu_{13} = \frac{p_{14} + p_{41}}{p_{13} \cdot p_{41}} \tag{3.18}$$

$$\mu_{31} = \frac{p_{23} + p_{32}}{p_{23} \cdot p_{31}} \tag{3.19}$$

The average durations t_b and t_g associated with the sojourns in the 'good' and 'bad' states can be obtained from the applied packet size T_p (ms):

$$t_g = \mu_{13} \cdot T_p \tag{3.20}$$

$$t_b = \mu_{31} \cdot T_p. \tag{3.21}$$

The overall loss rates in the 'good' and 'bad' states p_g and p_b can be calculated from Equations (A.6)–(A.10) as

$$p_g = \frac{P_4}{P_1 + P_4} = \frac{p_{14}}{p_{14} + p_{41}} \tag{3.22}$$

$$p_b = \frac{P_3}{P_2 + P_3} = \frac{p_{23}}{p_{23} + p_{32}}, \tag{3.23}$$

where P_1 to P_4 are the *state probabilities* for states 1–4, quantifying the probability that the chain is in the respective state.

In Chapter 4, the parameters presented in this section will be exploited for investigating speech quality under different types of dependent packet loss.

---

[13]From the overall ten parameters, four can be deduced from the other six, since probabilities departing from one state add up to 1.

3.3.5.3 Packet Loss Recovery

Different techniques exist to recover the speech information lost with a packet (Perkins *et al.*, 1998, for an overview). These techniques can be distinguished as methods that are applied only at the receiver side, and methods that are based on sending additional data used for the recovery by the receiver. In actual VoIP networks, combinations of different approaches or hybrid approaches are often employed.

- Packet Loss Concealment (PLC): No additional data is sent and the lost packets are compensated for at the receiver side, for example, based on (coding-) information from previously received packets. The minimal measure taken is the insertion of a silence interval corresponding to the interval of consecutively lost frames to yield timely playback. Depending on the type of phoneme that has been subject to loss (i.e. whether it is voiced or unvoiced), noise or pure tones were found to lead to an impression of a continuous signal, with a noise or tone sequence perceived in addition (phonemic restoration, see Section 1.1.2). Assuming that a lost frame and the previous frame are part of one phoneme, a simple frame repetition can be used for concealment. Although intelligibility can be enhanced with such a simple approach, speech quality is reduced compared with the error-free signal. More sophisticated approaches like insertion of frames interpolated from coding information of previous frames are applied in codecs typical of VoIP (e.g. ITU–T Rec. G.723.1, 1996; ITU–T Rec. G.729, 1996). Since PLC does not require additional data to be sent and thus to be added to later or earlier packets, it does not add a considerable processing delay.

- Forward Error Correction (FEC): An additional, duplicate and thus redundant version of the coded speech data is transmitted by the sender. It can be used at the receiver side to restore the lost information (IETF RFC 2733, 1999). This method yields relatively high quality even in case of bursty loss or of loss with higher overall rates (Jiang and Schulzrinne, 2002; Rosenberg *et al.*, 2000). If a loss can be recovered, it is recovered in a bit-exact manner. During packetization, different techniques exist for the distribution of the redundant data relative to the actual speech data. Examples are the usage of parity check codes (IETF RFC 2733, 1999) or Reed–Solomon Codes (Jiang and Schulzrinne, 2002; Rizzo, 1997).

- Low-Bitrate Redundancy (LBR): A redundant version of the speech signal that is coded at lower bitrate is sent in addition to the transmitted speech data. It can be used at receive side for the replacement of lost packets (Hardman *et al.*, 1995; IETF RFC 2198, 1997). This method leads to improved quality compared with PLC only, but is less effective than FEC, due to the fact that recovery is not bit-exact (see e.g. Jiang and Schulzrinne, 2002). The packetization of the additional LBR data can be handled similar to the FEC case described above.

Both FEC and LBR lead to additional data that has to be processed, added to earlier or later packets, and sent across the network. Thus, both techniques increase the network delay and network load.

3.3.5.4 Packet Size

The packet size T_p refers to the payload contained in a packet, and is measured in ms. Other identifiers of the same object are payload size (bytes), packet interval or packet length (ms), or number of codec frames per packet. Additional header information is added to the speech payload by the different protocols during packetization (see Section 3.1). To reduce the overhead data transmitted with the speech data, it is desirable to limit the number of sent packets, i.e. to pack several codec frames into one packet. In turn, larger packets lead to an increase of transmission delay, and potentially lower speech quality in case of packet loss. As a compromise, current VoIP services use packets containing a small number of frames, with packet sizes in the range from 10–60 ms[14].

3.3.5.5 Macroscopic versus Microscopic Packet Loss

From the considerations on different packet loss models it follows directly that packet losses may show both long-term and short-term dependencies. This is reflected in speech quality perception of time-varying distortions: While the short-term behavior of packet loss dictates how subjects perceive speech quality instantaneously, that is, during a certain time window around particular loss events, the long-term behavior of packet loss relates to variations of this instantaneous impression.

   This differentiation has been substantiated by extensive studies carried out by Gros and Chateau (Gros, 2001; Gros and Chateau, 2001, 2002). As an alternative to the model-oriented view on packet loss distributions presented above, they took a more perception-oriented approach. In their studies, Gros and Chateau used various loss profiles consisting of sections of different 'stationary'[15] levels of random packet loss P_R, extending over periods in the range of several seconds (see Figure 3.8). While the packet traces necessary to

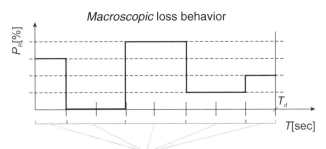

Periods of different *Microscopic* Loss Behavior

Figure 3.8: Example for loss profiles used by Gros and Chateau (2001). The durations T_d used in the tests were taken from the set [45 s, 90 s, 120 s, 190 s]. The levels of random packet loss (in %: P_R) chosen for different sections were taken from the set [0, 2.5, 10, 20, 30]. The dichotomy distinguishing *macroscopic* and *microscopic* loss behavior as it is proposed in this book is illustrated as well (see text).

---

[14]Excessive header data can further be reduced using header compression techniques (IETF RFC 3095, 2001, e.g.).

[15]Here, the term 'stationary' relates to the behavior of the loss rate over time. The corresponding perceptual degradation is time-varying by nature.

produce such profiles do not accurately reflect actual packet networks, they lead to strongly time-varying degradations of speech quality.

The approach taken by Gros and Chateau can easily be generalized to the distributions of packet loss encountered in real VoIP networks: Obviously, loss profiles can be conceived, where the different periods of 'stationary' loss behavior are described using Markov chain models of a low number of states. For example, instead of using a random model with packet-independent loss during these periods, a 2-state Markov model may be employed. This way, all conceivable packet traces can principally be modeled (a corresponding approach for network monitoring purposes has been suggested by Clark, 2001).

According to the notion of time-varying and instantaneous, 'stationary' speech quality, a classification of the loss behavior of packet networks is proposed in this book:

Microscopic **loss behavior** refers to a type of degradation that is clearly perceived as time-varying on a feature level, without leading to time-varying quality, i.e. considerable level-changes in instantaneous quality judgments. For example, some PLC algorithms are affected more by consecutively lost coding frames than others. Hence, speech quality is a function of the underlying *microscopic* loss behavior.

Macroscopic **loss behavior** refers to time-varying distortions where the quality perceived instantaneously is time-varying over longer periods of time (as in Gros and Chateau, 2001). Accordingly, *macroscopic* refers to a loss behavior, which is associated with clearly recognized changes in speech quality. In auditory speech quality tests, these changes can be observed also from an adaptation of the subjects rating behavior to an altered quality level. Changes in the perceived speech quality level can result only when both the previous and the present period of a particular *microscopic* loss behavior are of sufficient duration. The two different quality levels can then be recognized. The duration thresholds for this condition to be fulfilled are not precisely defined at this point.

The generalization of this taxonomy for other types of speech transmission channels is directly related to the notion of quality features:

Microscopic channel behavior is bound to a stationary magnitude of the perceived features.

Macroscopic channel behavior is characterized by time-varying feature magnitudes that may lead to time-varying quality.

3.3.5.6 Microscopic Loss Behavior: Quality Elements

The packet loss distribution, the employed repair methods like FEC, the packet size, and the PLC applied in the decoder all impact the decoded speech signal, as is illustrated in Figure 3.9.

Packet loss recovery mechanisms effective on the packet-level, like FEC, enable complete restoration of some of the packets lost at network level. In turn, additional packets may be lost due to discard by the jitter buffer, if they arrive too late for playback. The

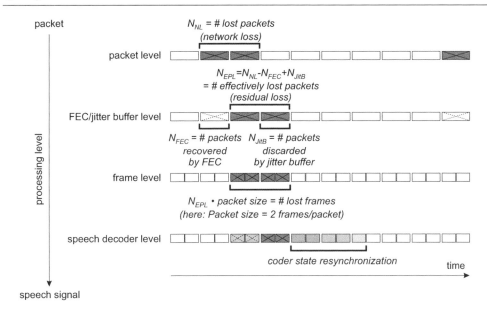

Figure 3.9: Illustration of packet loss at packet and frame level: Influence of network loss, packet size, jitter buffering and repair mechanisms prior to decoding (like Forward Error Correction, FEC) and during decoding (PLC).

residual loss after FEC can be considered as the *effective* packet loss the decoder and related PLC algorithm are faced with[16].

At the frame level, the packet size determines how much information gets lost with a packet. Its influence on quality is strongly related to the applied PLC: some algorithms are more sensitive to consecutive loss of codec frames than others (see Section 3.3.5.7). In the scenario depicted in Figure 3.9, two frames were used per packet. The last level of the figure shows the effect of PLC mechanism applied at the decoder level: some frames may be restored based on previously received information. However, some artifacts may still be perceivable, as the restoration is not bit-exact (as opposed to FEC), which is indicated by the gray shading. The additional degradation introduced by some decoders due to state-desynchronization under packet loss is indicated by the gray shading of frames arriving *after* the actual frame loss.

In this context, the 'convergence time' is defined as the time until the difference between the error-free and the degraded signal falls below a certain threshold. For example, in two similar studies the convergence time was determined instrumentally: In Rosenberg (2001), it was measured purely on the signal level for the G.729 codec and its built-in PLC, based on frame-wise energy differences. In Sun *et al.* (2001), the convergence time was measured for the G.729 as well as for other codecs applying the same approach as in Rosenberg (2001), but additionally using signal-based speech quality measures like PSQM (perceptual speech quality measure) (ITU–T Rec. P.861, 1996).

---

[16]For a list of studies reporting analytical expression to relate the network packet loss to the packet loss after FEC, see Appendix A, Section A.2.1. More details on the impact of jitter and the employed jitter buffer on packet loss can be found in Section A.2.2.

The resulting convergence time showed to be relatively less dependent on the number of frames lost in a row. In turn, it was found to depend mainly on the lost speech information: Lost voiced speech sounds were found to yield convergence times of up to 300–400 ms[17], while lost unvoiced speech sounds did not lead to any considerable state desynchronization (Sun *et al.*, 2001).

In their study, Hammer *et al.* (2004b) have backed up that the exact loss location determines the effect of packet loss on an underlying speech signal. They employed the signal-based measure PESQ (perpetual evaluation of speech quality) (ITU–T Rec. P.862, 2001) for quality estimation. Hammer *et al.* found considerable differences in estimated quality for a shift of the applied loss pattern by only one packet relative to the speech signal (up to 0.6 points on the PESQ quality scale).

3.3.5.7 Microscopic Loss Behavior: Quality Features

For speech quality perception, the importance of the packet loss location found with signal-based analysis could be confirmed by Bernex and Barriac (2002). Their analytical listening tests were aimed at the multidimensional analysis of the feature space related to packet loss. They found that both the perceived features and speech quality depend considerably on the position of a (three packets long) loss sequence in a given (six seconds long) speech sample.

As stated in the previous text, the packet size or number of consecutively lost frames do not influence the degree of state desynchronization encountered in some decoders in case of packet loss. However, several utilitarian studies have reported an effect of packet size (T_p) on speech transmission quality under packet loss, depending on the applied PLC algorithm:

If silence insertion is chosen as concealment, the resulting signal is the same as if the signal was interrupted for the duration of the lost packets. Most studies reported in the literature on silence insertion did not show a statistically significant effect of packet sizes relevant for VoIP (10 to 60 ms, see: ETSI Tiphon 11 Temporary Document 64, 1999; ITU–T Contribution COM 12–28, 2001; ITU–T Delayed Contribution D.110, 1999; Watson, 2001; Yamamoto and Beerends, 1997)[18]. However, a minor tendency for better quality with smaller packet sizes was found in some of the above studies even for $T_p \in [10\,ms, 60\,ms]$ (and in the study on random interruptions by Gruber and Strawczynski, 1985). It is interesting to note that in this range packet sizes are smaller than the average duration of a phoneme (e.g. approximately 80 ms for American English, see Umeda, 1975, 1977).

If more sophisticated strategies for loss concealment are applied in a decoder, the performance in case of packet loss can significantly be improved (see e.g. Perkins *et al.*, 1998, for an overview). In many cases, however, these algorithms lead to a more critical impact of the packet size chosen for a connection: often, they are designed for a specific

---

[17]The impact of a lost frame was most severe during the onset of a voiced speech sound. In this case, a PLC algorithm may replace the lost information by information retained from a previous, unvoiced frame (e.g. using a wrong excitation signal and vocal tract filter coefficient estimates). Interestingly, Sun *et al.* (2001) found the convergence time to almost correspond to the time remaining of the voiced sound after the loss event.

[18]The effect of packet size can be compared to the effect of a particular *frame* loss distribution showing consecutive frame loss. However, the resulting run lengths of lost and found codec frames are not geometrically distributed, so that the parameters derived in the previous section for the Markov-model cannot easily be applied to this problem.

packet size, for example corresponding to the codec's frame size (one frame per packet). The quality may then degrade more rapidly, the more frames are lost in a row (assuming the same average packet loss rate).

An effect of packet size on speech quality under packet loss was reported for different PLC algorithms (e.g. Hardman *et al.*, 1995; ITU–T Contribution COM 12–28, 2001; ITU–T Delayed Contribution D.110, 1999; Watson, 2001)[19]. The results obtained for the packet sizes relevant for VoIP (10–80 ms) can be summarized as follows:

- For packet repetition as loss concealment strategy, larger packet sizes lead to a decrease in quality (Watson, 2001). For packet sizes around 80 ms, some of the repeated signal units go beyond the borders between adjacent phonemes, leading to perceivable artifacts.

- The PLC employed in the GSM-EFR and the App. I-PLC of the G.711 were found to be sensitive to the choice of packet sizes (ITU–T Delayed Contribution D.068, 2002; ITU–T Delayed Contribution D.110, 1999).

- A minor effect of packet size was reported for the G.723.1 and its built-in PLC (ITU–T Contribution COM 12–28, 2001; ITU–T Delayed Contribution D.068, 2002; Voran, 2003).

- No effect of packet size could be found for the G.729 and its native PLC (ITU–T Contribution COM 12–28, 2001; ITU–T Delayed Contribution D.068, 2002; ITU–T Delayed Contribution D.110, 1999).

The number of consecutively lost frames is also impacted in case of dependent packet loss, as is described, for example, by a 2-state Markov model. However, only very few auditory tests are reported in the literature on speech quality under other types of microscopic loss behavior than random or periodic loss. The only two sufficiently documented tests of this type indicate a more important impact of packet loss on speech quality, when the number of consecutively lost packets was increased (Jiang and Schulzrinne, 2002; Sun and Ifeachor, 2002b)[20].

In Section 4, the results summarized in this section are complemented by additional speech quality tests, and are employed as the basis for a parametric model of speech quality under microscopic packet loss.

3.3.5.8 Macroscopic Loss Behavior: Quality Elements

In this book, we have defined changes in the microscopic loss behavior as *macroscopic loss behavior*. Such a macroscopic loss behavior is linked to the particular property of packet-based transmission across a network that is dynamically used by a number of users

---

[19]In all of these tests, the samples were degraded using random loss, except from ITU–T Delayed Contribution D.068 (2002), where periodic loss was used.

[20]The numerous studies reported in the literature that were carried out using signal-based quality measures such as PESQ (e.g. Hammer *et al.*, 2004b; Hoene *et al.*, 2003; Sun and Ifeachor, 2002a) are not addressed here, since the validity of these measures for nonrandom packet loss may be questionable (Pennock, 2002). Moreover, using quality estimates from one model to develop an enhancement of another model – one of the ultimate goals of this book – appears to be a dangerous approach.

requiring different amounts of network resources at different times. Moreover, many networks are configured to adapt their behavior, such as the jitter buffer length, to the current network load, which in turn leads to a progressive change in delay and loss behavior. As stated earlier in this book (see Section 3.3.5), Markov state models allow the macroscopic packet loss behavior to be described similar to the underlying microscopic loss behavior. For example, an overall, long-term loss behavior can be described by an 'outer' Markov model, where states of different microscopic loss behavior communicate among each other. These states – in turn – can themselves be described by simple state models representing the microscopic loss behavior.

3.3.5.9 Macroscopic Loss Behavior: Quality Features

The most relevant auditory tests with regard to longer-term behavior of packet loss were carried out by Gros and Chateau (Gros, 2001; Gros and Chateau, 2001, 2002). For their tests, they adopted the rating procedure described in ITU–R BT-500-8 (1998) for instantaneous judgments of speech quality degraded by different packet loss profiles. In Section 2.1.1.2, the method is described in more detail. In the tests by Gros and Chateau, subjects listened to long samples of speech (between 45 and 190 s long) and judged the quality they perceived instantaneously using a slider. In addition, subjects were asked for an integral quality judgment at the end of each sample. The comparison between the instantaneous judgment profiles and integral quality judgments allows conclusions to be drawn on time-varying quality perception.

In their tests, Gros and Chateau used quality profiles consisting of different levels of random packet loss (from the set $[0\%, 2.5\%, 10\%, 20\%, 30\%]$; see Figure 3.8 for an illustration). The main findings from their tests can be summarized as follows:

- Stabilization of the quality judgments:

  - After a change of the random packet loss level subjects took some time to stabilize their judgments. The instantaneous rating reaches the new quality level following some type of exponential decay curve.

  - The time necessary for stabilization is asymmetric with regard to the direction of a change in quality. In case of quality improvements, the corresponding average exponential decay time constant was found to be 14.3 s, and in case of a decrease in quality of 9 s (Gros, 2001, p. 89).

  - These reaction times are different from the ones obtained for instantaneous quality ratings of speech samples degraded by signal-correlated-noise of time-varying signal-to-correlated-noise ratio (Gros, 2001; Hansen and Kollmeier, 1999). For the latter type of degradation, stabilization was found to be symmetric for improvements and degradations, and also much faster (1–2 s reaction time). Similarly, short reaction times were observed from the instantaneous loudness judgments obtained for time-varying levels of nonspeech sounds by Susini *et al.* (2002).

- Integral quality ratings: The temporal average over the instantaneous ratings leads to an estimate of integral quality. The estimate is highly correlated with the integral quality ratings obtained in the test.

- Recency effect:

  - Periods of low or high quality positioned at the end of a speech sample had a stronger influence on the integral quality judgments than when such periods were positioned in the beginning of the sample.

  - The recency effect found by Gros and Chateau differs from the classical recency effect reported for verbal recall experiments (see Section 1.1.2, and Cowan, 1984; Crowder and Morton, 1969), and also from that found in experiments on loudness perception (Susini *et al.*, 2002). In all of these tests, the time thresholds for recall were in the range from 2 to 20 s. The recall of time-varying quality events, however, extends over durations between 1 and 2 minutes (Gros, 2001). Gros discusses this observation in the light of attention-driven processing related to the concept of working memory (Baddeley, 1997; Gros, 2001). According to this view, higher cognitive processing levels seem to be involved that enable longer storage.

  - A recency effect was found also in ITU–T Delayed Contribution D.064 (1998) for 60 s long-speech samples and 8 s long-burst segments (degraded by noise or interruptions). In the study described in ITU–T Contribution COM 12–28 (2001), a statistically significant recency effect was not observed. However, the packet loss bursts and stimulus durations applied here were shorter than in the tests by Gros and Chateau (2001) and ITU–T Delayed Contribution D.064 (1998) (3 s and 48 s, respectively). No recency but a primacy effect (see Section 1.1.2.3) was observed in ITU–T Contribution COM 12–21 (1997), where fading profiles typical of mobile telecommunication channels were used. Consequently, profiles showing strong degradation in earlier parts were rated worse than profiles showing strong degradation mainly in later parts.

- For a given profile, the integral quality judgment obtained in a listening test is comparable to that obtained in a conversation test. However, no recency effect could be observed in the conversation test carried out by Gros and Chateau (see Section 3.6).

Gros and Chateau interpreted the considerable reaction times and the asymmetry in the reactions on quality improvements and degradations with the rating strategy applied by the subjects. This strategy is thought to involve a temporal integration over impairment events, basically accounting for the number of perceived impairments per time unit. If quality improvements occur, the integration process takes longer (as fewer, or less important impairment events are perceived), and hence stabilization of the quality impression is achieved only after some time. In case of a decrease in quality, the intensity or rate of the impairment events is increased, permitting a faster adaptation.

This interpretation provides an explanation for the differences between the instantaneous rating of time-varying quality under packet loss on one hand, and under time-varying signal-to-correlated-noise ratios on the other hand. These differences (the stabilization duration and asymmetry) can be ascribed to the different natures of the two impairment types: The perceptual mechanisms involved in the instantaneous quality evaluation based on speech-signal-to-correlated-noise levels are thought to be more peripheral than the mechanisms involved when integrating over certain impairment events (Gros, 2001). A similar argument holds for the instantaneous judgment of time-varying loudness[21], which can also be performed based on low-level auditory information (e.g. Susini *et al.*, 2002).

None of the studies on macroscopic channel behavior reported in ITU–T Contribution COM 12–21 (1997), ITU–T Delayed Contribution D.064 (1998), Jekosch (2000) and ITU–T Contribution COM 12–28 (2001) involved instantaneous judgment. Hence, a comparison of the integral quality ratings obtained in these tests with corresponding time-averages over instantaneous ratings cannot be made. However, Hollier *et al.* (ITU–T Contribution COM 12–21, 1997), Rosenbluth (ITU–T Delayed Contribution D.064, 1998), and Jekosch (2000) compared judgments obtained for short speech passages to judgments obtained for longer stimuli consisting of combinations of the short passages. All these clearly indicate that the simple average over the ratings obtained for individual segments cannot be used as a predictor for the integral quality measured for the concatenation of the segments. Obviously, some weighting of impairment events is involved, both with regard to the position within the sample and with regard to the intensity of the degradation. For example, in the work of Gros and Chateau (2001) an implicit weighting is assumed that is reflected in the reaction times of the instantaneous judgment of time-varying speech transmission quality. Aspects of impairment weighting are further discussed in Section 4.2.

3.3.6 Bit Errors

Bit errors may occur due to wireline access technology such as Digital Subscriber Line (DSL), or wireless radio networks. Here, bit errors are caused by signal interference on the radio link, which are commonly expressed by the carrier-to-interference ratio C/I [dB]. How such bit errors may affect the speech packet stream largely depends on the employed access technology. First, we have to differentiate circuit-switched from packet-switched networks. Also, packet-switched transmission over circuit-switched networks is conceivable, as it is employed, for example, by GPRS (over GSM). In all cases, however, bit errors are reduced by the employed channel coding techniques.

In case of circuit-switched radio access as, for example, with GSM, different mechanisms are applied to recover from bit errors (channel coding, i.e. block codes and convolutional codes; interleaving). In addition, the employed codec classifies the bits in each frame according to their estimated relevance for intelligibility. For example, the GSM-FR codec distinguishes three classes, Class 1a, Class 1b, and Class 2 (ETSI GSM 06.10, 1988). In case of unrecovered bit errors, speech codec frames are discarded only when the Class 1a bits are affected. Otherwise, the corrupted frame is used as is in the decoding process.

For CDMA-based technologies such as UMTS, the situation depends on the actual implementation: For the circuit-switched mode, channel coding and other bit error recovery mechanisms are comparable to GSM, although somewhat more elaborate. Similar to the GSM-codecs, the AMR-codec (ETSI TS 126 071, 2002) used in UMTS provides unequal error protection (UEP), subdividing the bits of each frame according to their estimated perceptual relevance into three classes. Consequently, codec frames are discarded only when the most relevant bits were effected. In case of packet-based implementations of UMTS

---

[21]Or of discomfort due to the noise perceived in a bus, which is rated mainly based on loudness (Parizet *et al.*, 2003).

for speech delivery, similar measures are taken to ensure a comparable error-resistance as in circuit-switched mode.

Instead of transmitting speech in the dedicated voice mode, packet-based radio networks can also be employed in their data-mode for actual VoIP (e.g. UMTS, GPRS). In case of the simple assumption that the delivery system does not differentiate speech from nonspeech data, a bit error that is not recovered by lower layer protocols and channel protection schemes imply a dropping of the effected packet, for example, if the UDP checksum fails (note that, for example, the IP checksum only protects its header, not the IP-payload). The same holds true in case of VoIP over wire-line access networks involving DSL technologies. Corresponding considerations on speech quality of VoIP over UMTS can be found in the study by Kwitt *et al.* (2006), who have used an E-model based approach for quantifying speech quality. For VoIP over any type of access technology exposed to bit errors, the study by Hammer *et al.* (2004c, 2003) has shown that keeping all the packets that are corrupted by bit errors leads to higher quality than both discarding only packets where relevant information is effected, and discarding all degraded packets. However, the latter approach is automatically employed, if packets are dropped in case of an unsuccessful UDP checksum.

3.3.7 Talker Echo

A transmitted speech signal may be reflected at some point in the speech path from the mouth of the talker to the ear of the listener, and be transmitted back to the talker. The reflections necessary for talker echo to occur may arise both in the network and at the point of the user interface (see Section 3.2): In the past, where wireline handset telephones were generally used, reflections occurred at network level, especially due to imperfect coupling at the 2–4 wire junctions (hybrids) between the analog telephone and the network end office. Today, reflections are typically acoustical ones, due to the microphone part of the user interface picking up the speech sounds from the hearing part (e.g. in hands-free terminals).

Talker echo is perceived when for a given level of reflection the delay of the reflected speech signal exceeds a certain delay threshold of approximately 30 ms (Appel and Beerends, 2002; ETSI Technical Report ETR 250, 1996, p. 50). In some limits, the delay threshold is level-dependent, due to masking of the echo by the current speech (due to bone conduction, air-path conduction and talker sidetone). Below the echo threshold, the reflected signal is perceived as (delayed and thus hollow) sidetone, which may mask the speech coming from the other interlocutor. In case of round-trip echo delays larger than 150–200 ms, i.e. in the duration range of average syllables (see Section 1.1.1), difficulties in talking may arise (e.g. Appel and Beerends, 2002; Lee, 1950). Due to masking, talker echo during periods of double talk is perceived as far less annoying than during periods of single talk (e.g. ITU–T Contribution COM 12–102, 1999). Owing to the increasing echo degradation with increasing delay, talker echo is especially important in case of network technologies that introduce considerable network delay, like VoIP or GSM. For more details see (Möller, 2000, pp. 30–32 and pp. 164–166).

On the quality element side, talker echo is typically described using two parameters, namely, the (frequency weighted) echo attenuation and the one-way echo delay ($T[ms]$). In case of network echo, the echo attenuation is expressed as 'Echo Loss' (EL) (ETSI Technical Report ETR 250, 1996); the attenuation on an acoustic echo path is expressed as weighted 'Terminal Coupling Loss' $TCLw$ (ITU–T Rec. P.342, 2000). The overall echo attenuation called the Talker Echo Loudness Rating ($TELR$) results from the reflection loss, and the attenuation introduced by the user interfaces in send and receive direction (depending on the actual echo-path, see Sections 2.2.2.2 and 3.3.10).

3.3.8 Listener Echo

Listener echo results from multiple reflections of a transmitted speech signal within the network or the employed user interfaces. For overall echo delays in the range of some milliseconds, the echo is not perceived as such, but as a 'hollowness' of the signal (see ETSI Technical Report ETR 250, 1996). In today's networks, listener echo has become less and less important, as multiple reflections without echo compensation are rare, even when the potential usage of echo-prone user interfaces such as hands-free terminals is considered. Hence, listener echo will not be addressed in greater detail in this section. A discussion of the tests described in Bodden and Jekosch (1996) can be found in Möller (2000), p. 165, and ITU–T Contribution COM 12–37 (1997).

3.3.9 Echo Cancellers

Echo cancellers (ECs)[22] are applied to reduce the effect of far-end reflections of the delayed speech signal stemming from the talker at the near end (talker echo). ECs – like other signal-processing components – are increasingly integrated in the user interfaces, as nowadays the main cause for echoes lies in acoustic coupling between loudspeaker and microphone[23].

Different implementations of ECs were proposed (for an overview see Vary *et al.*, 1998, Vary and Martin, 2006, pp. 505–567.). The basic principles are illustrated in Figure 3.10.

In this example, the cancellation of acoustic echo is assumed. In Figure 3.10, the configuration composed of the loudspeaker, the room, and the microphone – the system across which the far end speech signal $x(k)$ is transferred into the talker echo signal $\tilde{x}(k)$ – is referred to as Loudspeaker-Enclosure-Microphone-system (LEM). It is characterized by the impulse response $g(k)$. This impulse response is considered to be time-varying, since, for example, the speaker in the room may change position. The signal $y(k) = s(k) + n(k) + \tilde{x}(k)$ transmitted back to the far end talker is composed of the wanted speech signal $s(k)$, the background noise signal $n(k)$, and the echo signal $\tilde{x}(k)$. The actual EC consists of an adaptive filter with impulse response $h(k)$, which estimates the LEM impulse response $g(k)$. If $g(k)$ exactly matches $h(k)$, no residual echo $e(k) = \tilde{x}(k) - \hat{x}(k)$ will remain after cancellation, and $\hat{s}(k) = y(k) - \hat{x}(k) = s(k) + n(k)$. Many ECs work reliably unless the LEM response changes. In periods of adaptation, a considerable amount of residual echo $e(k)$ may be perceived. Note that the echo estimate $\hat{x}(k)$ is an additive signal. Consequently, in case of large mismatch between the real echo and the estimate, the additive signal may be perceived as additional noise. To suppress the residual echo, nonlinear postprocessing is typically applied. Three types of postprocessing approaches can be distinguished:

- Center clipping

- Level switching

- (Nonlinear) postfiltering.

---

[22]Note that the term EC is used here to refer to the entire system consisting of the actual EC and additional postfilter methods applied for residual echo suppression.

[23]For example, owing to the short distance between loudspeaker and microphone in increasingly small mobile phones, or the loudspeaker level of hands-free terminals

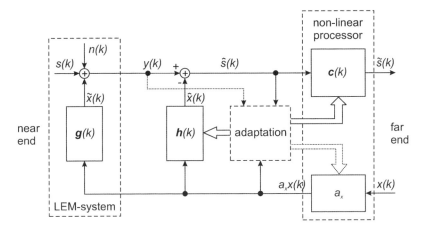

Figure 3.10: Schematic drawing of an EC employed to cancel the echo stemming from a given combination of loudspeaker, room (enclosure) and microphone (adopted from Vary *et al.*, 1998, Loudspeaker enclosure microphone, LEM; see text for details).

A center-clipper cuts out all portions of the transmitted signal $\hat{s}(k)$ that lie below a certain threshold A. If the threshold is set too low, stronger echo-signals may still be perceivable. In turn, if it is set too high, portions of the wanted speech stemming from the near end may be clipped as well, leading to perceivable impairment (clipping/switching).

Alternatively, a level-switching device may separate the necessary attenuation into two parts, one in the receive and one in the send direction (viewed from the near end). The corresponding attenuations $\mathbf{c}(k) = a_y$ of the send path and a_x of the receive path (see Figure 3.10) can be set according to the conversation situation, and the degree of adaptation of the impulse response $\mathbf{h}(k)$ (Vary *et al.*, 1998, pp. 453–454). Particularly critical in this respect are periods of double talk (Gierlich, 1996; ITU–T Contribution COM 12–42, 1997). For example, when the speaker at the near end tries to interrupt the active speaker at the far end, the coefficient a_y in sending direction may still be set to yield the typical target attenuation of 40 dB. In this case, the level-switching device cuts off portions of the near-end utterance (switching/front-end clipping). A badly adjusted level-switching device may cause perceivable switching effects both in send and receive direction (Gierlich, 1996).

As a last example, adaptive nonlinear post filtering can be applied, such as the frequency-selective damping proposed by Martin (1995). In this approach, the spectrum of $\hat{s}(k)$ is time- and frequency weighted based on the momentary spectral properties of the wanted signal $s(k)$, the residual echo $e(k)$, and the noise signal $n(k)$. Under unfavorable conditions like double talk, such postprocessing may lead to a transient, nonlinear distortion of the signal in send-direction.

The effects of echo cancellation on speech quality are summarized in Table 3.5. Note that owing to the dependency on the adaptation of the entire system, and also to transmission delay varying from talkspurt to talkspurt (Section 3.3.3), the effects may be time-varying by nature. Generally, postfiltering devices may introduce additional overall transmission delay. Auditory test methods to measure the impact of a given EC on integral quality or on selected quality features are described in ITU–T Rec. P.831 (1998); ITU–T Rec. P.832 (2000).

Table 3.5: Summary of echo cancellation effects on speech quality. For more details see also ITU–T Rec. G.165 (1993); ITU–T Rec. G.167 (1993).

Feature	Dependency on Conversation	Direction
Residual echo	Both single and double talk, however, lower importance during double talk (masking)	Send
Nonlinear distortion	Both single and double talk, but especially double talk	Send
Switching/Clipping	Especially double talk	Send and Receive
Delay	Both single and double talk	Send

Up to now, no standardized parametric description of ECs is available. However, a number of measurable characteristics of ECs are described, for example, in ITU–T Rec. G.165 (1993); ITU–T Rec. G.167 (1993); ITU–T Rec. G.168 (2002); ITU–T Rec. P.342 (2000). A possible approach to describe the impact of ECs on speech quality could consist in ascribing the introduced effects to those generic quality elements that have already been defined in a parametric way, such as the residual echo (considered as talker echo, see Sections 2.2.2.2 and 3.3.7) and the delay introduced by the EC (see Section 3.3.4). The time-varying degradation of the transmitted speech, for example, owing to center clipping or level switching may then be handled similar to the degradation due to packet loss (see Chapter 4).

However, ECs introduce a new paradigm with regard to speech quality in an actual conversation, namely, the distinction between single and double talk (see Table 3.5). The impact on speech quality of degradations that occur predominantly during periods of double talk as compared to the impact of degradations that occur both during single and double talk have to be studied in more detail, before a coherent mapping between elements and features can be derived. Therefore, the proportions of single and double talk to be found in an average telephone conversation have to be quantified. Moreover, the quality-related weightings of degradations occurring during either single or double talk or both have to be quantified. A study on the amounts of single and double talk as a function of delay[24] is described in ITU–T Delayed Contribution D.214 (2004). It is shown that double talk increases with increasing delay, as the conversation partners are increasingly interrupting each other (see also Kitawaki and Itoh, 1991). A first simplified attempt to quantify the weighting for degradations that occur only during periods of double talk has been proposed by Johannesson (1996; see also ITU–T Delayed Contribution D.012, 1997), in the framework of activities on the extension of the ITU–T's quality prediction model for network planning (the E-model, ITU–T Rec. G.107, 2005, see Section 2.2.2.2, pp. 43).

3.3.10 Loudness

In telephonometry, the loss in perceived loudness due to transmission is typically expressed as the *loudness rating* of the link, or of respective parts of it. Therefore, the loss induced by the system under consideration is compared to the loss induced by a reference system (in case of narrowband telephony the reference is the Intermediate Reference System (IRS),

---

[24]Delay being the requirement for echo to occur.

see ITU–T Rec. P.48, 1989; ITU–T Rec. P.830, 1996, Annex D). The underlying loudness principle is based on the findings by Fletcher and Munson (1937) on critical bands and masking, applied according to Richards (1974).

A technical description of how loudness ratings are calculated in telephony based on amplitude spectra measured in the ISO-preferred 1/3-octave bands can be found in ITU–T Rec. P.79 (1999). For each band, the contribution to loudness is calculated from the part of the speech energy per band, which is above the hearing threshold for continuous spectrum sounds – assuming a generalized speech spectrum at the channel's input. The same is done for the reference channel (instead of this step being performed each time, suitable coefficients corresponding to the reference channel and input speech spectrum have been defined in ITU–T Rec. P.79, 1999). In both cases, the loudness can be calculated as the frequency-weighted sum of the contributions of each frequency band. Then, the *loudness rating* corresponds to the level of overall amplification in dB necessary to adjust the loudness associated with the test channel to the loudness associated with the reference channel. An instructive, more detailed derivation of loudness ratings is also given in Möller (2000; pp. 19–26). Examples of the loudness ratings defined in telephonometry are listed in Section 2.2.2.2 as input parameters to the E-model (ETSI Technical Report ETR 250, 1996; ITU–T Rec. G.107, 2005, see).

It has to be noted that the loudness rating algorithm was initially defined for NB telephony. In case of WB speech (e.g. 50–7000 Hz), the method can be retained, however, by exchanging the reference system for a WB one. Corresponding WB loudness rating coefficients are given in ITU–T Rec. P.79, Annex G (2001), which were adjusted to yield a loudness rating of 0 dB, when the system under test is the NB IRS (IRS; ITU–T Rec. P.48, 1989, and ITU–T Rec. P.830, 1996, Annex D).

The focus of this book is not on the effect of loudness on speech quality. However, loudness plays a crucial role for almost all covered quality elements and features. It is the dominating effect in most cases, as a sufficiently high speech level is the prerequisite for an efficient communication. Moreover, loudness loss is directly related to several other quality elements, for example:

- Together with the noise level on the line, it determines the signal-to-noise ratio.

- In combination with the echo return loss, it determines the amount of echo perceived by a user.

- Excessive loudness may lead to annoyance.

- A speech level still sufficiently high for communication but lower than the optimum may diminish the degradation due to nonlinear distortions (see Chapter 4, Section 4.4).

3.3.11   Noise

The types of noise encountered in telecommunications can be classified as

- Line Noise
- Signal-correlated Noise
- Background Noise.

Owing to the signal loss on long analog lines, noise was one of the major problems in traditional, analog circuit-switched telephony. After its almost complete extinction on all-digital circuit-switched lines, it regained importance with the advent of mobile communications, and bandwidth-critical transmission techniques. In mobile communications, radio-signal variations may cause impulsive noise-like events. In mobile and other networks, where transmission bandwidth may be sparse in periods of high- network traffic, comfort noise may be inserted. Comfort noise may also be perceived as degradation, since it may show a time-varying behavior and can be quite different from the original noise at send side. Noise can generally be characterized by its level, spectral shape and amplitude distribution. While the loudness of WB stationary circuit noise can be captured by the 'psophometrically'-weighted level measurement according to ITU–T Rec. O.41 (1994), no comparable standardized procedure is available for impulsive noise.

Signal-correlated noise is not additive but multiplicative to the speech signal. In telephone networks, signal-correlated noise is typically introduced by quantization processes (i.e. is the quantization noise). A method for synthetic generation of signal-correlated noise is described in ITU–T Rec. P.810 (1996). Until recently, this degradation type was used as a reference degradation in tests on the auditory evaluation of low bit-rate codecs (see Section 3.3.2, and Appendix D.1).

Background or ambient noise is different from line noise in that it may carry useful information for the listener. If the room noise is present at receive side, and a handset telephone is used, the impact of low background noise levels is typically low. If the user at receive side concentrates on the telephone conversation, central processes account for an exclusion of the background noise. As was shown in tests carried out by Bodden and Jekosch (1996), loud noise of significance or emotional impact for the listener, however, lead to a stronger impact on speech quality (for more details on these unpublished and further tests see Möller, 2000, pp. 33–34, and pp. 160–172)

Background noise at send side may have several effects:

- It may carry information on the environment of the conversation partner at send side.

- It impacts the speech signal because of potential masking, lowering speech intelligibility. This impact is partly compensated for by the Lombard effect, which causes speakers to adopt their speaking behavior to the noisy environment, for example, by speaking up and stressing speech sounds differently (see Section 1.1.2; see Möller, 2000, pp. 33–34, and pp. 159–166, for more details on the impact of Lombard speech on conversational speech quality).

- Depending on the type and level of the noise, it may cause annoyance for the listener at receive side.

- Degradations on the transmission path like those due to low bit-rate coding or packet loss may considerably impact background noise transmission, potentially leading to artifacts ultimately making the background noise even more annoying (e.g. when music is coded with a low bit-rate speech codec, or when packet loss degrades an initially pleasant background noise).

3.3.12   Noise Suppression

Two main challenges are related to noise suppression: The unwanted signal is to be suppressed as far as possible, however, affecting the wanted speech signal as little as possible. Technical details on the variety of approaches can be found in (Vary *et al.*, 1998, pp. 377–428) or (Vary and Martin, 2006, pp. 389–504). In principle, the approaches can be distinguished based on the number of channels (i.e. microphones) employed, the method for noise statistics estimation and the approach for 'subtraction' from the wanted signal.

The quality improvement achieved by noise suppression may be counterbalanced by the additional degradations it can introduce. The quality degradation caused by noise suppression algorithms can be due to the effect on the speech signal, due to an audible residual noise – with possibly more bothering characteristics than before suppression – or due to a combination of these effects. For example, residual noise degradations may decrease the stationarity in the noise signal, both in the time- and in the frequency-domain. A prominent example of residual-noise degradations is sometimes referred to as 'waterfall effect' or 'musical tones', which is due to unpleasant sounds created by spectral variations. The 'degraded' residual noise is often audible during speech activity as well as speech pauses. The degradations on the speech signal may show both overall as well as temporally more located effects. For example, unvoiced phonemes may be misinterpreted as noise. Consequently, even speech intelligibility and listening effort, and thus to a large extent speech quality may be degraded, instead of being improved with the achieved increase in signal-to-noise ratio.

The two main types of features related to noise suppression have been accounted for in the standardization of a new method for assessing the resulting speech quality: In ITU–T Rec. P.835 (2003), a three-fold rating procedure is recommended. According to this method, judgments of integral quality are collected (on the 5-point ACR [absolute category rating] scale, the MOS scale), together with two additional separate judgments on the distortion of the speech signal and of the intrusiveness associated with the noise signal. Parametric descriptions of the (signal-processing) elements underlying the different features associated with noise suppression are not yet available.

3.3.13   Linear Distortion and Wideband Transmission

In this section, the transmission link will be regarded as a linear time-invariant (LTI) system. Under this assumption, the spectral shape of a mouth-to-ear speech transmission channel depends mainly on the user interfaces at the send and receive side, and the channel bandpass. In traditional NB wireline telephony, the channel bandwidth is confined to the range 300–3400 Hz, as a result of the channel filter defined in ITU–T Rec. G.712 (1992)[25].

For auditory tests and for telephonometric measurements, the acoustic-electric transfer functions at send and receive side are often adjusted to a generalized model of handset telephones (the Intermediate Reference System (IRS), ITU–T Rec. P.48, 1989; ITU–T Rec. P.830, 1996, Annex D). The IRS-send filter is restricted to the typical 300–3400 Hz telephone band and is characterized by a positive slope of approximately 3 dB/octave. The IRS-receive filter is flat over the entire telephone transmission band. For WB terminals, a corresponding reference system has not yet been defined (for more details on the impact

---

[25]Depending on the exact filter implementation, the lower cut-off frequency may be below 300 Hz.

of user interfaces see Section 5.4.1). A channel bandpass for WB transmission is recommended in ITU–T Rec. G.722 (1988). It has a flat frequency response within its band-limits ranging from 50–7000 Hz.

3.3.13.1 Quality Features: Coloration and Speech Sound Quality

Before the perceptual features related with the quality element *linear distortion* are discussed, a working definition of a term is provided that identifies the quality of linearly distorted speech: speech sound quality.

Linear distortion mainly affects the *timbre* of a transmitted speech sound. In this subsection, the term *speech sound quality* is introduced as a measure of the speech quality[26] that is associated with the effect of a transmission system on the timbre of the transmitted speech.

Timbre and Coloration

A sound event is characterized by its loudness, pitch, perceived duration, timbre and spaciousness (see Section 1.2.1; Letowski, 1989). Of these perceptual features, loudness, pitch, duration and localization can directly be associated with certain physical properties of the sound signal (see e.g. Blauert, 1997; Zwicker and Fastl, 1999). As this book deals with speech communication links with only one channel per direction, aspects of spatial impression and localization are of minor importance for this research.

Principally, all of the remaining features may be affected by a VoIP link. The loudness loss due to the transmission is dealt with in Section 3.3.10. The perceived duration of speech units like phones or syllables may be impacted by the replacement of lost codec frames using repetitions of previously received frames. Time-varying distortions like packet loss are *generally* linked to duration: Duration of degraded passages, of clean, undistorted passages between degradation events, and so on. Aspects of frame or packet loss are dealt with in Sections 3.3.5 and Chapter 4.

Pitch is the psychoacoustic dimension on which two sounds can be ordered according to a 'high-low' criterion. The dimension has ratio-scale properties, so that one sound can be said to be twice as high as another (Zwicker and Fastl, 1999). As discussed e.g. in Brüggen (2001), pp. 7–14, pitch cannot always clearly be separated from timbre. This is linked to the fact that timbre cannot easily be defined directly. Instead, it is typically defined *indirectly* as the criterion based on which a listener can distinguish two sounds that are similarly presented, without using their pitch, loudness or duration as criteria (Pratt and Doak, 1976). Timbre is related to the identity of the sound source (Letowski, 1989). It is of multidimensional nature. Thus, the timbre associated with a particular sound can be described by its position in a multidimensional perceptual space[27].

Any change in timbre, i.e. modification of the position in the feature space, can be described by the vector connecting the initial position (corresponding to the reference timbre) with the new position, the modified timbre. Often, the reference timbre is not

---

[26]The speech transmission quality, if one-way quality is concerned.

[27]In this book it is assumed that noise signals like background and line noise as well as echo signals are considered as [unwanted] additional sounds by subjects, which do not form part of the wanted speech sound. Due to the human ability of auditory scene analysis, this assumption seems valid for a wide range of additional degrading sounds (Bregman, 1990). The timbre related to the unwanted sounds does not form part of the timbre of the wanted sound. However, it is clear that effects like masking due to additional, unwanted sounds may affect the timbre of a given speech signal (e.g. Roberts and Moore, 1990).

explicitly known. However, a listener typically associates an auditory event with a reference point in the timbre space, which reflects a desirable set of features. The directed difference between the reference point and the new position has been defined as *coloration* in the literature (see e.g. Brüggen, 2001, p. 7). Since the desired timbre is application-dependent, the degree of coloration is application-dependent, too.

Speech Sound Quality

Even if a listener does not know a particular sound source directly, for example in case of an unknown speaker, the listening experience still allows certain degrees of coloration to be perceived. Depending on the application, the degree of coloration determines a certain level of quality. For example, for a concert hall a controlled level of coloration due to the room acoustics is appreciated (Blauert and Lindemann, 1986), while for speech communication coloration by the room degrades intelligibility, and is thus less desirable (Houtgast and Steeneken, 1985).

Like timbre, coloration represents a multidimensional quantity, and the reference timbre can be viewed as its origin. For technical applications, however, it is often desirable to have a single-valued quality index at hand. Hence, in the framework of this book, the term *speech sound quality* is introduced, with the following definition:

Def.: **Speech Sound Quality** is the speech quality associated with a particular coloration.

Speech Sound Quality under Linear Distortion

Several studies have been reported in the literature on the quality assessment of linearly (spectrally) distorted speech, for telecommunication purposes as well as for other applications such as hearing aids (Gabrielsson *et al.*, 1985; Gleiss, 1970; Moore and Tan, 2003; Pascal and Boyer, 1990; Voran, 1997).

Owing to the ample context provided in telephone conversations, *comprehension* under telephone bandwidth limitations is very robust (Stickney and Assmann, 2001, and see Section 1.1.2). In a series of paired comparison tests by Gleiss (1970), perceived *naturalness* of bandpass filtered speech showed improvement to the same extent, when the lower 300 Hz band limit was widened by 100 Hz, as when the upper band limit of approximately 3000 Hz was widened by an amount of 5–10 times that step size (i.e. 500–1000 Hz). Interestingly, the relative difference in frequency band increments is consistent with that between the corresponding critical bandwidths (Section 1.1.2). In principle, naturalness of bandpass-filtered speech was found to be independent of loudness and additive noise (Gleiss, 1970). In turn, naturalness varies with different talkers (Gleiss, 1970; Pascal, 1988), since certain spectral characteristics of the connection can be perceived only, when the transmitted signal has appropriate properties to make the channel characteristics audible. Consequently, the widening of the upper limit of the telephone bandwidth was found to increase speech quality more for female talkers, while the widening of the lower frequency limit increased it for male talkers (Pascal, 1988).

An interesting observation was made by Lawson and Chial (1982) from the results of a speech quality magnitude estimation test for different bandpass filters: When the ratings were displayed logarithmically, they showed a (close to) linear behavior over log-frequency. Such a relation was proposed by Stevens (1957) with his power law as a fundamental descriptor of psychophysical phenomena.

The study by Bücklein (1962) has shown that peaks in the frequency response of WB speech are much more audible and also more detrimental to *intelligibility* than corresponding holes are. Flohrer (1968) carried out a test on the *naturalness* of speech degraded by holes in the frequency spectrum. If one hole is introduced into the spectrum at frequency f, the effect on naturalness becomes audible, when the relative bandwidth $\Delta f/f$ exceeds a value of approximately 0.2. The shape of this threshold $\Delta f = 0.2 \cdot f$ plotted over frequency is similar to the dependency of critical bandwidth on frequency. Flohrer (1968) interpreted this observation in the following way: As long as some part of the basilar-membrane that belongs to the critical band under consideration is still being excited by the speech signal, no change in naturalness can be perceived. When the hole in the spectrum exceeds the detectability threshold, this condition is no longer fulfilled. Flohrer (1968) also cites observations by Schäfer (1938), who determined a threshold curve for the detectability of changes in naturalness of bandpass-filtered speech, when the lower or upper limit of the pass-band was varied. Here, too, some similarities to the relation between critical bandwidth and frequency were observed.

The influence of the slope of the channel's frequency response on speech quality can be examined from the overview given by Gleiss (1989). For narrow transmission bands, a slope between 0 and 3 dB/octave was found to be preferred, with the upper end corresponding to the IRS-send slope. Speech intelligibility is improved with a slope of 4–6 dB in the relevant frequency region, i.e. 300 to 4000 Hz (e.g. Gabrielsson *et al.*, 1985). In turn, depending on the source signal, this measure may increase the 'sharpness' of the presentation to an unfavorable level (Gabrielsson and Sjogren, 1979; Gabrielsson *et al.*, 1985). Consequently, the (NB) IRS-send filter can be viewed as a compromise between intelligibility and speech sound quality. For 'Hi-fi band' speech transmission (up to 16 kHz), positive and negative slopes extending over a wide frequency area were both found to be detrimental to naturalness; when positioned in a higher frequency area, positive and negative slopes did not affect naturalness (Moore and Tan, 2003). In their tests, Moore and Tan also studied the effect of spectral ripple, which was found to be most critical for naturalness in the lower frequency region, and of far less effect for higher frequencies.

Another user interface-related factor comes into play when hands-free terminals are employed: The perceived features and quality depend on the room acoustics at send an receive side. In a third-party listening test described in ITU–T Contribution COM 12–42 (1997), different hands-free terminals were assessed in different room-environments. The sound impression ratings (which may have been influenced by additional features like switching or nonlinear distortion in this case) show a clear dependency on the hands-free terminals, but a relatively small impact of the rooms. In turn, a similar test is described in Gleiss (1989) for a hands-free terminal used in two rooms (anechoic [A] and reverberant [R]), which were combined in all possible ways as the rooms at send and receive side (Send → Receive: [A] → [A], [A] → [R], [R] → [A], [R] → [R]). The slope of the transmission band (200–4000 Hz) was either flat or showed a preemphasis of 6 dB/octave. Naturalness as well as 'intelligibility' was assessed by absolute category rating. In this case, the strongest effect clearly stems from the choice of the room at send side: Changing from [A] → [A] to [R] → [A] leads to both a naturalness and 'intelligibility' drop of approximately 1.2 points on a 1–5 point scale (similar to the MOS scale, see Section 2.1). In turn, a change from [A] → [A] to [A] → [R] did not result in any significant naturalness change, and only in a minor decrease of 'intelligibility'. In turn, preemphasis lead to a decrease in naturalness, and interestingly not to an increase in intelligibility. Using a semantic differential method, Brüggen (2001) identified two main factors determining the amount of coloration due to the room at send side: The amount of spectral coloration and the amount of 'diffusity'. The former is related to the early reflections of the room, the latter is related to late reflections, i.e. the reverberation of the room.

A question, which cannot be unequivocally resolved at this point, is whether a bandpass centered at a particular, possibly bandwidth-dependent frequency is preferred over a pure low-pass filter. The tests carried out by Moore and Tan (2003) to evaluate the naturalness of sound reproduction systems imply that this – too – may depend on the source signal: In case of music, a lower cut-off frequency of around 55 Hz was preferred to any other lower band limit, while naturalness of speech was consistently rated higher with a lower limit of 123 Hz (regardless of the upper limit, which was varied between different settings from ≈3500 to 17000 Hz). In a listening test carried out for different combinations of lower and higher pass-band limits, Krebber (1995) found quality to be highest for WB speech ranging from 0 to 11 kHz. Similarly, in the listening test by Voran (1997), quality increased with decreasing lower band limit (for similar bandwidth). One cause for these discrepancies may be the recording procedure of the source stimuli, where certain combinations of the distance from the microphone with the microphone type may result in the proximity effect (Josephson, 1999), amplifying lower frequency components.

Another factor for a possible quality preference in case of a bandpass that attenuates lower frequency components could be the presentation method: If monotic, i.e. one-ear listening is employed, lower frequency components may appear implausible, since in everyday life these components are typically perceived with both ears. Level differences due to head shadow (i.e. interaural level differences, ILDs) consider only higher frequency components, for which diffraction around the head is reduced.

Wideband versus Narrowband

After the definition of a 64 kbit/s WB codec in the mid-eighties, the extension of the narrow telephone bandwidth to WB was under active discussion (ITU–T Rec. G.722, 1988; Modena *et al.*, 1986). According to Pascal (1988), the fundamental requirements for a telephone service to be acceptable are a sufficient loudness and intelligibility, and with less importance its 'pleasantness'. The main accomplishment expected from WB telephony was that of fidelity or naturalness. One can only speculate on the restrictions that held back WB telephony in the infancy of ISDN. One potential cause may be the lack of appropriate user interfaces. Owing to the acoustic leakage between the listening part of a handset telephone and the user's ear, significantly attenuating lower frequency components (see Section 3.3.14), the advantage of a wider frequency range, extending especially to lower frequencies, is not easily reached at low costs. For example, ITU–T Contribution COM 12–11 (1993) cites results from ITU–T Contribution COM 12–63 (1990), that the '[…] tests indicated little difference in quality between NB and WB speech and music signals as listened to on WB handset receivers[…]'. Today, with headsets and hands-free terminals becoming wide-spread, the range of user interfaces is a larger one.

The quality advantage of WB over NB is beyond doubt. Different sources provide different numbers, but the advantage can be said to be around 1.3–1.5 points when expressed on the MOS scale (ITU–T Contribution COM 12–11, 1993; Krebber, 1995; Pascal, 1988). In the tests conducted by Moore and Tan (2003), the naturalness difference between

313–3547 Hz and 123–6987 Hz was as high as 4 points on a 10-point category rating scale. It has to be noted that the perceived quality advantage of WB over NB may depend on the test set-up, and the additional conditions presented in the test: For example, studies by McGee suggest that the presence or absence of low-frequency components may lead subjects to switch between two modes of perception: a WB- and NB-mode, as a result of their switching expectations (McGee, 1964, 1965).

For WB speech, attribute ratings were related to properties of the frequency response by Gabrielsson and Sjogren (1979) and Gabrielsson *et al.* (1985). For example, 'clearness' was found to be related to a broad frequency range with a flat frequency response and low nonlinear distortion (Gabrielsson and Sjogren, 1979), but can be increased with a small positive slope for higher frequencies (> 1 kHz, see above; Gabrielsson *et al.*, 1985). In turn, a reduction in low-frequency components (< 1 kHz) reduced the perceived 'fullness', and the higher frequency bands (> 1 kHz) determined the degree of 'brightness'. According to Gabrielsson *et al.* (1985), 'nearness' is assumed to be mainly related to low-frequency components, but was found to increase with small positive slopes in the higher frequency range as well.

3.3.14   User Interfaces

Instead of traditional handset telephones, a variety of user interfaces may nowadays be employed, such as:

- Headsets,

- Hands-free terminals,

- Mobile phones,

- Cordless phones.

Hence, a user of a telecommunications service is rarely confronted with a transmission path of an IRS-shaped spectrum (Section 3.3.13). The geometry, especially of mobile or cordless phones leads to considerable spectral deviations from the IRS-shape. In receive direction, the acoustic leakage between the handset and the listening-ear leads to a signal loss in the region of lower frequency bands (up to 700 Hz, depending on the used phone). The form and shape of the inevitable gap between handset and ear – and thus the low-frequency loss associated with it – depend on the applied force with which the telephone is pressed against the ear (Krebber, 1995, pp. 40–59). For the send-direction, a different effect has to be mentioned, since handset telephones are getting shorter and shorter, especially in mobile telephony: The length of a telephone and the angular position it is held in, both lead to a uniform attenuation or amplification of the entire transmission spectrum, without introducing considerable linear distortion. In both cases, an optimal region exists, where the microphone is closest to the mouth, and the level of the picked-up speech is highest. Consequently, for the overall shape of the frequency response the handset position relative to the ear is of much higher importance than the position relative to the mouth (Krebber, 1995, pp. 19–20). Further sources of deviation from the IRS-shape are, for example, the differences in the chosen microphone- or loudspeaker technology.

Other types of user interfaces may likewise have a considerable impact on the overall frequency response: In different tests, headsets were characterized in send-direction (ITU–T Contribution COM 12–47, 2002). Here, some changes to the characteristics imposed by the flat channel bandpass may occur, depending on the manufacturer. In receive direction, the frequency response of headsets was shown to be much more critical (ITU–T Contribution COM 12–54, 2003). For some manufacturers, the bandwidth in receive direction was confined to a range as narrow as 1000–3000 Hz. Hands-free terminals introduce another factor, namely, the influence of the room (both at send and at receive side, see Section 3.3.13).

In addition to the influence on linear distortions, user interfaces like hands-free terminals are typically equipped with signal-processing components like ECs or noise suppression, which are bound to additional degradations (see Sections 3.3.9 and 3.3.12).

Apart from the influence on the transmission channel, the user interface is also related to a particular expectation, i.e. internal reference of the user. For example, when headsets are employed, the expected quality is typically higher than in case of a telephone handset (Möller, 2000, pp. 133–141). This aspect is discussed in some more detail in Section 3.7.

3.3.14.1 Sidetone

Different sidetone paths exist across which the talking user perceives his own voice as well as ambient noise. In telephonometry, a distinction is made between ambient noise perception (listener sidetone) and own voice perception (talker sidetone). If the user employs a handset telephone, a direct air path exists to the free ear. Owing to central auditory processing, for low and medium levels of noise this path is typically deactivated during a telephone conversation. Several sidetone paths exist to the ear that is covered by the handset: An acoustic air path is due to the acoustic leak between the receiver and the ear. Another acoustic path results from bone conduction (e.g. Pörschmann, 2001). Finally, sound can also be borne by the mechanical parts of the receiver. An additional (electrical) sidetone path is typically employed to compensate for the insertion loss in hearing one's own voice, which is caused by user interfaces shielding one or both ears. Moreover, the presence of electrical sidetone enables the user to verify whether the user interface is operational. When devices such as hands-free terminals are used, no shielding occurs, and hence no artificial sidetone has to be presented.

In case of additional ambient noise, the resulting listener sidetone may be detrimental to quality. The sidetone pick-up of user interfaces is thus described by both a parameter quantifying the talker sidetone attenuation and one quantifying the listener sidetone attenuation (sidetone masking rating *STMR*, and listener sidetone rating *LSTR* respectively, which are based on the loudness rating concept described in Section 3.3.10). If – assuming a typical talker sidetone level – a total delay beyond 4–10 ms is added to the sidetone path, the talker sidetone starts to degrade the talking quality; when the delay exceeds approximately 30 ms, a distinct talker echo is perceived (Appel and Beerends, 2002; ETSI Technical Report ETR 250, 1996, p. 50; see also Section 3.3.7). Some more detailed considerations on sidetone are summarized in Möller (2000, pp. 28–30).

3.4 Quality Dimensions

Like many other perceptual magnitudes, one-way speech transmission quality is of multidimensional nature: In multidimensional scaling experiments it was confirmed that a speech sample transmitted across a particular transmission link can be associated with a particular position in a multidimensional feature space (see e.g. Mattila, 2001 for an overview of multidimensional scaling applied to different contexts of perception). Prominent individual examples from the domain of speech quality evaluation are McGee (1964), McDermott (1969), Gabrielsson and Sjogren (1979), Hall (2001), Mattila (2002a), Bappert and Blauert (1994), and Wältermann (2005). Some of these tests are summarized in Table 3.6[28].

Four of the test series are of particular interest for this book (and others will be referred to where appropriate):

1. Hall (2001) has delivered a proof based on multidimensional scaling that signal-correlated noise is perceptually different from low bit-rate coded speech: The three dimensions he determined and the respective loadings for different conditions show that signal-correlated noise is the only degradation in the test strongly loading on 'noisiness'. This is of relevance, as until recently, signal-correlated noise of different levels has typically been used as the reference condition for the evaluation of low bit-rate codecs (see also Sections 3.3.2, and D.1). Hall's data indicate that the replacement of this procedure by a method using reference stimuli taken from a set of other, known codecs is appropriate, as these can be assumed to be related to the same feature space as potential codecs under test[29] (the new method is defined in ITU–T Rec. P.833, 2001, see also Appendix D). The three dimensions Hall (2001) found could be mapped to quality ratings obtained on the MOS scale. Therefore, Hall used a multivariate regression on the linear combination of the dimensions (implying a vector model of preference, see Section 2.1.2.4). The result is a simple formula of the form:
$$\text{MOS} = a \cdot dim_1 + b \cdot dim_2 + c \cdot dim_3 + d,$$

   where dimension 1 showed by far the highest impact on quality, followed by dimension 2 ($dim_1 \equiv$ 'naturalness', $dim_2 \equiv$ 'noisiness', $dim_3 \equiv$ 'amount of low-frequency content').

2. The test series described in Bernex and Barriac (2002; see also ITU–T Contribution COM 12–23, 2000) is the first of the only two multidimensional analyses reported in the literature on the features related to VoIP packet loss (the other study is described in ITU–T Delayed Contribution D.071, 2005; Wältermann, 2005). For their study, Bernex and Barriac degraded two speech samples, one uttered by a female, one by a male, with consecutive packet loss of three packets. Coding algorithms were the G.723.1 at both 5.3 and 6.3 kbit/s, and the G.729B (with the option of including voice activity detection). For each stimulus, they displaced the loss pattern relative to the speech samples and thus collected 70 different samples per condition. All in all, 12 conditions were studied, 10 of which differed only in one parameter, namely, the codec, the gender of the speaker or whether voice activity detection was used or not, and two conditions representing combinations of these parameters.

---

[28]It was chosen to distinguish different dimensions and to combine others based on the descriptions of the dimensions given in the respective publications. Especially in tests employing a multitude of different types of degradations, the identified dimensions can be considered as 'composite dimensions'. The dimensions obtained from tests where similar types of degradations were studied (e.g. Brüggen, 2001) may be considered as 'subdimensions' of such composite ones. As a consequence, this display of results should be interpreted with some caution.

[29]The impact on quality of using signal-correlated noise as reference in low bit-rate codec evaluation tests, and of applying the corresponding test data normalization according to the 'equivalent-Q' method (ITU–T Rec. G.113, 1996), was studied by Möller based on absolute category rating listening and conversation as well as isopreference tests (Möller, 2000, pp. 121–129). He found that the MNRU-method may lead to an erroneous test data transformation.

Based on a free categorization task, the subjects were asked to group stimuli that they thought sounded similar. After the creation of groups, the subjects were required to verbally describe the groups they had created. The output data were converted into similarity matrices and then analyzed using multidimensional scaling on one hand,

Table 3.6: Summary of tests for the multidimensional analysis of speech transmission. Abbreviations: 'MDS' ≡ Multidimensional Scaling; 'SD' ≡ Semantic Differential; 'Pref. Similar.' ≡ Preference similarity scaling; 'Pref.' ≡ Preference ratings; 'ROD' ≡ Free verbal description of per dimension rank-ordered stimuli (see text for details); 'ME' ≡ Magnitude Estimation of quality; 'MDPS' ≡ Multidimensional Preference Scaling; 'Free Cat.' ≡ Free Categorization. The different studies are: 'Pascal' ≡ Pascal (1988); 'McGee' ≡ McGee (1964, 1965); 'Gabrielsson' ≡ Gabrielsson and Sjogren (1979); Gabrielsson *et al.* (1985); 'Bruggen' ≡ Brüggen (2001); 'Hall' ≡ Hall (2001); 'Bappert' ≡ Bappert and Blauert (1994); 'McDermott' ≡ McDermott (1969); 'Mattila' ≡ Mattila (2001); 'Bernex' ≡ Bernex and Barriac (2002); ITU–T Contribution COM 12–23 (2000); 'Wältermann' ≡ ITU–T Delayed Contribution D.071 (2005); Wältermann (2005).

System	Study	Methods	Intelligibility/Clarity	Naturalness	Noisiness	Amount low frequencies/Band limitation/Fullness	Amount high frequencies/Sharpness	Bubbling	Color of sound/coloration	Smoothness/interrupted	Diffusity/Directness	Signal versus background distortion	loudness
bandpass (9)	Pascal	MDS & SD				X	X						
bandpass (15)	McGee	SD & Pref. Similar.			X	X							
bandpass (div.)	Gabrielsson	SD & Pref.			X	X	X	X					
rooms (14)	Brüggen	SD								X	X		
codecs (10)	Hall	MDS & ROD & Pref.			X	X	X						
codecs (12)	Bappert	MDS & SD & ME			X			X					
PSTN (22)	McDermott	MDS & MDPS			X						X	X	
GSM (clean: 44)	Mattila	MDS & SD & Pref.			X	X	X	X	X				
GSM (noisy: 44)	"	MDS & SD & Pref.			X	X	X	X					
packet loss	Bernex	MDS & Free Cat.				X		X	X				
VoIP/PSTN (14)	Wältermann	MDS & SD			X	X		X	X				

and a tree analysis method on the other hand. The resulting dimensions describe different features related to packet loss, expressing the dependency of the features on the loss location. A fourth 'dimension' not displayed in Table 3.6 was that of 'no perceived impairment'. An analysis of the results revealed that some subjects had used the criterion of 'equal location of degradation' in the similarity categorization process, while others judged based on different perceptual features caused by the packet loss. Another interesting result of their study is the negligible perceptual difference between the codecs G.723.1 and G.729 under packet loss. It confirmed the assumption that the similar PLC algorithms applied in these codecs lead to similar perceptual effects.

3. Mattila (2001) conducted a broad analysis of the perceptual dimensions underlying speech transmission quality in mobile communications. A large number of signal processing conditions were used (44) such as, different codecs with and without transmission errors, center clipping, noise reduction systems, different user interfaces at send and receive side and some artificial linear filtering and nonlinear distortion conditions. Since the perceptual effect due to packet loss is similar to that of transmission errors on a mobile network radio link (that are partly translated into frame-erasures at receive side)[30], the test obviously allows conclusions to be drawn also on VoIP-transmission quality. The test was carried out separately for two different background noise situations ('clean' and 'noisy', where the latter was a car noise), yielding an overall of $2 \cdot 44 = 88$ conditions. The test methodology employed a special screening and training of the test subjects prior to the actual analytic test (Mattila, 2001; Mattila and Zacharov, 2001, pp. 105–120). Moreover, the descriptive language used in the semantic differentiation task was specifically developed together with the selected listening panels. This way, inter- or intraindividual inconsistencies in the formation of the perceptual space could be avoided (e.g. in Pascal, 1988, the 'naive' listeners did not differentiate between higher quality conditions). In the Individual Difference Scaling (INDSCAL) analysis of the similarity data, five dimensions were found for clean speech, and four dimensions for noisy speech. The dimensions are listed both in Tables 3.6 and 3.7, using the labels Mattila determined in his attribute scaling tests (Mattila, 2001, 2002a, pp. 169–202). It has to be noted that the terms depicted in Table 3.7 are the ones chosen by Mattila (2001), which were rephrased for Table 3.6 to achieve a coherent framework applicable to different test results. Similar to Bernex and Barriac (2002), Mattila found an individual dimension linked to time-varying distortions such as bit errors in GSM and resulting frame loss ('smooth – fluctuating – interrupted', in Table 3.6 referred to as 'smoothness').

Mattila used two different external preference mapping models in order to relate the perceptual dimensions to perceived quality (see Section 2.1.2.4). In Mattila (2001), a vector model of preference is employed, that assumes a linear relation between quality and the perceived quality dimensions. With this approach, the integral quality judgments he had obtained can be predicted with a multiple correlation between test results and predictions of 0.95, and a low root mean squared error (i.e. of 8.5 on the employed quality line scale ranging from 0 to 100). In Mattila (2002a), he has extended the preference mapping by using an ideal-point model of preference

---

[30]This is pointed out also by the similar dimensions 'interrupted'/'smoothness' found by Bernex and Barriac (2002) and Mattila (2001), respectively.

as the basis of a multivariate regression. This choice was made as the dimension 'dark–bright' found for clean speech (in Table 3.6 subsumed under 'band limitation') is likely to have an ideal point of preference. In other words, neither of the two extremes dark and bright are expected to be infinitely desirable (see Section 3.3.13.1). A similar approach was chosen for noisy speech. Here, the dimension 'low–high' had been found in correspondence to the 'dark–bright' dimension for clean speech[31]. The multivariate regression resulted in a set of weights relating the dimensions, interactions between dimensions, and squared terms of the dimensions to integral quality (see Section 2.1.2.4). Both for clean and for noisy speech certain interactions and squared terms were found to be less important for quality than others. Hence, in a simplification and refinement of the model, Mattila (2002a, 2003) reduced the set of variables used for the preference mapping. The resulting equation allows the speech transmission quality of a given system to be estimated based on its position in the multidimensional feature space. In Table 3.7, the relative weights of different relevant dimensions, interactions and squared terms are listed. With the refined model, the multiple correlation between predictions and test results was slightly increased to 0.97 (as

Table 3.7: Relative weights for quality obtained by Mattila for different dimensions and combinations thereof: clean (Mattila, 2002a) and noisy speech (Mattila, 2003), multivariate regression of multidimensional scaling data.

	Variable	Relative weight [%]
Clean speech	Synthetic – natural	14.6
	Dark – bright	4.6
	Smooth – fluctuating – interrupted	23.1
	Bubbling	10.4
	Noisy	1.0
	Synthetic – natural * bubbling	0.7
	Dark – bright * bubbling	10.5
	Smooth – fluctuating – interrupted * bubbling	22.2
	Smooth – fluctuating – interrupted * noisy	8.3
	Synthetic – natural * synthetic – natural	1.3
	Bubbling * bubbling	1.4
	Noisy * noisy	1.9
Noisy speech	Low – High	13.6
	Synthetic – natural; boiling	21.1
	Smooth – fluctuating – interrupted	46.1
	Noisy	3.9
	Smooth – fluctuating – interrupted * noisy	10.0
	Noisy * noisy	5.3

[31] It is interesting to note that the subjects detected the band limitation underlying this dimension mainly based on the background noise in case of noisy speech, and based on the speech signal in case of clean speech.

opposed to 0.95 in case of the vector model). Correspondingly, the root mean squared error was reduced from 8.5 to 6.4.

Both for noisy and clean speech the 'smooth – fluctuating – interrupted' dimension proofed to be most relevant for integral quality. For clean speech, by far the most important interaction was found to be the one between 'smooth – fluctuating – interrupted' (i.e. 'smoothness') and 'bubbling'. Here, 'bubbling' is a dimension related to a characteristic that varies fast relative to the speech signal, for example, due to LPC (linear predictive coding) distortions. The interaction with 'smoothness', which corresponds to a characteristic varying relatively slowly along the speech signal, can be illustrated as follows (Mattila, 2002a): When speech is strongly interrupted, 'bubbling' hardly has any effect on quality. In turn, when speech is smooth, 'bubbling' clearly degrades quality. Obviously, 'smoothness' is a dominant effect that works like a switch with regard to 'bubbling'.

The interaction between 'noisiness' and 'smoothness' found for the noisy speech works just the other way round: Only when 'noisiness' is low, increasing 'smoothness' leads to a considerable increase of quality. This finding can be explained as follows: Firstly, additional noise may lead to phonemic restoration, partly counterbalancing the effect of lacking smoothness, i.e. of perceivable interruptions. Secondly, additional noise may to some extent mask the artifacts introduced by the decoder in case parts of the transmitted speech are missing. Thirdly, noise is a more obvious degradation than the artifacts or small interruptions impacting the smoothness-dimension, so that the users' attention is directed to the 'noisiness' rather than the 'smoothness'.

Other interactions like the 'dark – bright * bubbling' one can be explained with the fact that in the judgments of similarity different fine-grained dimensions were grouped to form overall dimensions, which were used by the subjects to distinguish the stimuli (as was revealed by the comparison with the attribute scaling data; Mattila, 2002b).

4. Wältermann has conducted three different tests: an absolute category rating test on speech quality, a multidimensional scaling experiment and an attribute scaling test using a semantic differential method (ITU–T Delayed Contribution D.071, 2005). He studied 14 different network conditions typical of VoIP/PSTN networks, including time-varying distortions such as packet loss, different codecs, background noise, noise-suppression algorithms and user interfaces such as hands-free terminals. A total of 20 subjects took part in each of the three test runs. The analysis of the multidimensional scaling (MDS) data leads to a stable configuration with four dimensions: 'noisiness', 'interruptedness', 'frequency content' and 'directness'. The semantic differential data yielded a reduction of the dimensionality to three instead of four, merging the dimensions 'frequency content' and 'directness' to one dimension ('directness/frequency content'). For the investigated samples, Wältermann found the highest impact on quality for 'interruptedness', followed by 'noisiness' and 'directness/frequency content' (see Table 3.8). To determine a predictor of quality that is based on the loadings of the underlying quality dimensions, Wältermann employed a vector model of preference. Using a multivariate, linear regression, he determined the (normalized) regression coefficients shown in Table 3.8. With the resulting quality

Table 3.8: Z-score normalized coefficients of the linear regression carried out by Wältermann to map the observed dimensions to quality, based on a vector model of preference (ITU–T Delayed Contribution D.071, 2005). Note that a negative sign indicates a decrease of quality with an increasing amount of the respective dimension, while a positive sign indicates an increase of quality with an increase of the respective dimension.

Variable	Weight
interruptedness	−0.698
noisiness	−0.472
directness/frequency content	0.457

prediction formula, again in the form

$$\text{MOS} = a \cdot dim_1 + b \cdot dim_2 + c \cdot dim_3 + d,$$

as in case of the results obtained by Hall (2001), Wältermann could explain about 90% of the total quality variance.

Wältermann's results are of particular interest for the modeling of quality of VoIP networks:

- According to his study, 'interruptedness' plays the major role with regard to quality. It results from degradations such as packet loss, but may also stem from signal-processing components such as noise suppressors. While packet loss is covered by current speech quality models such as the E-model (ITU–T Rec. G.107, 2005), signal processing components like noise suppressors or ECs are not.

- The dimension of 'noisiness' appears to be less important for quality than 'interruptedness', but still plays a considerable role. Since it is the most fundamental type of degradation, noise is typically quite well considered by most quality models (for the E-model, this has been shown by Wältermann in ITU–T Delayed Contribution D.071, 2005).

- The aspect of frequency distortion as it is introduced by user interfaces such as hands-free terminals is not covered by most quality models, although its role is comparable to that of the two other dimensions identified by Wältermann. It could be argued that the markedness of certain degradations studied in a given test may not reflect the markedness of these effects in real networks. However, it becomes clear from Wältermann's tests that user-interface related effects – the internal signal processing as well as the electro-acoustic properties – have to be studied in more detail, if quality models are to be developed that can safely be applied to VoIP networks. Obviously, the 'speech sound quality' introduced earlier in this book (Section 3.3.13) has to be specifically addressed in this case.

The author's research on this topic and an approach for speech quality prediction under bandwidth restrictions of WB speech are discussed in Chapter 5.

3.5 Combined Elements and Combined Features

Up to this point, only individual quality elements have been discussed. However, in real networks combinations of different quality elements are typically encountered. In such cases, an interaction between impairments may already be effective at the signal level: For example, the loudness rating of the line has an impact on a possible echo level as well. The situation becomes more complex, when a direct interaction on the signal level is not to be expected, but different impairments interact on a perceptual level. For example, how does a user perceive quality, when s/he is confronted with packet loss and talker echo?

In the literature, only few studies on combined degradations are reported. For example, in perception tests performed by Allnatt and coworkers on the degradation of video-pictures (Allnatt, 1975; Lewis and Allnatt, 1965), it was found that an additivity of the studied impairments held, if an appropriate perceptual scale was used[32]. In their case, independent impairments were additive when expressed as a perceptual magnitude on a ratio-scale. The discussion of multidimensional scaling implies that the degradations used here represent orthogonal features, which were both bound to a vector-type preference relation (see also Section 2.1.2.4). Other tests were conducted by Bodden and Jekosch (1996) and Möller (2000), mainly to evaluate the additivity assumption of the E-model (see Möller, 2000, pp. 167–172). Our auditory tests on combined elements and features and implications for the validity of the additivity principle of the E-model are discussed in Chapter 4, Section 4.4.

3.6 Listening and Conversational Features

The auditory features related to a telephone link can coarsely be divided into listening and conversational features. While listening features are perceived already in a listening-only situation, conversational features such as delay or talker echo require a conversational situation. This section briefly discusses issues associated with the perception of listening features in a conversational situation.

An aspect relevant for both listening and conversational features, but bound to a conversational situation is that of asymmetry (Richards, 1973, pp. 311–313). For example, one direction of a two-way communication link may suffer from considerable loudness loss. Hence, speech of one conversation partner (A) may be strongly attenuated, while the other's speech signal is not affected (B). As a result, quality as perceived by A, too, is lower than in case of optimal speech level at both sides. This is due to an adaptation of B's behavior to the quality she/he perceives from A: B assumes symmetry, i.e. 'shares' the perceived difficulty with his partner. With today's increased variety of user interfaces and the potential routing of speech packets across different network paths – possibly leading to very different loss and delay conditions for individual directions – asymmetry is of particular importance for VoIP networks.

---

[32]It is recalled that these studies formed the basis of the impairment factor principle underlying the E-model, which assumes additivity of individual impairment factors (see Section 2.2.2.2).

Conversation tests represent the most realistic framework for speech quality assessment in telecommunications (see Section 2.1.1.4). In turn, they are related to considerable effort and costs. Hence, most tests carried out to evaluate a network component that impacts listening quality are listening tests. The general question is whether the impact of listening features is judged similarly in a conversational situation as it is done in a listening situation. Several tests were carried out by Möller (2000; pp. 129–133) to answer this question. They showed that for stationary listening degradations like signal-correlated noise (MNRU), low bit-rate coding and different bandwidth-limitations, listening-only and conversation tests lead to comparable results. In some cases, though, a tendency for rating more critically in a listening test than in a conversation test could be observed. A similar observation was made by Gros and Chateau (2002) for quality degradation due to packet loss. Depending on the test type, the subjects' attention may be focused on different criteria. In a conversation test, the content, i.e. what is said, is of much higher relevance than in a listening test. Consequently, subjects tend to be more tolerant towards small degradations in this case. In turn, in a listening test subjects typically focus more on the form of a stimulus, and are more critical with its degradation.

Another effect was found by Gros and Chateau (2002) when comparing conversation and listening test results for the integral quality of speech degraded by different packet loss profiles: A recency effect could clearly be found for the listening situation, but not in the conversation test. A possible explanation is related to the nature of the recency effect: Due to the conversational mode, a subject can associate strong changes in quality with certain instances of the conversation, and thus more readily remember them. According to this interpretation, the increased focus on semantic aspects may enable a recoding of the perceptual degradation into 'place-in-conversation' information.

3.7 Desired Nature

So far, this chapter has mainly dealt with the impact of the actual transmission system – including the electro-acoustic interfaces – on the features and the quality perceived by users. However, quality results from a comparison of the perceived features with *desired* features (see Chapter 1, Section 1.2).

This section focuses on the question 'what *impacts* the desired features?', rather than on the question 'what *are* the desired features?'. Answers to the latter question can be obtained from the relation between multidimensional scaling data and corresponding quality ratings: The dimensional configuration yielding optimum quality provides a good guess of what the desired features might be (see Section 3.4, 2.1.2.4 and 5.4).

The desired features may depend on several factors, which can be external, or internal (*modifying factors* and *personal factors*, respectively; see Chapter 1, Section 1.2). The following list provides examples of such factors, which may also be interrelated[33]:

- The properties of the acoustic signal itself may be associated with a particular category of perception, for example, telephone speech puts the listener into a different perception 'mode' than radio-quality speech (see Section 5.4).

- The content, for example, of what is said by a conversation partner (see Section 5.4).

---

[33]In Möller (2000, pp. 133–141), many of these factors are summarized into three more general groups.

- The mobility, for example, whether a mobile or wireline service is used.

- The environment, for example, whether an interlocutor is situated in a bus or in a private living-room.

- The type and nonacoustic properties of the user-interface, (e.g. whether it is a handset telephone or a headset).

- Additional or alternative service 'features'[34].

- The cost of the particular service.

- The group, the particular user belongs to.

- The call motivation, for example, whether a professional call or a private call is made.

- The specific task associated with the call.

- The long-term experience, i.e. the reference the user has stored for certain features (e.g. representing his everyday telephone connections).

- The 'midterm' experience, for example, the features the user was confronted with during more recent calls.

- The user's current emotional state.

The cost factor is of particular interest for VoIP networks. In principle (and in contrast with classical wire-line telephony), VoIP allows a trade-off between a certain level of quality and the corresponding cost (e.g. per minute) to be made: Certain quality elements can be negotiated during call set-up. Depending on factors like their motivation and the task to be completed, users may choose more expensive, high-quality connections or less expensive, lower quality ones. In case of such a pricing scheme, quality is typically controlled by the choice of codec (i.e. the choice of bit-rate), the choice of sender-based loss concealment (requiring more data to be sent), the delay and jitter requirements to be met by the network and the recipient, and the tolerable packet loss rate (the latter two are both achieved e.g. by bandwidth allocation). With these options, the overall bandwidth necessary for a particular call can be reduced, and the overall call-capacity of the network be increased.

The user interface factor too, is of special relevance for VoIP: Due to the convergence of different network types (see Figure 3.2), a user may have different choices of user interfaces to access the network. However, the transmission technology applied in the network may be quite different from what the user interface typically is associated with. For example, when an analog phone is used, but the core network involves VoIP, the perceived features may be unexpected. Obviously, the long-term experience plays a role as well (see Duncanson, 1969, for a study on long-term vs. short-term experience with telephone calls).

Another important aspect is the potential *improvement* of VoIP speech quality over PSTN speech quality, due to the potential of transmitting WB speech. Due to the fact that wideband *telephone* speech is not yet in service for a long time and at larger scale, it cannot have led to any long-term experience as yet. Instead, the usage of a headset and the perception of WB speech features may evoke reference to other 'schemas' of speech

---

[34]Here in the sense of service option, or of a service like SMS (Short Message System), which is additional to the main telephony service.

perception, like that of radio speech. In turn, the desired features related to traditional *narrowband* telephone speech may be affected by the introduction of WB telephony (see Section 5.4).

Some of the influencing factors presented in the above list – external as well as personal – have been studied in the literature (see also Möller, 2000, pp. 115–145).

3.7.1 Context

The context of use was studied in a conversation test by NORTEL (ITU–T Delayed Contribution D.009, 1997). In this test, an attempt was made to quantify the advantage of using a mobile phone over using a wireline phone in terms of an equivalent signal-to-noise ratio. The results show that the larger the relative mobility compared to the fixed wireline scenario, the greater the tolerance towards lower signal-to-noise ratios, i.e. the larger the 'advantage' measured in circuit noise level.

Complementary observations were made in a conversation test conducted by Möller (2000; pp. 137–141). The test subjects had to assess 14 different telephone lines degraded by 'stationary' impairments (signal-correlated noise, low bit-rate codecs or talker echo). One subject of each test pair was using a portable phone (i.e. a cordless phone connected to the network via a cable) and was situated on a balcony outside the laboratory building. The other subject used a traditional, wireline handset telephone, and was seated inside an office-type room. While the ratings averaged over subjects did not reveal any considerable difference between the two situations, a more detailed analysis did: Experienced mobile phone users rated the mobile network access by more than 0.5 points MOS higher than the unexperienced ones. In contrast, the wireline network access was rated almost identically by both user groups. This observation indicates an interaction between the user group factor and the context factor. Another conclusion can be drawn from the comments of the subjects after the test: Many of the subjects, with prior mobile phone experience, argued that the auditory features perceived during the test did not match the (time-varying) features they knew from real mobile networks. This finding points to an interaction between factors such as the user group, the long-term experience and the context of use.

3.7.2 Cost

MOS ratings were not the only judgments asked from the subjects by Möller. After each call, they also had to indicate the price they were willing to pay for the connection they had been using (Möller, 2000, pp. 141–145). Interestingly, a clear offset in prices could be observed between the mobile and the wireline access, directly from the average over all subjects. Regardless of the quality level, subjects were willing to pay more for the mobile than for the wireline access. In this case, the cost factor interacted with the context factor, however, without affecting the quality ratings.

The cost factor also played a role in the trade-off experiment conducted by Watson (2001). Her study reflects the pricing schemes made possible with VoIP, namely, of choosing a particular trade-off between the quality level and the cost of a call. In Watson's conversation test, quality could be increased by the subjects *during* a call at the expense of some budget, which was available to them for the entire call (Watson, 2001, pp. 145–170). The higher the quality, the faster the budget was spent. The call quality was controlled by the subjects using a slider method similar to the one applied by Gros and Chateau (2001) for quality assessment, in this case used for quality control. The subjects could position the slider at a certain position of their choice that was linked to a certain packet loss level, and to a certain budget (less budget spent for higher loss levels). The budget clearly impacted the threshold of quality tolerated by the subjects: In the 'budget' case, the first change to the slider position was made at an average loss rate of $\approx 26\%$, while in the no budget case, the loss rate was decreased by the subjects at an average of $\approx 12.5\%$. The mean loss rate tolerated in the budget case was of $\approx 28\%$, and in the case of no budget of $\approx 20\%$. These results underline that the desired features (i.e. the threshold of tolerance) may be modified by the cost of a service.

Another series of tests regarding VoIP quality and pricing was conducted in the framework of the M3I-project (Hands *et al.*, 2001). The EU-funded project was aimed at finding an efficient and yet acceptable, dynamic pricing scheme for Internet services. In one of the experiments, three groups of subjects were formed. Each group was assigned a virtual pricing scheme. The groups were labeled according to the relative (virtual) amount they were assigned to pay for their service, as 'gold', 'silver' and 'bronze' users. Furthermore, it was explained to the subjects that the tariff-group was linked to a certain level of quality. In a series of conversation tests carried out at BT labs, users were asked to assess the speech quality of different lines degraded by random packet loss (0–10%). In case of increasing packet loss, the acceptability ratings obtained from the 'bronze' group were least affected by the degradations, followed by the silver and the gold group. Obviously, expectations were higher, the higher the (virtual) user group level assigned to the subjects.

3.7.3 Service Options

A different type of trade-off was investigated in the study reported in T1A1.1/ 2000-014 (2000). In this case, the test subjects had to trade-off an additional service option but with reduced speech quality ('Package B') with a clean connection but without additional service option ('Package A'). The additional service consisted in 'unified messaging' i.e. a service including e-mail, voice and fax retrieval by PC or telephone, with a text-to-speech tool for e-mail and fax headers. The messaging option was not actually used in the test, but its functionality was demonstrated prior to the test. During each conversation, the subjects could chose between the two different packages (A and B), which they selected depending on the additional degradation introduced in B (packet loss and delay). Obviously, the subjects accepted quite considerable amounts of delay (730 ms) or packet loss (3.41%) in return for the additional messaging option.

3.8   Open Questions

As can be retained from the earlier discussion of elements and features, a broad range of unresolved questions can be formulated. These questions are related to two different points of view on speech quality research:

- *How* do users perceive the speech quality associated with the features of VoIP? This point of view corresponds to a subject-oriented perspective, with a particular focus on quality features.

- *What* is the speech quality associated with a particular VoIP network set-up? Here, an object-oriented perspective is taken, seeking to relate quality elements to quality.

In the remainder of this book, three main topics are addressed in more detail:

(a) Time-varying distortion: How do users perceive integral speech quality in case of time-varying distortion? What is the quality impact of longer-term, that is *macroscopic* variations of the shorter-term (*microscopic*) behavior of the channel? The quality element addressed is packet loss.

(b) Linear distortion: How do users perceive the quality of linearly distorted speech? How do users perceive the quality of WB telephone speech? What is the impact of the related quality elements on speech quality? The quality elements addressed are the transmission bandwidth, the employed speech codec, and user interfaces. A special emphasis is laid on the comparison between WB and NB speech, and on factors impacting the expectation of the user (i.e. the desired nature).

(c) How do users perceive quality under combinations of (a) or (b) with better studied degradations like echo, noise or delay?

This choice is focused on the intersection between pure research and a pragmatic, application-oriented approach: VoIP is an application of today's network and signal-processing capacities to speech transmission over an existing data network-infrastructure. In turn, the combination of speech on one hand and packet-based transmission on the other leads to novelties both on the level of the quality elements, and on the level of the associated quality features. From the different elements, time-varying distortion and WB speech transmission are the ones related to the most prominent quality features character-istic of VoIP. These new features turn speech quality of VoIP networks into an object of perception-oriented research.

In turn, any progress in the understanding of the relation of VoIP speech quality elements to VoIP quality satisfies the application-oriented necessity to soon dispose of a planning model that enables quality predictions based on the quality elements typical of VoIP. It is the aim of this book to establish a link between VoIP relevant, parametrically describable quality elements and the associated speech quality.

3.9 From Elements to Features: Modeling VoIP Speech Quality

After a review of the literature on VoIP speech quality elements and features, we will now briefly rediscuss different approaches to instrumentally measuring or predicting the speech quality of telephone networks or services, in the light of the quality elements particularly characteristic of VoIP. Therefore, we employ the differentiation of speech quality models into signal-based measures, network planning models and monitoring models introduced in Chapter 2, Section 2.2 (see Möller and Raake, 2002). Moreover, we will restrict ourselves to those modeling approaches that are either the ones most frequently used or the most VoIP-relevant representatives of their kind. As it was handled in previous chapters, our main focus will be on network planning models, which may in principle also be applied to quality monitoring.

3.9.1   Signal-based Measures and VoIP

The core unit of many VoIP quality measurement tools are signal-based measures such as PESQ (see Section 2.2 ITU–T Rec. P.862, 2001). For example, using intrusive service monitoring approaches, the quality resulting from the speech transmission between two points A and B in the network can be estimated using the speech signals at A and B as input to the signal-based measure. If desired, additional parameters such as delay can be collected, for example, from RTP packet header information. An integral speech quality estimate can then be obtained, for example, using an E-model-based approach deriving the required equipment impairment factor from the output of the signal-based measure ITU–T Rec. P.834 (2002).

For NB services, PESQ works quite reliably with valid predictions, in case the application scenarios are within the range of what the model has been designed for. However, two types of problems principally may be related to measurement tools such as PESQ:

(a) A general, systematic deviation between the MOS-scores obtained in auditory listening tests (MOS-LQS: 'subjective' Listening Quality, see Section D.4), and those derived from the signal-based measure (MOS-LQO: 'objective' Listening Quality).

(b) Deviations for particular types of degradations or measurement configurations.

For PESQ, ITU–T's Study Group 12 has resolved many of the reported problems of type (a) by defining an appropriate mapping function between raw PESQ-scores and MOS-LQO (ITU–T Rec. P.862.1, 2003).

Problems of type (b) were reported for:

- Packet loss distributions strongly deviating from the random loss case, i.e. showing more runs of several packets lost consecutively than in case of independent ('random') loss (Pennock, 2002). The general discrepancy between the test data and PESQ estimates observed by Pennock (2002) would have improved by the above mentioned mapping function between raw PESQ scores and MOS-LQO (ITU–T Rec. P.862.1, 2003). In turn, this mapping does not resolve the issue that PESQ does not distinguish between 'random' and burst (dependent) loss observed by Pennock (2002). Note that Pennock identifies the linear correlation – a measure often used to quantify the performance of PESQ as compared to MOS results obtained in listening tests – as misleading, since it may hide systematic deviations between estimates and auditory test data.

- In case of background noise, PESQ works better when a noisy signal is used as the reference instead of a clean reference signal (ITU–T Delayed Contribution D.061, 2005). In turn, such an approach may fail when noise reduction algorithms are employed. However, PESQ was not designed for estimating speech transmission quality of circuits involving noise reduction, as mentioned in ITU–T Rec. P.862 (2001).

- In case of severe linear (frequency) distortions, especially spectral limitations below the NB bandwidth of 300–3400 Hz, PESQ underestimates the impact on speech transmission quality. Möller (2000, pp. 176–183) has compared listening test results and model predictions under different bandwidth restrictions for other signal-based measures than PESQ. The results show that PESQ's predecessor PSQM (ITU–T Rec.

P.861, 1996) overestimates the effect of linear distortions, and Telekom Objective Speech Quality Measure (TOSQA) underestimates it (Berger, 1998).

In the light of VoIP transmission, PESQ has been extended to cover WB transmission (ITU–T Rec. P.862.2, 2005). The corresponding extensions consist in:

- An exchanged input-filter, showing a flat frequency response above 100 Hz, and a slight roll-off below this limit to reflect headphone-listening.

- An appropriate mapping function between the raw WB-PESQ estimates and a WB MOS-LQO, which differs from the NB-mapping function defined by ITU–T Rec. P.862.1 (2003).

The WB version of PESQ applies to quality estimates in WB and mixed NB/WB scenarios. However, due to the different mappings to MOS-LQO, outputs from NB-PESQ (ITU–T Rec. P.862, 2001) and WB-PESQ (ITU–T Rec. P.862.2, 2005) cannot be compared.

The reliability of WB-PESQ – as of other WB speech quality models – is limited by the available training test databases, which cannot yet compare with what is known of NB (telephone) speech quality. However, the proximity of the current WB-PESQ approach to the well established 'classical' PESQ ITU–T Rec. P.862 (2001) may also lower its validity and reliability. For example, similar problems of WB-PESQ as for its NB-version were observed with regard to background noise transmission – especially involving noise suppression algorithms (ITU–T Delayed Contribution D.061, 2005). Another known problem of WB-PESQ is the mapping between the raw WB-PESQ-score and an MOS-LQO, which seems to be language- and codec-dependent (for example, the G.722 codec shows a different correlation line with raw WB-PESQ estimates than the G.722.1 and G.722.2; see ITU–T Delayed Contribution D.067, 2005). This and other issues are addressed in the NB- and WB-PESQ application guide ITU–T Rec. P.862.3 (2005).

Like NB-PESQ, WB-PESQ underestimates the impairment in case of strong linear-frequency distortions, especially below 300–3400 Hz. With regard to linear frequency distortions, ITU–T Contribution COM 12–20 (2005) proposes an improved version of both WB- and NB-PESQ as PESQ+, which is based on a diagnostic signal-based approach, see ITU–T Contribution COM 12–04 (2004).

3.9.2 Parametric Models and VoIP

In this section, we will focus on the E-model, since it provides a good basis for establishing a link between VoIP relevant, parametrically describable quality elements and the associated speech quality, which is the main goal of this book. The E-model currently is the recommended tool for network planning, and has thoroughly been evaluated for all degradations typical of analog or digital PSTN networks (Möller, 2000; Möller and Raake, 2002). Consequently, it was chosen as the starting point for the parametric quality impairment models presented in this book. Note, that in the meantime, the model proposed in this book for the impairment due to packet loss has led to an update of the E-model (ITU–T Rec. G.107, 2005).

Table 3.9 shows a summary of the quality elements relevant for VoIP. The table indicates which elements are or can be related to the input parameters of the E-model as it was

Table 3.9: Quality Elements of VoIP and their coverage by the initial E-model (ITU–T Rec. G.107, 2000). Also indicated is what we have addressed in the original research presented in Chapters 4 and 5, as well as the corresponding new input parameters.

Quality Element	Covered?	E-model Parameters	New Parameters	Previously Unresolved Features	Covered in this book?
Packet Loss	(✓)	Ie	Ppl, 2-state / 4-state parameters, $BurstR$	PLC, $f(codec)$?, $f(t)|_{t\ big/small}$, packet size	X
Jitter	–	(Ta)	Ppl, others?	–	(X)
Transmission delay	✓	Ta	–	Trade-off against other degradation types; impact on *perceived* quality	X
Talker echo	✓	$TELR, T$	–	–	
Listener echo	✓	$WEPL, Tr$	–	–	
Loudness	✓	SLR, RLR, OLR	–	WB loudness ratings suitable?	
Noise	✓	$Q = f(qdu)$	–	–	
"		$Nc, Nfor$	–	–	
"	Partly	Ps, Pr	–	Background noise transmission	
Linear distortion	–	–	f_c, z_{bw}	Not covered	X
Wideband	–	–	–	Not covered	X
Echo canceller	–	$Ta, TELR, T$	See packet loss?	Switching, nonlinear distortion send & receive, echo distortion	
Noise suppression	–	Nc (equivalent)	–	Noise and signal distortion	
VAD	–	see packet loss	–	Front-end clipping	
CNG	–	–	–	Noise switching	
Codecs (NB)	(✓)	Ie	Tandeming	–	
Codecs (WB)	–	Ie	–	Not covered	X
Talker & listener sidetone	✓	$STMR, LSTR$	–	–	
Combinations	Some	–	–	For example packet loss, linear distortion + others	X

formulated in its first releases (until the version from 2000). The last column indicates the effects we addressed in the original research described in this book.

In spite of the restrictions summarized in Table 3.9, the old version of the E-model – including tabulated data for the impairment due to packet loss – has already been widely employed in the context of VoIP (Clark, 2001; Cole and Rosenbluth, 2001; Janssen *et al.*, 2002; Jiang and Schulzrinne, 2002; Klimo, 1999; Markopoulou *et al.*, 2002; Sun and Ifeachor, 2003). For example, it was used for speech quality monitoring based on protocol information: Parameters such as delay and packet loss characteristics can serve as direct input parameters to the model, if certain simplifications and assumptions are made.

At least one of the following problems are related to all of the approaches described in the literature:

- The impact of different types of packet loss distributions on speech quality is not addressed systematically. The model either employs the tabulated data provided in the now withdrawn ITU–T Rec. G.113 Appendix I (2001), or is based on a larger number of curve-fitting parameters, and make the model applicable only to random, i.e. independent packet loss (for more details on packet loss models see Chapter 4).

- The approach does not question the additivity assumption underlying the E-model, although the model was developed for 'stationary' degradations and not for time-varying ones like packet loss.

- The approach postulates that the only types of degradations to be expected in VoIP systems are delay and codecs under packet loss. However, this claim stands in contrast to how VoIP is applied in real networks: A common usage scenario is PSTN–VoIP–PSTN. Consequently, it has to be proven, whether the impairment due to packet loss is additive to that due to other types of degradations, before the model is applied to such configurations. The necessity for such an investigation is underlined by a study carried out by Watson and Sasse (2000). Here, factors like the characteristics of the user interface or echo were found to be decisive for the quality of internet speech. Even if the described approach considers the impact of other impairments like echo or noise, it intends to quantify only the effects of packet loss, jitter and delay on speech quality. Because of the doubts in the additivity assumption of the E-model, it is questionable whether even relative quality evaluation is possible with such an approach: for example, degradations like packet loss may amplify the impairment due to background noise.

- User interfaces other than handset telephones or other types of linear distortions are not covered by the current version of the E-model. Moreover, assumptions on these and other, not IP-related, characteristics made from packet-trace data may not reflect reality. Some increase of validity of the quality estimates may result from the usage of additional, quality-related data exchanged by the two endpoints, which can inform, for example, about jitter-buffer configurations and corresponding packet discard (RTCP XR, see IETF RFC 3611, 2003). However, neither information on the user interfaces nor on the related echo-problems are provided by these RTCP 'extended reports'.

- In many cases, some of the model parameters themselves are estimated using signal-based models like PESQ. While this approach may be valid when no transmission

errors like packet loss occur, that is in case of elements well covered by these measures, it is questionable especially with regard to bursty packet loss (as stated earlier,
the accuracy of measures like PESQ for quality under dependent packet loss was
challenged, for example, by Pennock, 2002).

Moreover, no parametric model is available, which would enable realistic predictions of
quality in case of WB speech, or of linear distortions introduced by different user interfaces
or channel band limits. The initial PSTN version of the E-model only covers the loudness
impact of linear distortion. As we have detailed earlier with regard to signal-based measures,
all standard versions of the quality models addressed in this section are of limited validity
in case of linear distortions.

In this book, several of the shortcomings of the older E-model versions (up to ITU–T
Rec. G.107, 2000) mentioned are addressed, and several new formulae are proposed to
overcome them (see Table 3.9). As stated earlier, the algorithms developed to include the
impairment due to packet loss have recently been standardized in an update of the model
(ITU–T Rec. G.107, 2005). The extensions we propose are developed in Chapters 4 and 5,
and are summarized in Chapter 6.

3.10   Quality Elements and Quality Features of VoIP: Summary

This chapter has given an overview of the main quality elements typical of VoIP, and
of the perceptual features and the speech quality that are associated with these elements.
The available literature on the relation between the quality elements and perceived speech
quality is summarized. Based on this summary, open research questions with regard to VoIP
speech quality perception and prediction are pointed out.

On the one hand, the quality elements addressed are the *overall* elements to be encountered in VoIP networks, that is, the elements that may be contributed to by different network
components or factors related to VoIP transmission. Examples of these overall elements are
delay jitter, the overall transmission delay, packet loss, noise and the linear distortion
introduced by the system. On the other hand, some selected individual quality elements
of particular relevance for VoIP speech transmission are considered: The employed user
interfaces, the codecs, the noise suppression algorithms, the ECs, and the additional VAD.

From the addressed quality elements, packet loss is identified as the one leading to the
most characteristic quality features of packet-based transmission: the 'smoothness' of the
transmitted voice and time-varying quality. Hence, packet loss receives particular attention
in this chapter: On the quality element side, a detailed description of typical packet loss
distributions is provided. On the quality feature side, a quality-perception oriented view on
packet loss distributions is presented, distinguishing between *microscopic* and *macroscopic*
loss. According to this new framework, macroscopic loss describes the loss behavior that
leads to time-varying quality, that is to perceivable changes between different quality levels. It reflects the longer-term loss characteristics. Microscopic loss, in turn, is introduced
as the property of the loss distribution that effects the behavior of the involved signal-
processing components, like the decoder or PLC algorithms. Thus, by definition a given
microscopic loss behavior is linked to a certain level of speech quality. A sequence of

different microscopic loss passages compose a particular macroscopic loss behavior, which leads to perceivable speech quality changes.

For the sake of completion, this chapter further addresses other generic quality elements also encountered in non-VoIP telephone networks, like transmission delay, noise and echo. In this respect, the chapter summarizes how these elements are described in telephony, and how they affect speech quality. Moreover, more recent developments are reflected by addressing the measures taken to compensate for some of these traditional types of degradations: The chapter attempts to condensate the knowledge on how noise suppression systems or ECs may effect speech quality. For example, ECs may be described by overall quality elements that are already considered in other contexts: The introduced switching or clipping may be represented as some type of packet loss, the residual echo as talker echo and the delay introduced by such systems be added to the transmission delay.

Packet loss or echo are clearly linked to a *degradation* of speech quality. As stated in the introductory parts of this book, however, VoIP comes along with a considerable potential of *improvement* of NB telephone speech quality: A far greater variety of user interfaces may be applied, the NB transmission range may be extended to WB – especially due to the flexible employment of corresponding low-bitrate speech compression algorithms – and new types of (multimedia-) services can easily be implemented, exploiting the convergence enabled by the unified application of IP. Taking this novelty of VoIP into consideration, this chapter summarizes the existing knowledge on how the spectral characteristics of the transmission channel as well as the speech codec impact the perceived speech quality.

As outlined in Chapter 1, quality results from a comparison of perceived and expected features (Jekosch, 2005b). The notion of quality features is related to the notion of quality dimensions: Speech quality is multidimensional by nature. The dimensions represent the perceived features that e.g. could allow two telephone connections to be distinguished. This chapter summarizes different studies reported in the literature in order to identify the quality dimensions characteristic of VoIP connections. It is shown that from the effects encountered in VoIP systems, packet loss is the one most strongly affecting quality. In turn, it is also shown that features not typically handled by speech quality prediction models play an important role for quality perception, such as the speech sound quality related to user interfaces other than handset telephones. The quality that is ultimately related to the perceived features depends on the expected features, and on the way in which a given feature is mapped to quality. In this chapter, it is concluded that the main quality features of VoIP, namely those related to packet loss and the spectral behavior of the system, are mapped to quality in different ways: The feature related to packet loss – the 'smoothness' of speech – pertains to a vector-type preference behavior, with a 'the less, the better' type of preference relation. The features related to linear distortion ('fullness', 'sharpness'), in turn, correspond to an ideal-point preference model, where 'some point is ideal' (Carroll, 1972; Mattila, 2001).

Where this ideal point lies in case of, for example, linear distortion or WB speech, and how badly packet loss is sanctioned by a user in terms of quality, depends on the user's expectations. In order to reflect this situation, the chapter gives an overview of studies on factors that have an impact on the desired features of VoIP connections, and on how the desired features affect speech quality. For example, additional service features like the enhanced mobility provided by mobile services or messaging features clearly make subjects more tolerant towards degradations. On the other hand, the impact of these factors depends on whether the users have prior experience with the presented services. Higher costs make users more critical, and additional service capabilities like messaging more tolerant for VoIP-typical degradations.

To complete the picture, this chapter addresses the perceived quality when a given connection is characterized by combined quality elements. Here, different perceptual features – although possibly representing orthogonal dimensions – may dominate or mask each other. From the point of view of an actual telephone conversation, quality features cannot only be distinguished according to the notion of quality dimensions, but also according to whether they are perceived only during an actual conversation, or are pure listening features. In this respect, the chapter provides considerations on quality perception of listening features in a conversation situation.

In summary, the chapter forms the basis for the more indepth and original analysis of the two major quality features of VoIP: time-varying distortion and linear distortion or WB transmission. With its additional information on other quality elements and features, it serves as a source of information to better understand the perceived speech quality of VoIP systems.

4

Time-Varying Distortion: Quality Features and Modeling

The first class of overall elements addressed in this book, both with the author's own auditory tests and by parametric modeling, are time-varying distortions. They are the type of degradations most characteristic of Voice over Internet Protocol (VoIP) (Chapter 3). Packet loss[1] is the most crucial of the possible time-varying distortions, since it (i) degrades speech quality considerably, and (ii) is directly linked to the packet-based transmission technique.

Since time-varying distortions are perceptually different from stationary distortions, quality perception under time-varying distortions is different, too. The first subject-oriented studies on the impact of time-varying distortion, such as, interruptions on speech perception date back to the fifties (Miller and Licklider, 1950). With the transmission technologies made possible more recently by the advances in digital signal processing, which have lead to mobile networks and VoIP, an object-oriented[2] assessment of speech quality under packet and frame loss has been initiated (Coleman *et al.*, 1988; Hardman *et al.*, 1995).

Considerable insight into speech quality perception under time-varying distortions has lately been gained regarding two aspects (Sections 3.3.5.7, 3.3.5.5):

- The quality features related to time-varying distortions like frame or packet loss were identified, measured and embedded into the framework of other features of transmitted speech (for details see Sections 3.3.5.7 and 3.4; Bernex and Barriac, 2002; Mattila, 2001; Wältermann, 2005). It was found that this type of time-varying distortions is different from stationary degradations in at least one perceptual dimension (referred to as 'continuity' or 'smoothness of speech'; Mattila, 2001).

- The instantaneous perception of time-varying quality was investigated, for two different underlying quality elements (signal-correlated noise and packet loss, respectively;

---

[1]The term packet loss refers to the combined effect of network packet loss, packet discard by the jitter buffer, and frame discard by the decoder (e.g. in case of wireless links).

[2]Object-oriented vs. subject-oriented: The system is the goal of the assessment, not the user's perception, see Section 1.2.2.

Speech Quality of VoIP: Assessment and Prediction   Alexander Raake
© 2006 John Wiley & Sons, Ltd

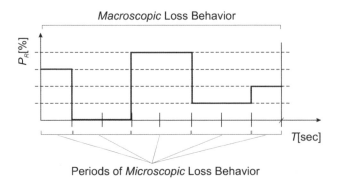

Figure 4.1: Representation of the overall *macroscopic* packet loss behavior as a concatenation of periods of *microscopic* loss behavior. Here, P_R [%] is the average packet loss percentage during passages of given microscopic loss behavior.

Gros and Chateau, 2001; Hansen and Kollmeier, 1999). The studies by Gros and Chateau (2001) and similar approaches have related the integral speech quality of longer passages to speech quality ratings obtained instantaneously, or obtained for shorter passages (ITU–T Contribution COM 12–21, 1997; ITU–T Delayed Contribution D.064, 1998; Jekosch, 2000).

Starting from these findings, two perception-based timescales for describing a time-varying channel behavior have been defined in this book: *Microscopic* and *macroscopic* channel behavior (Sections 3.3.5.7; 3.3.5.5). At the feature level, the *microscopic* behavior determines the quality a user perceives momentarily, or during short periods of a given connection. In case of packet loss, the microscopic behavior is related to factors such as the number of packets lost in a row, and the applied loss concealment algorithm (packet loss concealment [PLC]). For example, the PLC may be more or less robust with regard to consecutive packet loss, or may lead to particular perceptual features (see Section 3.3.5.7). Based on the findings by Gros and Chateau (2001), Rosenbluth (ITU–T Delayed Contribution D.064, 1998), Jekosch (2000) and Hansen and Kollmeier (1999), it can be assumed that the integral quality, i.e. the quality rated at the end of a connection, is related to a *weighted* time-average over the rating of passages of different microscopic channel behavior. The sequence of such passages of a given microscopic behavior is what was defined as *macroscopic* channel behavior in Section 3.3.5.5. The situation is depicted in Figure 4.1.

These considerations are reflected in the research questions addressed in the present chapter:

(a) How can the impact of the *microscopic* packet loss behavior on speech quality be quantified parametrically? (Section 4.1)

(b) How do individual segments of particular *microscopic* loss behavior have to be weighted to determine the integral quality associated with a given *macroscopic* loss behavior? (Section 4.2)

(c) How can the impact of the macroscopic loss behavior on integral speech quality be quantified parametrically? (Sections 4.1, 4.2)

(d) How can the impact of packet loss and other degradations on communicability, i.e. on conversational aspects of speech quality be assessed in a way that reflects the actual usage of telephone networks? (Section 4.3)

(e) Is the basic assumption of the E-model, namely, that impairments properly transformed onto the psychological impairment scale are additive, valid for time-varying distortions like packet loss? (Sections 4.4)

(f) What is the integral speech quality perceived when packet loss of 'stationary' macroscopic behavior is combined with other types of degradations? (Section 4.4.2)

(g) What is the integral speech quality perceived when packet loss of 'nonstationary' macroscopic behavior is combined with other types of degradations? (Section 4.4.3)

In Chapter 3, Section 3.7, it was pointed out that the *desired* nature or composition of a perceptual event (i.e. the user's expectation) determines the quality the user associates with the *perceived* features. In case of VoIP, the users' expectations may be quite different from what is perceived, because the user interface no longer guarantees a particular set of perceptual features. To account for this situation, the investigations described in the following take the applied user interface and aspects of expectation implicitly into consideration. Therefore, the telephone situation is mimicked to a maximal extent in a laboratory environment. In addition, the auditory tests were predominantly carried out as conversation tests, which represent the most realistic method for telephone speech-quality assessment (see Section 2.1). To reduce the variation of the subjects' expectation that may be induced by the user interface, all tests on speech quality under packet loss were carried out with standard handset telephones.

4.1 Microscopic Loss Behavior

This section deals with the question, as how the impact of the *microscopic* packet loss behavior on speech quality can be quantified parametrically. To answer this question, different microscopic loss distributions are addressed, namely, random loss, and dependent i.e. 2-state Markov loss. In this case, the *macroscopic* loss behavior is characterized by a constant *microscopic* loss behavior, as depicted in Figure 4.2.

4.1.1 Random Packet Loss

Random packet loss is the simplest case of loss behavior. Its effect on speech quality has extensively been studied in the literature. It can be described by only one parameter, the overall packet loss percentage *Ppl* (Raake, 2004).

Almost all auditory tests described in the literature indicate a similar behavior of speech quality when plotted over random packet loss percentage, regardless on the loss concealment (PLC) strategy applied by the decoder, or the packet size chosen for the connection: The curve-shape is comparable. An example for this behavior taken from the author's own tests on speech quality under random interruptions[3] is depicted in Figure 4.3.

---

[3]Silence insertion interrupts the speech signal for the duration of the missing packets.

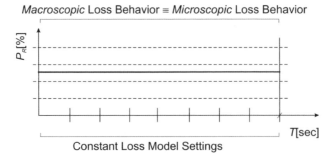

Figure 4.2: Loss profile in case that only *microscopic* packet loss occurs that does not lead to considerable variation of speech quality over time.

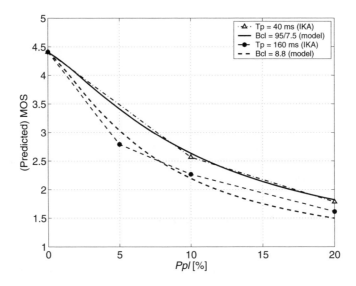

Figure 4.3: Speech quality impaired by random interruptions as obtained in the author's own experiments (Raake, 2004; Raake and Möller, 1999). Abscissa: Interruption percentage Ppl; ordinate: MOS-ratings. Also shown are the predictions by the parametric model described in the text, with two parameter settings chosen according to the interruption lengths.

The results were obtained on the mean opinion score (MOS) scale in one listening-only and one conversation test (referred to as 'LOT-GSM' and 'SCT-GSM' in the following). The interruptions were inserted at random, with two different interruption durations of 40 ms and 160 ms. As the results of the conversation and the listening test are very similar, the ratings averaged over the two tests are presented here (for details see Appendix G, Section G.1).

Johannesson, the editor of the E-model (Section 2.2.2.2), has proposed a parametric model of the quality impairment due to temporal clipping, i.e. interruptions introduced by DCME (Digital Circuit Multiplication Equipment; ITU–T Delayed Contribution D.071, 1995).

$$Icl = \frac{7.5 \cdot Cl}{1 + 7.5 \cdot Cl/95} \qquad (4.1)$$

Here, Icl is an additional impairment factor for speech clipping to be added to Equation (2.5), with Cl as the amount of clipped speech in percent:

$$R = Ro - Is - Id - Ie - Icl + A \qquad (4.2)$$

As depicted in Figure 4.3, the predictions obtained with this extended E-model are in good agreement with the test results for speech quality in case of interruptions of $Tp = 40$ ms duration.

To use Equation (4.1) later with other PLC techniques, it can be rewritten as

$$Icl = 95 \frac{Cl}{Cl + \dfrac{95}{7.5}} = 95 \frac{Cl}{Cl + Bcl}, \qquad (4.3)$$

where $Bcl = \frac{95}{7.5} \approx 12.7$ is a measure of the robustness of the system (or in this case of the user) against clipping. The factor 95 corresponds to the maximum possible impairment.

With this modified formula, all listening and conversation test data reported in the literature on speech quality under random interruptions or silence insertion can be approximated by using appropriate settings for the 'robustness' Bcl.

This observation is in line with what was discussed in Section 3.3.5.7: At a constant percentage of interrupted (or lost) speech, quality may depend on the duration of the interrupted or lost segments. To further investigate this effect, an analysis of variance (ANOVA) was carried out for the MOS-results of the listening test ('LOT-GSM'), using the interruption length and percentage as well as the speaker as fixed factors (see Appendix G.1, Table G.3). The resulting F-ratios show a statistically significant effect of the interruption percentage ($F = 66.3$, $p < 0.001$), and a weak effect of the interruption length ($F = 5.6$, $p < 0.05$). Note that the measurable impact of interruption duration cannot directly be generalized to VoIP with G.711 and silence insertion, since an interruption length of 160 ms does not reflect the packet sizes typical of VoIP (10–60 ms, see Section 3.3.5.4).

In the more general form of Equation (4.3), the clipping impairment formula is applicable to packet loss of log-PCM (pulse code modulation) coded speech with silence insertion as PLC. Here, we replace the clipping percentage Cl in Equation (4.3) by the random packet loss percentage Ppl, and the 'clipping robustness' by the 'packet loss robustness' Bpl. In Figure 4.4, the predictions derived with Equation (4.3) are compared with the results from auditory listening tests. These have been obtained by different institutions for G.711 with silence insertion under random packet loss $Ppl \equiv Cl$, for different packet sizes Tp (ETSI Tiphon 11 Temporary Document 64, 1999; ITU–T Contribution COM 12–28, 2001; ITU–T Delayed Contribution D.110, 1999). For comparability, we have normalized all results according to the procedure discussed in Appendix D, Section D.3.

As can be seen from the curves, the value of $Bcl = 8.8$ seems to yield a good fit of all test results, except for the tests indicated by the dashed and dotted curve (denoted by 'D.110'). In this case, Bcl in Equation (4.3) has to be set to $Bcl = 4.3$. It has to be mentioned that in this test an uncommon procedure was applied to introduce packet loss, guaranteeing that only packets containing speech get lost. If random packet loss is introduced without particular control *where* loss occurs, the overall random loss rate corresponds to the portion of lost *speech* packets ($p_{pl} = p_s$), regardless of whether voice activity detection is used or not: It is equally probable to lose a speech packet as it is to lose a silence packet. Hence, for the test 'D.110', the actual portion of lost speech packets $p_{s,D.110}$ was higher than for all other studies:

$$p_{s,D.110} = \frac{\#lost\ speech\ packets}{\#speech\ packets} = \frac{\#lost\ packets}{\#speech\ packets} \tag{4.4}$$

$$> \frac{\#lost\ packets}{\#speech\ packets + \#silence\ packets} = \frac{\#lost\ packets}{\#packets} = p_{pl} = p_s$$

Although it reflects a special test case, the clipping robustness factor of $Bcl = 4.3$ has been adopted by the ITU–T for PCM-coding with silence insertion (as Bpl; ITU–T Rec. G.113 Appendix I, 2002, Appendix I). A more realistic setting for today's VoIP systems with G.711 and silence insertion appears to be $Bcl = Bpl = 8.8$.

Figure 4.4 also shows that all test results for G.711 with silence insertion under packet loss are lower than the prediction with the initial setting of $Bcl \approx 12.7$, and the results obtained in the author's tests for 40 ms-interruptions (Figure 4.3). The latter can be attributed to how the interruptions were inserted in our tests: In contrast with the sharp interruptions caused by packet loss in conjunction with silence insertion, a cos^2-switch was used with a rise- and fall-time of 2 ms, leading to a smoother transition and hence less degradation than for actual packet loss.

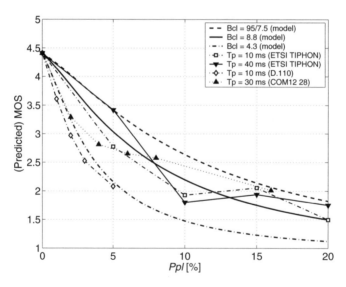

Figure 4.4: Speech quality impaired by random packet loss using G.711 with silence insertion as PLC (adopted from Raake, 2004). Abscissa: Packet loss percentage Ppl; ordinate: MOS-ratings. The depicted results are taken from various auditory tests and predictions from Equation (4.3) with different Bcl settings ('ETSI TIPHON': ETSI Tiphon 11 Temporary Document 64 (1999); 'D.110': ITU–T Delayed Contribution D.110 (1999); 'COM-12 28': ITU–T Contribution COM 12–28 (2001)).

The above findings can be summarized as follows:

- All test results for speech quality under random interruptions, speech clipping or packet loss with silence insertion as PLC show a similar dependency on interruption or loss rate.

- The relation between interruption percentage and speech quality can be modeled using Equation (4.3), with two parameters: The loss or interruption rate Cl, and the robustness against this type of loss Bcl.

- From the available test results we can conclude that a common Bcl-value of $Bcl \approx 8.8$ is applicable to all packet sizes relevant to VoIP (in the range from 10 to 60ms).

- Safe predictions can be achieved setting $Bcl = 4.3$.

- If silence insertion is assumed to be the least efficient means of 'concealing' lost packets, this setting reflects a worst-case scenario.

4.1.1.1 IP Codecs and Packet Loss Concealment

Based on listening test results reported in the literature, we have modified Equation (4.3) to be applicable to log-PCM with silence insertion as well as other codec/PLC combinations. For current codecs and their native PLC strategies, curve fitting of the test data with an additional parameter revealed that replacing the multiplicative constant 95 used in Equation (4.3) by $(95 - Ie)$, Ie for the equipment impairment under error-free coding conditions, leads to more accurate predictions. Following this approach, we estimate the Equipment Impairment Factor for a codec under random packet loss according to equation (4.5) (ITU–T Delayed Contribution D.044, 2001):

$$Ie,eff = Ie + (95 - Ie)\frac{Ppl}{Ppl + Bpl}. \tag{4.5}$$

Here, we have replaced Bcl in Equation (4.3) by Bpl, the *Packet Loss Robustness Factor* mentioned earlier in this book (Section 2.2.2.2). Obviously, the higher the value for Bpl, the smaller the effect of packet loss. Ie,eff is the *Effective Equipment Impairment Factor* covering both the impairment due to low bit-rate coding and that introduced by additional (random) packet loss. At 0% packet loss, Ie,eff equals the equipment impairment factor Ie.

   In Equation (4.5), it is implied that the impairment caused by the speech-coding algorithm and that caused by the PLC under packet loss are not independent. This is expressed by the factor $(95 - Ie)$: A smaller step of impairment is expected, if the quality of the codec without packet loss is already low.

   The codecs and PLC schemes covered by the model are those that are widely used in VoIP-transmission. Equation (4.5) also applies to log-PCM coded speech under random packet loss, when silence insertion is employed as PLC (see the preceding text): With $Ie = 0$, Equation (4.5) is equivalent to Equation (4.3). Table 4.1 gives values for Ie and Bpl recommended in ITU–T Rec. G.113 Appendix I (2002)[4] are given. In addition, it

---

[4]We have derived these Bpl values by least-square fitting of tabulated data. This data was used with the E-model for the equipment impairment factor under packet loss, before the model presented in this section was included in the E-model (for the old tabulated data see ITU–T Rec. G.113 Appendix I, 2001, which is now withdrawn). For more details on impairment factors, see Section 3.3.2.

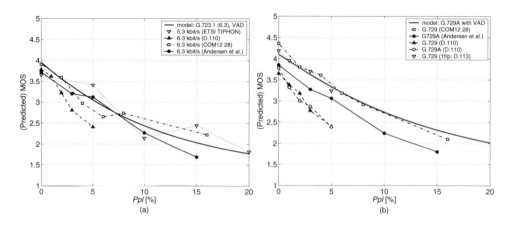

Figure 4.5: Speech quality impaired by random packet loss as determined in auditory tests on the MOS scale, and the predictions of Equation (4.5) with appropriate settings for the Equipment Impairment Factor Ie and the Packet Loss Robustness Factor Bpl (adopted from Raake, 2004). Abscissa: Packet loss percentage Ppl; ordinate MOS-rating. (a) G.723.1 (5.3 and 6.3 kbit/s). (b) G.729/G.729A. 'ETSI TIPHON': ETSI Tiphon 11 Temporary Document 64 (1999); 'D.110': ITU–T Delayed Contribution D.110 (1999); 'COM-12 28': ITU–T Contribution COM 12–28 (2001); 'Andersen *et al.*': Andersen *et al.* (2002); 'D.113': ITU–T Delayed Contribution D.113 (1999).

shows a Bpl estimate for the Internet low bit-rate codec (iLBC), which we derived from the listening test data provided in Andersen *et al.* (2002). For the latter, we estimated an equipment impairment factor of $Ie = 11$, which is equal to that for the G.729A.

Whether the provided settings are limited to the packet size listed in Table 4.1 depends on the sensitivity of the codec PLC toward consecutive frame loss, For example, the settings for the G.729A, and for the G.711 with silence insertion can also be used for other packet sizes below 60 ms, while the settings for the G.711 with the Appendix I PLC should be used solely with 10 ms (see Section 3.3.5.7).

Figure 4.5 depicts comparisons between predictions from Equation (4.5) and results from auditory listening tests, for the G.723.1 [30 ms packet size: figure 4.5(a)] and for the G.729 codec [20 ms packet size, if not indicated otherwise; $1 f/p \equiv 1 frame/packet \equiv 10 ms$: figure 4.5(b)][5]. Note that all results were normalized following the linear normalization procedure described in Appendix D.3.

As can be seen from Figure 4.5, the model predictions are in relatively good agreement with the listening test results. Only in case of the test curve denoted by 'D.110' (ITU–T Delayed Contribution D.110, 1999), the model overestimates speech quality. This is due to the uncommon packet-loss insertion procedure described in the preceding text, see Equation (4.4). If these results are discarded, a correlation between the model predictions and the mean test results of $\rho = 0.963$ is obtained for the G.729 codec, and of $\rho = 0.944$ for the G.723.1.

In Figure 4.6, the performance estimation for the different codecs and PLC algorithms listed in Table 4.1 is illustrated based on model predictions. The slope of each curve shows

---

[5]The G.729A is the low-complexity version of the G.729 codec, yielding slightly lower speech quality (ITU–T Rec. G.729, 1996, Annex A). The VAD used in some of the tests is that native to the G.729.

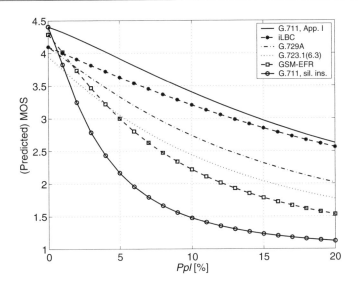

Figure 4.6: Prediction of speech quality (in MOS) impaired by random packet loss (*Ppl* [%]) for different codecs according to the settings for *Ie* and *Bpl* given in Table 4.1 (Raake, 2004).

Table 4.1: Settings for the Equipment Impairment Factor *Ie* (codec) and the Packet Loss Robustness Factor *Bpl* for different codecs and corresponding PLC algorithms. The packet length *Tp* and a flag indicating whether VAD was used are provided in addition. All settings, apart from those for the iLBC, are now included in ITU–T Rec. G.113 Appendix I (2002). For the listening test results underlying the iLBC parameter-estimates see Andersen *et al.* (2002).

Codec (Bit-rate)	Tp [ms]	VAD	PLC	Ie	*Bpl* [%]
G.711 (64 kbps)	10	–	Sil. Ins.	0	4.3
G.711 (64 kbps)	10	–	App. I	0	25.1
G.723.1 (6.3 kbps)	30	Annex A	Native	15	16.1
G.729A (8 kbps)	20	Annex B	Native	11	19.0
GSM-EFR (13 kbps)	20	–	Native	5	10.0
iLBC (13.867 kbps)	30	–	Native	11	32.0

the robustness of the codec–PLC combination against packet loss; the distance from $R = 93.1$ on the R-scale at $Ppl = 0\%$ shows the codec's performance in the loss-free case.

More recent tests have shown that the approach described here for modeling the impairment due to narrowband speech codecs under packet loss may also be applied to wideband codecs: ITU–T Delayed Contribution D.064 (2005) shows a first preliminary estimate of corresponding (wideband) packet loss robustness factors Bpl (for details on wideband speech quality, see Chapter 5).

4.1.1.2 Conversational Quality

Most results presented so far were collected in listening tests. A comparison of the random loss impairment model to speech quality judgments obtained in the author's short conversation tests is shown in Figure 4.7 (Raake, 2004).

The short conversation tests were conducted to assess speech quality under combinations of random packet loss with other degradations, and are discussed in more detail in Section 4.4.2. Here, we will restrict the discussion to the test results obtained for packet loss as the only degradation type presented to the subjects. The codec tested under packet loss was the G.729A (20 ms packet size), without voice activity detection.

As can be seen from the figure, the test results are in good agreement with the model predictions, showing a correlation of $\rho = 0.91$, and a root mean squared error of

$$RMSE = \sqrt{\frac{\sum_{i=1}^{n} \left(MOS_i - \widehat{MOS_i}\right)^2}{n}} = 0.32, \tag{4.6}$$

where MOS_i is the mean opinion score for connection i, and $\widehat{MOS_i}$ is the corresponding E-model prediction.

It has to be noted that in two of the four tests, speech quality under *low* rates of random loss ($Ppl \leq 3\text{--}5\%$) was found to be relatively less affected than predicted by the model (Figure 4.8). In other words, a 'step' can be observed for loss rates around 3–5%, separating the loss range into two regimes of lower and higher importance to quality. This may be due to the conversation mode, where subjects pay less attention to degradations of the form (due to small impairments), and more attention to degradations affecting intelligibility, that is, the comprehension of what is said (due to larger impairments; Sections 1.1.2 and 2.1).

We observed this effect only for the conversation tests SCT11 and SCT12, where we have used noise as additional degradation in some conditions. Thus, the observed step may also be due to a shared attention of the test subjects between the two types of listening degradation (i.e. packet loss and noise). In the tests on packet loss and talker echo (SCT31), and packet loss and delay (SCT21), we could not find such a step.

A similar but more important step (or threshold) was observed in a listening test on speech transmission quality of natural and synthetic speech prompts to be used as the speech output of a smart home system operated over the phone (see Figure 4.8, curve LOT (INSPIRE); Möller *et al.*, 2004). The test involved a particular kind of task: The subjects first had to mark on a test sheet which home appliance a particular speech prompt had referred to; this way, it was assured that the subjects focus on the contents of the prompts,

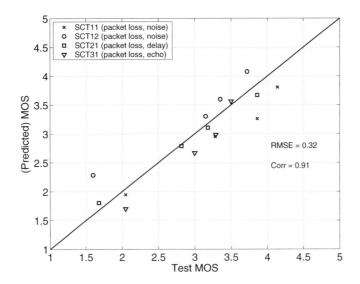

Figure 4.7: Comparison of the author's own conversation test results (abscissa) and E-model predictions (ordinate; Raake, 2004). The employed codec was the G.729A, with 20 ms packet size and its native PLC. The random loss rates used in the tests were 0%, 3%, 5% and 15%. See text for more information.

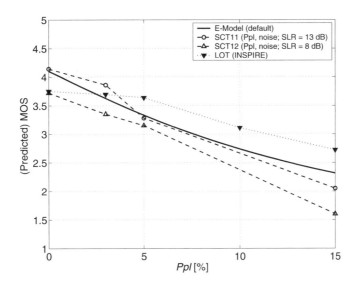

Figure 4.8: Behavior of speech quality over packet loss for SCTs on packet loss and line noise (SCT11: 'o'; SCT12: '△'). For comparison, the figure presents the results obtained in a listening test on speech prompts of a smart home system operated via a spoken dialogue system over telephone (LOT (INSPIRE); Möller *et al.*, 2004).

and not only on their surface form. In this case, loss rates below 5% had almost no effect on the quality ratings. In the test, however, only a few combinations of packet-loss patterns and underlying speech patterns were tested per condition. Thus, further tests are needed to exclude the possibility of a coincidental combination of the speech and loss patterns. Moreover, synthetic and natural utterances were presented in one test run in the INSPIRE test. Hence, the threshold can also indicate a packet-loss level, where an 'unnaturalness' dimension common to both synthetic speech and natural speech under low packet-loss rates is complemented by an 'interruptedness' or 'smoothness' dimension (for more details on quality dimensions see Section 3.4).

4.1.1.3 Comparison to Other Approaches

If compared with other approaches, Equation (4.5) has the advantage of deploying only two additional parameters, namely, the overall loss percentage Ppl and the packet loss robustness Bpl. For example, the formulae suggested by Cole and Rosenbluth (2001) or by Janssen *et al.* (2002) need three such additional parameters, and assume independence of codec and packet loss impairments. For example, the equation by Cole and Rosenbluth (2001) is of the form:

$$I_{loss} = x_1 \cdot ln(1 + x_2 \cdot Ppl), \qquad (4.7)$$

where I_{loss} is the additional impairment due to packet loss, and x_1 and x_2 are two additional codec-specific parameters.

An entirely different approach is suggested by Klimo (1999). Here, the impairment due to packet loss is included in the Simultaneous Impairment Factor Is (see equation 2.5), by modeling lost packets as an additional, signal-correlated noise, as it could result from quantization processes (i.e. 'masking' the lost speech information). Although this approach has the advantage of primarily being applicable to different loss distributions, it is questionable, whether a perceptually different degradation can be used to model the effect due to packet loss: According to the multidimensional analysis carried out by Hall (2001), speech impaired by signal-correlated noise differs considerably from coded speech in at least one of the three quality dimensions he found (dimension 2, referred to as 'noisiness'; see Section 3.4). As discussed in Section 3.3.5.7, a perceptual difference can also be assumed between signal-correlated noise and coding under packet loss (see also Gros and Chateau, 2001; Hansen and Kollmeier, 1999).

Using the limited number of codec-specific parameters proposed here, the amount of tabulated data necessary to cover different codec/PLC combinations can be reduced considerably, without making reference to perceptually different degradations such as signal-correlated noise. Moreover, corresponding parameters for new codecs, as it was described for the iLBC, can easily be added by determining only one codec-specific value. This additional robustness factor Bpl provides a direct indication on the expected performance of the codec/PLC combination under random packet loss.

4.1.1.4 Summary Random Packet Loss

- We presented a formula for parametrically quantifying the impairment due to random packet loss. The formula is an extension of the E-model (Equation (4.5)), and is included in the recommended model algorithm (ITU–T Rec. G.107, 2005).

- Only two additional input parameters to the E-model are therefore required: The packet loss percentage *Ppl* and the packet loss robustness factor *Bpl*. The latter quantifies the codec/PLC robustness toward random packet loss, and thus is a helpful indication for the choice of a particular codec. We have derived corresponding *Bpl* values for different codec/PLC combinations (Table 4.1), which are recommended in ITU–T Rec. G.113 Appendix I (2002).

- The model predictions were shown to be in good agreement with auditory test data compiled from the literature and determined in the author's own tests.

- The approach was compared with other, E-model-based modeling approaches, and its advantages were pointed out.

4.1.2 Dependent Packet Loss

So far, the discussions were focused on speech quality under random packet loss, which does not reflect the loss distributions typically encountered in real networks (see Section 3.3.5.1). In this section, the question will be addressed, how speech quality depends on other types of microscopic loss behavior, that is, macroscopically nonvarying packet loss that shows a certain degree of dependence between loss events.

Therefore, an extensive listening test was conducted. It focused on the comparison between speech transmission quality under random packet loss and under packet loss distributed according to a 2-state Markov model (Section 3.3.5.1, Figure 3.5). For many network connections that do not show an expressed *macroscopic* loss behavior, this relatively simple loss description already leads to a satisfactory modeling of the observed packet traces (see Section 3.3.5.1; Sanneck and Carle, 2000; Yajnik *et al.*, 1999).

4.1.2.1 Test Set-Up and Procedure

For the packet loss conditions presented to the subjects in this test, different overall loss rates were chosen, and different average numbers of consecutively lost packets. The latter was controlled by the 2-state model settings, based on the average sojourn time in the bad state ('1') that can be calculated as $\mu_{10} = 1/q = 1/(1 - pc)$ (see Section 3.3.5.1, Equation (3.5)). Here, $pc = 1 - q$ is the conditional loss probability, which expresses the probability of losing the next packet, when the current packet was lost.

The longest pause between packet-loss events to be expected in the listening test determines the suitability of a particular speech sample duration. The longest pause can be estimated as follows: Given an overall loss probability *ppl* and a conditional loss probability *pc*, the average number of received packets between loss events μ_{01} is derived from Equations (3.5), (3.6) and (3.7) as:

$$\mu_{01} = \frac{1 - ppl}{(1 - pc) \cdot ppl}. \tag{4.8}$$

Obviously, the smaller the overall loss rate *ppl*, and the higher the conditional loss probability *pc*, the longer is the pause between loss events. Based on the largest conditional loss probability in the test of $pc = 0.5$, the smallest overall loss rate of $ppl = 0.015$, and the applied packet size of 20 ms, the longest average pauses between loss events to be expected in our test is

$$T_{pause,max} = \mu_{01} \cdot 20\,ms \approx 2.62\ s. \tag{4.9}$$

To assure a minimum of seven to eight occurrences of loss events, and that the test samples were long enough to allow stable judgments, samples of approximately 20 s duration were used. Forty German sentences were recorded from four speakers (2f, 2m). The sentences were taken from the EUROM language-material (Gibbon, 1992; Gibbon et al., 1997). In total, we applied 30 different transmission channels, across which the source sentences were transmitted. With four speakers per condition, 120 samples had to be judged by the listeners.

Of the 30 conditions, 11 conditions were used as reference conditions, reflecting the idea of the procedure described in ITU–T Rec. P.833 (2001; see also Appendix D.2). To determine equipment impairment factors for codecs under packet loss to be employed in conjunction with the E-model, ITU–T Rec. P.833 recommends the usage of 14 codec and codec tandeming conditions (plus a number of conditions for checking the impairment factor estimate for tandems involving the codec under test) and 10 additional reference conditions including transmission errors. Since the focus of the tests described here was on several different settings of a 2-state Markov model, not all 24 or more reference conditions could be applied. Instead, six reference conditions without transmission errors (codecs in single-operation), and five reference conditions with the G.729A under random packet loss were used. Here, we assume that the impairment due to G.729A coding under random packet loss is correctly predicted by the random loss model of the E-model, Equation (4.5). The 11 reference conditions and 19 test conditions are depicted in Table 4.2.

For all test conditions, the G.729A was used, with overall loss percentages Ppl from the set [0%, 1.5%, 3%, 5%, 10%, 15%] and conditional loss probabilities $pc = 1 - q$ from the set [0.15, 0.3, 0.5]. In case of $pc = 0.15$, the distribution obtained for $Ppl = 15\%$ corresponds to the random case, which is already included in the reference conditions. Hence, we employed $5 \cdot 3 - 1 = 14$ test conditions of this type. All other parameter settings of the line simulation model were set to the E-model default values listed in Table E.1.

In addition, we used five conditions reflecting a realistic network setup. The corresponding traces were collected by Florian Hammer at the Telecommunications Research Centre Vienna, Austria (Forschungszentrum Telekommunikation Wien, .ftw). He used a simulation of internet traffic of different numbers of users affecting the simulated RTP traffic between two VoIP-users connected to the same network. The test-bed employed for data collection is described in Hammer et al. (2004b, 2003). Traces of one-hour duration were recorded, from which appropriate 20 s long sections were selected fulfilling the requirement of a fixed overall loss percentage Ppl (also from the value set used in this test, i.e. [0%, 1.5%, 3%, 5%, 10%, 15%]). A total of 19 test conditions result from this choice.

As mentioned in the previous text, source material from four talkers was employed for each condition. To avoid that we investigate a particular trace in each condition instead of a class of traces corresponding to a particular 2-state loss-model setting, we used a different trace for each talker in each condition. Furthermore, two sets of test files were created with two alternative versions of the four traces, yielding a total of eight different traces per condition. This is different from the procedure applied in other tests on 2-state loss, such as the study described by Jiang and Schulzrinne (2002). According to a personal

Table 4.2: Test setup for the listening test on speech transmission quality under 2-state Markov loss. The effective equipment impairment factor values Ie,eff for the reference conditions that are listed in the last column are taken from ITU–T Rec. G.113 Appendix I (2002) and are used for the test data normalization based on ITU–T Rec. P.833 (2001).

No.	Codec	$Ppl = \dfrac{p \cdot 100}{p + q}[\%]$	$pc = 1 - q$	Ie,eff
1	G.711	–	–	0
2	G.726(32)	–	–	7
3	G.729A	–	–	11
4	GSM-FR	–	–	20
5	G.726(24)	–	–	25
6	G.726(16)	–	–	50
7	G.729A	1.5	Random: 0.015	17.1
8	G.729A	3	Random: 0.03	22.5
9	G.729A	5	Random: 0.05	28.5
10	G.729A	10	Random: 0.1	40
11	G.729A	15	Random: 0.15	48.1
12	G.729A	1.5	0.15	
13	G.729A	3	0.15	
14	G.729A	5	0.15	
15	G.729A	10	0.15	
16	G.729A	1.5	0.3	
17	G.729A	3	0.3	
18	G.729A	5	0.3	
19	G.729A	10	0.3	
20	G.729A	15	0.3	
21	G.729A	1.5	0.5	
22	G.729A	3	0.5	
23	G.729A	5	0.5	
24	G.729A	10	0.5	
25	G.729A	15	0.5	
26	G.729A	1.5	Real trace	
27	G.729A	3	Real trace	
28	G.729A	5	Real trace	
29	G.729A	10	Real trace	
30	G.729A	15	Real trace	

communication from Jiang, that study used only 1–2 traces and 1–2 different talkers per condition, so that the obtained results may be the effect of a particular sample trace rather than the effect of a particular loss model setting.

As stated previously, 40 sentences built the source material of the tests. To avoid too frequent repetitions of the sentences, we used each sentence three times in the test

$(3 \cdot 40 = 120)$, and different sentences for the four talkers in each condition. To avoid subject fatigue, the test was separated into two sessions held on two days. Each session was further subdivided into two parts, with a short break in between. The data preparation was carried out using the online-tool described in Appendix B. The subsystems were combined in a way that reflects a VoIP-typical usage scenario, corresponding to the configuration PSTN–VoIP–PSTN. We introduced packet loss using the network simulation NetDisturb (see Appendix B.3). The processed samples were presented in different randomized orders via normal handset telephones. The play lists as well as the quality ratings were controlled using a computer program.

Twenty-eight subjects participated in the tests (12 f, 16 m), covering an age range from 18 to 36 years (with an average of 25.9 years). They were, to their own account, normal hearing. They were asked to rate the overall impression (quality) of the link they had just used on the MOS scale (Figure 2.1), and the degradation of the line on the CR-10 scale (Borg, 1982; ITU–T Rec. P.833, 2001, Appendix I).

4.1.2.2 Test Results

We have normalized the test results according to the P.833-methodology (Appendix D.2), however, using the reduced set of only 11 reference conditions. The linear normalization line determined by least square fitting is shown in Figure 4.9.

The transformed MOSs are shown in Figure 4.10(a). For clarity, the ratings obtained for $pc = 0.15$ have been omitted, as well as those for the real packet traces. The latter and

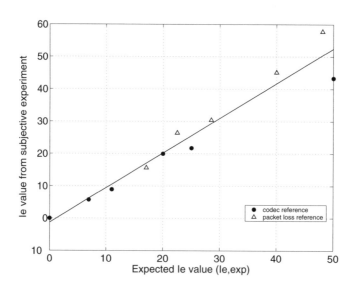

Figure 4.9: Normalization line for listening test data normalization according to the P.833-methodology (see Section D.2; ITU–T Rec. P.833, 2001). Abscissa: Expected impairment factor values; ordinate: Impairment factor values determined from listening test results. '•': Error-free codec reference conditions (conditions #1–#6, Table 4.2); '△': Reference conditions with G.729A under random packet loss (conditions #7–#11, Table 4.2).

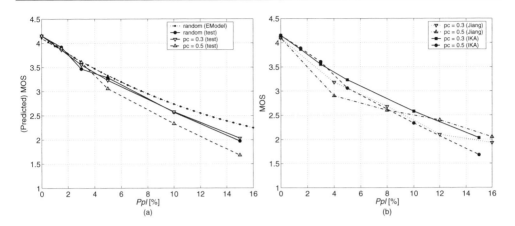

Figure 4.10: Listening test results for G.729A under 2-state Markov packet loss. (a) Author's own results. (b) Comparison to results from Jiang and Schulzrinne (2002). Abscissa: Packet loss percentage *Ppl*. Ordinate: Transformed MOS-scores.

more details on the results can be found in ITU–T Delayed Contribution D.222 (2004). All transformed ratings, including those for random packet loss, are lower than the prediction by the E-model. This may be due to the large number of packet-loss conditions presented in the test, which may have drawn the subjects' attention to this degradation type, to a larger extent than in tests with more varied degradation types. As it is also revealed from the curves, the results for random loss are practically identical to those collected for $pc = 0.3$. Only with a higher conditional loss of $pc = 0.5$, and thus an average of two consecutively lost packets, a tendency for lower quality can be observed. However, the differences are not statistically significant.

For further analysis, an analysis of variance (ANOVA) was carried out for the results of the actual test conditions, using the speaker, the packet loss percentage *Ppl* and the conditional-loss probability pc as fixed factors (Appendix G.2, Figure F.1). The conditional loss only showed a very minor but statistically significant impact on quality ($F = 5.450$, $p < 0.01$), while we found clearly the highest F-ratio for the packet loss percentage ($F = 419.51$, $p < 0.001$). Moreover, the speaker factor showed a much stronger effect on quality than the conditional loss, with $F = 42.805$, $p < 0.001$. A diagnostic analysis of the test data revealed that one of the female speakers was particularly disliked (on average over all conditions her samples were rated by ≈ 0.6 MOS worse than those of the other speakers). Excluding this speaker from the analysis, the influence of speaker and conditional-loss probability are found to be comparable ($F(Ppl) = 329.231$, $p < 0.001$; $F(pc) = 5.466$, $p < 0.01$; $F(speaker) = 5.777$, $p < 0.01$).

In Figure 4.10(b), the results for $pc = 0.3$ and $pc = 0.5$ are compared with the MOSs reported by Jiang and Schulzrinne (2002). As opposed to our test, their study showed neither a consistent difference between the quality ratings for the two different conditional loss settings, nor a difference that increases with the overall packet loss percentage. Since their tests were carried out with a reduced number of traces and speakers per condition, the results seem to reflect the impact of the particular sample-trace–speaker combination, rather than that of the underlying loss model. Hence, these results were not employed for quality modeling.

4.1.2.3 Parametric Model of Quality under 2-State Loss

In this section, we describe a generalization of the random-loss formula (Equation (4.5)), which extends the initial E-model to the case of the dependent, i.e. 2-state packet loss. Since no other appropriate auditory test data are available from the literature, the enhancement is based solely on the author's test results presented in the previous section. The effective equipment impairment factor *Ie,eff* that quantifies the impairment under coding and 2-state packet loss can be written as

$$Ie,eff = Ie + (95 - Ie)\frac{Ppl}{100 \cdot p + Bpl},$$ (4.10)

where we exploit the fact that random loss is a special case of 2-state loss (see Section 3.3.5.1), and thus

$$Ppl_{random} = ppl_{random} \cdot 100\% = p \cdot 100\%.$$

To include the conditional loss rate as additional parameter, we can further exploit that

$$p = \frac{q \cdot ppl}{1 - ppl} = \frac{(1 - pc) \cdot ppl}{1 - ppl} = \frac{(1 - pc) \cdot Ppl}{100 - Ppl}.$$ (4.11)

This leads to

$$Ie,eff = Ie + (95 - Ie)\frac{Ppl}{Ppl \cdot \dfrac{100 \cdot (1 - pc)}{100 - Ppl} + Bpl}.$$ (4.12)

The predictions derived with this rewritten formula are in astonishingly good agreement with the test results, as shown in Figure 4.11.

So far, we have restricted ourselves to 2-state Markov loss as the microscopic loss behavior. A simple generalization to arbitrary microscopic loss can be achieved, if we exploit that for 2-state loss (US Patent 6,931,017, 2005)

$$\frac{1 - pc}{1 - ppl} = BurstR.$$ (4.13)

Then,

$$Ie,eff = Ie + (95 - Ie)\frac{Ppl}{\dfrac{Ppl}{BurstR} + Bpl}.$$ (4.14)

In Equation (4.14), *BurstR* is the burst-ratio introduced in Section 3.3.5.1, Equation (3.14).

The simplicity of this new approach is remarkable, since it does not require any additional input-parameter to the E-model, apart from the conditional-loss rate *pc* necessary for a complete description of the 2-state loss. Moreover, Equation (4.14) does not change the E-model's predictions of quality under random packet loss, since *BurstR* = 1 in this case.

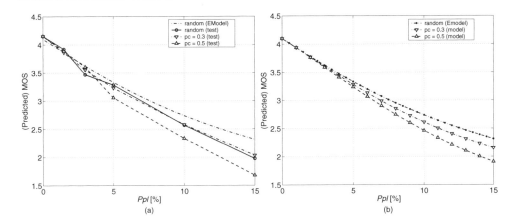

Figure 4.11: Listening test results for G.729A under 2-state Markov packet loss. Abscissa: Packet loss percentage *Ppl*. Ordinate: Predicted, or test MOS-scores. (a) Test results. (b) Model predictions using Equation (4.14) with the E-model.

To evaluate whether the approach is applicable to other codec/PLC combinations rather than the G.729A with its native PLC, further auditory tests are needed. The test described here is the only available one currently for quality under 2-state loss, that can be analyzed for modeling purposes. However, it seems reasonable to think that other codecs that use a state variable to capture the number of consecutively lost frames may as well be covered by this approach.

4.1.2.4 Summary of Dependent Loss

The following aspects were discussed in this section:

- We described a listening test carried out to understand the relation between dependent, microscopic loss and speech transmission quality for one example codec/ PLC combination.

- From the test results, we deduced a generalization of the parametric model of speech quality under random packet loss that is applicable to a more arbitrary microscopic loss behavior. It requires only one additional input-parameter to the E-model, which is necessary to describe the dependence of loss events (conditional loss probability *pc* or Burst Ratio *BurstR*). In case of random loss, the formula converges to the random loss model introduced in Section 4.1.1.

4.2   Macroscopic Loss Behavior

This section is concerned with two questions:

- How do individual segments of a particular *microscopic* loss behavior have to be weighted to determine the integral quality associated with a given *macroscopic* loss behavior?

- How can the impact of the macroscopic loss behavior on integral speech quality be quantified parametrically?

These questions are addressed in two ways. At first, a theoretical approach is chosen based on listening test data taken from the literature. Here, the emphasis lies on integral quality under adjacent periods of random packet loss (as the first type of microscopic loss behavior). In a second step, the results of a series of the author's conversation tests are presented, and analyzed with regard to the above questions. In this case, the focus is on subsequent periods of dependent loss i.e. 2-state Markov loss (as the second type of microscopic loss behavior).

4.2.1 Macroscopic Loss with Segments of Random Loss

In this section, adjacent segments of *microscopic*, random packet loss are assumed, that together form a particular profile of *macroscopic* loss behavior (see Figure 4.12).

The considerations presented in this section are based on the extensive auditory tests carried out by Gros and Chateau to establish the relation between instantaneous and integral quality (see Section 3.3.5.7, p. 76 ff.; Gros, 2001; Gros and Chateau, 2001; ITU–T Contribution COM 12–94, 1999; ITU–T Delayed Contribution D.139, 2000).

The test results serve to discuss, how integral quality can be predicted from knowledge of the packet loss rate and macroscopic loss distribution, by using the random packet-loss quality model outlined in Section 4.1.1. Two lines of thinking are followed:

1. The loss profiles employed by Gros and Chateau are concatenations of periods of 'stationary' random packet loss. Based on their results and the monitoring approach by Clark (2001), a set of assumptions is made.

   - The integral quality can be estimated by time averaging over instantaneous quality.

   - The quality for individual segments can be derived with the random-loss formula (Section 4.1.1).

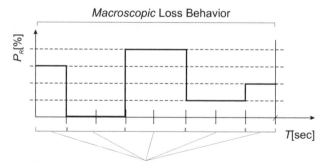

Figure 4.12: Loss profile showing a *macroscopic* loss behavior that is characterized by segments of different random packet loss as the *microscopic* loss behavior.

- During intersecting periods of quality improvements and degradations, instantaneous quality follows an exponential decay curve.

Starting from these assumptions, we determine the optimum settings for the exponential decay times that lead to the best fit of Gros and Chateau's integral quality data.

2. If an integration strategy is assumed for users establishing a quality impression of a link degraded by packet loss (Gros and Chateau, 2001), we investigate whether it is possible to predict integral quality simply on the basis of the integrated i.e. average loss percentage, for example, using the random loss model.

4.2.1.1 Estimating Integral Quality by Time-Averaging

Gros and Chateau used a large number of different loss profiles in their studies. These profiles of different durations (45–190 s) consist of different variations between two–five levels of random packet loss (0%, 2.5%, 10%, 20% and 30%). For all profiles, Gros and Chateau collected integral quality ratings from the test subjects. A simplified illustration of the instantaneous quality rating of a particular packet-loss profile is depicted in Figure 4.13.

In the study presented in this section, we used a subset of all profiles to derive optimum exponential decay times for the prediction of integral quality. The applied procedure was as follows:

(a) 35 packet loss profiles were selected from all profiles. Only those profiles were used, for which no expressed recency effect was observed, i.e. profiles without strong quality degradations or improvements concentrated in the beginning or at the end. Note that the selected profiles and the identification numbers given in Gros and Chateau's test descriptions are listed in Appendix H.1.1, Table H.1.

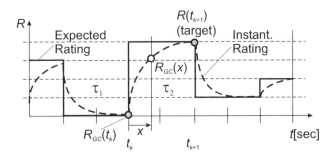

Figure 4.13: Illustration of the instantaneous rating behavior of subjects in case of a loss profile as used by Gros and Chateau (2001). The figure emphasizes the exponential decay during the stabilization of judgments between different segment borders, such as t_k and t_{k+1}. Note that in this illustration the judgments are depicted on the E-model R-scale, to explain some of the parameters used in the modeling approach discussed in the text. A corresponding explanation of the relative time x, and the values $R_{GC}(t_k)$, $R_{GC}(x)$ and $R(t_{k+1})$ is provided in the text, too.

(b) As an estimate of instantaneous quality, the Transmission Rating Factor $R_{GC}(x)$ for some instance x was calculated according to Equation (4.15). For an illustration of the different parameters see Figure 4.13:

$$R_{GC}(x) = R(t_{k+1}) - \left[R_{GC}(t_k) - R(t_{k+1})\right] \cdot e^{-\frac{x-t_k}{\tau_j}}. \tag{4.15}$$

Here, t_k and t_{k+1} identify the temporal borders of adjacent segments. $R(t_{k+1})$ is the transmission rating factor calculated for each target loss level at the temporal border (t_{k+1}) using the E-model with $BurstR = 1$ (i.e. the random-loss formula, Equation (4.5)). For the R-factor calculation, we set all input parameters of the E-model to their default values (Appendix E, Table E.1). Since Gros and Chateau employed the G.723.1 in their tests, the equipment impairment factor we used in the calculations was $Ie = 15$, and the packet loss robustness factor $Bpl = 16.1$ (see Equation (4.5); Table 4.1). For each segment, the packet loss percentage Ppl was set to the packet loss percentage corresponding to the period $[t_k, t_{k+1}]$. In an instantaneous judgment task, $R(t_{k+1})$ could stand for the quality level a subject aims at, while her/his judgment stabilizes after a quality change at segment border t_k. τ_j are the time constants for the exponential decay: τ_2 for improvement, and τ_1 for impoverishment.

The integral quality ratings were collected by Gros and Chateau (2001) on the MOS scale. The integral quality in MOS can be *estimated* by integrating Equation (4.15) over time (x) for each segment, dividing the result by the segment duration $(t_{k+1} - t_k)$, summing up the obtained quality estimates over segments, and converting the resulting average R into MOS. Note that Gros and Chateau (2001) have quantified the exponential rating behavior on the MOS scale. The algorithm developed by Clark (2001) for trace-data based VoIP quality monitoring is based on the averaging approach suggested by Gros and Chateau (2001), but is carried out on an impairment factor scale. In this book, we use the transmission rating factor R to express the quality behavior over time. In principle, this is not different from the other two approaches.

(c) The 35 profiles were divided into two groups of profiles.

- The first group of profiles (#1–#14) served for determining optimal settings for the time constants of the exponential decay. The optimization was done by finding those two parameters τ_1 and τ_2, for which the difference between subjective overall MOS and the MOS obtained from the estimated profile according to (b) were minimal in a least square sense, as it can be formulated by the minimization task shown in Equation (4.16).

$$\min_{\tau \in \mathbb{R}^2} \sum_{i=1}^{N} \left[MOS_{test_i} - MOS \left(\sum_k \frac{1}{t_{k+1} - t_k} \int_0^{t_{k+1} - t_k} R_{GC,i}(x_k, \tau) \, dx_k \right) \right]^2 \tag{4.16}$$

Moreover, the solution had to fulfill the condition that all individual absolute differences between the test results (MOS_{test_i}) and the estimated time-average are smaller than 0.35 points MOS:

$$\left| MOS_{test_i} - MOS \left(\sum_k \int_0^{t_{k+1}-t_k} R_{GC,i}(x_k, \tau) \, dx_k \right) \right| \leq 0.35 \qquad (4.17)$$

- The second group of profiles (#15–#35) was used in order to verify the determined time constants. This was done by calculating, whether the absolute of the difference between estimated overall MOS and integral (test) MOS was below 0.35 MOS.

The optimization of the exponential decay settings according to the processing steps (b) and (c) have lead to time constants of $\tau_1 = 9$ s and $\tau_2 = 22$ s. While the value for τ_1 – corresponding to a decrease of quality – is equal to Gros's findings ($\tau_{1,Gros} = 9$ s), the optimum setting determined for τ_2 – improvement of quality – is higher than that found by Gros ($\tau_{2,Gros} = 14.3$ s). In other words, the settings used here imply a slower reaction on quality improvements. This can be explained by the fact that Gros and Chateau determined the decay times to fit the *instantaneous* judgments. Instead, the values presented here were derived with regard to the *integral* quality judgments. For many profiles, the quality estimate Gros and Chateau obtained by averaging over instantaneous quality judgments is more optimistic than the integral rating delivered by the subjects. The longer time constant $\tau_2 = 22$ s for quality improvements we found gives more weight to periods of low quality. Consequently, the overall estimate is more pessimistic and closer to the integral test data.

The time constants are also different from the settings suggested in the monitoring approach by Clark (2001). Here, values of $\tau_{1,Clark} = 5$ s and $\tau_{2,Clark} = 15$ s are given. In the description of the approach, no information is provided on why these particular settings were chosen.

In Figure 4.14, the integral quality ratings obtained in Gros and Chateau's tests are compared with the predictions derived with the time-averaging approach outlined above (o; for details on the differences obtained for individual profiles see Table H.2, Appendix H.1.1.1). Also shown in the figure are the borders for which predicted quality lies within the range

$$\left| MOS_{pred.} - MOS_{test_i} \right| \leq 0.35,$$

according to the requirement set by item (c). As can be seen from the graph, there are only two outliers that do not fulfill this requirement. Since quality for other profiles similar to these outliers was well predicted, no principal doubt is raised with regard to the general prediction approach. The linear correlation between the test results and the time-averaging estimates is $r = 0.93$.

It can be concluded that in case of a *macroscopic* loss behavior showing adjacent segments of random packet loss, integral speech transmission quality can accurately be predicted by time-averaging over individual segments. Therefore, the quality levels associated with individual segments can be estimated using the random loss formula, Equation (4.5), i.e. setting $p \cdot 100 = Ppl$ in Equation 4.10 or $BurstR = 1$ in Equation 4.14. It has to be noted, however, that we could verify this finding only for the G.723.1 codec used in the studies by Gros and Chateau (2001).

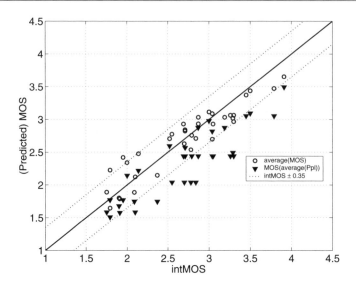

Figure 4.14: Comparison of integral quality ratings obtained in listening tests (abscissa), and predictions based on: 'o': segment-wise time-averaging of estimated instantaneous quality (ordinate: $\overline{MOS}(R)$); '▼': average packet loss (ordinate: $MOS(\overline{Ppl})$).

4.2.1.2 Estimating Integral Quality from Average Loss Rate

In this section, a study is presented that addresses the question, whether integral quality can be predicted from the average loss rate. To this aim, for each of the 35 loss profiles used in the previous section (Section 4.2.1.1), the *loss rate* was time-averaged over the constituting segments. The resulting average loss rates were then used as the input parameter Ppl to the E-model, to estimate integral quality (employing Equation (4.5), or Equation 4.10 with $p \cdot 100 = Ppl$).

The points denoted by '▼' in Figure 4.14 show the relation between integral quality ratings obtained by Gros and Chateau, and the predictions we derived with the E-model for the average loss percentages (for more details on the loss percentage of each profile and the related MOS-ratings – predicted as well as test-MOS – see Table H.2, Appendix H.1.1.1). The following observations can be made:

- For almost half of the profiles (17 out of 35), the quality estimate based on the average packet loss percentage yields acceptable agreement with the test data, that is,

$$\left| MOS_{pred.} - MOS_{test_i} \right| \leq 0.35.$$

- The overall correlation between the predicted MOS and the test data is of $\rho = 0.87$.

- In all relevant cases of deviation, the predictions are more *pessimistic* than the test results.

- Largest deviations occur, when the profiles show sparse passages of very different loss levels.

Thus, for quality prediction in the context of network planning, integral quality under a macroscopic loss behavior that shows adjacent periods of random packet loss can safely be estimated from the average loss rate of an entire profile. The limits of this approach are explored further in Appendix H, Section H.1.1.2. Also in this case, the generalization to other codecs than the G.723.1 used by Gros and Chateau should ideally be verified by further auditory tests.

4.2.2 Macroscopic Loss with Segments of Dependent Loss

In this section, we assume adjacent segments of dependent packet loss as the *microscopic* loss type, that together form a particular profile of *macroscopic* loss behavior (see Figure 4.15).

We conducted a series of three conversation tests at the Institute of Communication Acoustics (IKA) to address two main questions:

1. How can the relation between the macroscopic loss behavior of a network and the integral quality perceived by a user be quantified, if adjacent segments of microscopic behavior show dependent packet loss?

2. What is the integral speech quality perceived when packet loss of macroscopic behavior (that typically leads to time-varying speech quality) is combined with other types of degradations, that as such only lead to stationary speech quality?

We will discuss the second question in Section 4.4.3.

4.2.2.1 Loss Conditions

In the conversation tests, we introduced packet loss using a 3-state Markov model. This model is the limiting case of the 4-state model discussed in Section 3.3.5.1 (Figure 3.7), with: $p_{14} \to 0$, and $p_{44} \to 0$. The resulting loss model can be viewed as a 2-state model representing the 'bad state' that communicates with the 'good state', which now is represented by only one state, a found state (see Figure 4.16). On average, the resulting packet loss profile can thus be characterized as the concatenation of segments of 2-state Markov

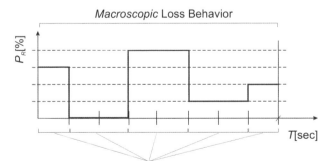

Periods of *Microscopic* Loss Behavior: 2-state Markov Loss

Figure 4.15: Loss profile showing a macroscopic loss behavior that is characterized by segments of different dependent loss (here 2-state Markov model type loss) as the microscopic loss behavior.

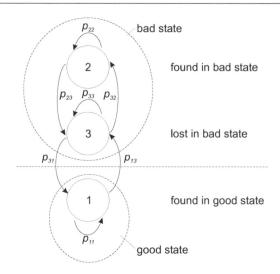

Figure 4.16: 3-state Markov model.

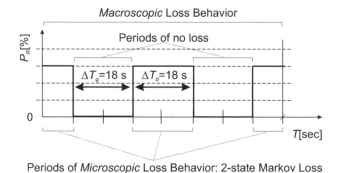

Figure 4.17: Loss profile that – on average – results from the employed 3-state Markov model (see text for details).

loss that represent the desired type of microscopic, i.e. dependent loss behavior. Together, their concatenation then shows the macroscopic packet loss behavior we aimed at in this section (see Figure 4.17).

We can reduce the 4-state transition probability matrix to the 3-state matrix by a corresponding choice of the respective transition probabilities given in Equation (A.1), yielding:

$$\mathcal{P}_{3-state} = \begin{pmatrix} p_{11} & 0 & p_{13} \\ 0 & p_{22} & p_{23} \\ p_{31} & p_{32} & p_{33} \end{pmatrix}. \tag{4.18}$$

All formulae derived in Section 3.3.5.1 to describe characteristics of the 4-state model can be applied to the 3-state model, if furthermore $p_{41} \to 1$ is assumed.

We chose this 3-state model to

(i) introduce loss in a parametric way,

(ii) create a strongly time-varying, i.e. macroscopic loss behavior with 'bad' and 'good' periods,

(iii) enable the control of the microscopic loss behavior during the 'bad' periods.

With these characteristics, the 3-state model represents a stringent combination of theoretical loss profiles such as those used by Gros and Chateau (2001), and a more realistic loss behavior as it is encountered in actual networks.

By an appropriate choice of the model parameters, the macroscopic and the microscopic loss behavior can be controlled. In case of the employed 3-state model, the *macroscopic* loss is characterized by the sojourn times in the 'good' and 'bad' state (μ_{13} and μ_{31}, respectively; see Equations (3.18) and (3.19)):

$$\mu_{13} = \frac{1}{p_{13}} \tag{4.19}$$

$$\mu_{31} = \frac{p_{23} + p_{32}}{p_{23} \cdot p_{31}} \tag{4.20}$$

The *microscopic* loss behavior is defined by the distribution of loss run-lengths, i.e. of the numbers of consecutively lost packets. In case of the employed 3-state model, no packets are lost in the 'good' state. Hence, the microscopic loss behavior is characterized by a 2-state Markov chain model in case of the 'bad' state, and a single no-loss-state for the 'good' state. As described in Section 3.3.5.1, the average number of consecutively lost packets can be calculated as the expectation of the geometrically distributed run length k (see Equations (3.4) and (3.5)):

$$E\{k\} = \frac{1}{1 - p_{33}} = \frac{1}{1 - pc}. \tag{4.21}$$

The overall loss percentage related to the 3-state model follows from Equation (A.11):

$$p_{pl} = P_3 = \frac{p_{13}\,p_{23}}{p_{13}\,p_{23} + p_{13}\,p_{32} + p_{23}\,p_{31}} \tag{4.22}$$

In total, four transition probabilities can be considered as the unknown variables of the 3-state model (e.g. p_{13}, p_{23}, p_{31} and p_{32}, see Equation (4.18); the dependent variables p_{11}, p_{22} and p_{33} result from the fact that the probabilities in one row of the transition probability matrix $\mathcal{P}$ sum up to 1). Starting from the applied packet size of 20 ms for the G.729A codec we used (two frames/packet), the four unknowns were derived from Equations (4.19) to (4.22) to meet a set of requirements (see Figure 4.17):

Macroscopic loss behavior

(1) Average sojourn time in the 'bad' state of $\Delta T_b = 18$ s (*burst duration*).

(2) Average sojourn time in the loss-free 'good' state of $\Delta T_g = 18$ s (*gap duration*).

Microscopic loss behavior

(3) On average, two packets are lost in a row.

Overall loss percentage

(4) $Ppl \in [0\%, 3\%, 5\%, 15\%]$.

(1)-(2): We considered the setting of 18 s for burst and gap duration to be a suitable choice of macroscopic loss behavior for the conversation tests, since 18 s is

- long enough to have subjects become aware of the presence or absence of the degradations, and partly stabilize their instantaneous impression of quality (reflecting the reaction times of 9–30 s found in the instantaneous judgment tests by Gros, 2001; Gros and Chateau, 2001),

- short enough to yield at least 2–4 interchanging sojourns in the 'good' and 'bad' states (considering a mean conversation duration of approximately 150 s for the employed short conversation test scenarios; Möller, 2000, p. 79),

and thus

- 'macroscopic' enough to assure time-varying quality perception.

(3): We chose dependent microscopic loss with an average of two packets lost in a row during bursts to complement the studies on time-varying random loss conducted by Gros and Chateau (see Section 4.2.1).

(4): The particular settings for overall loss rate were adopted for comparability to the author's short conversation tests on combinations of random packet loss with other degradation types (see Sections 4.1.1.2 and 4.4.2; Raake, 2004).

For more details on the test setup, test procedure and test subjects see Appendix G.4.

4.2.2.2 Test Results and Quality Prediction

We will restrict the following discussion to the results of two of the three conversation test series conducted on 3-state Markovian loss. In each test series, we not only presented conditions of packet loss alone, but also conditions of packet loss combined with another type of degradation to the test subjects (the latter to assess speech quality under combined degradations). The tests considered in this section are identified as 'CT-macro1' (packet loss and line noise) and 'CT-macro3' (packet loss and talker echo)[6].

Figure 4.18 shows a comparison of the MOS-results[7] obtained in the conversation tests on speech quality under random packet loss (Section 4.4.2; combinations of random packet loss and other degradations), and on speech quality under macroscopic loss (Section 4.4.3; combinations of 3-state packet loss and other degradations). Note that the figure solely presents the results for those test conditions, where packet loss was presented without additional degradations. Figure 4.18(a) depicts the results found in the test series on packet

---

[6]The conversation test series on packet loss and delay, 'CT-macro2', was conducted with particular conversation scenarios and at a higher line attenuation than the other two tests and is therefore not considered here.

[7]The results obtained on the CR-10 scale are very similar to the MOS-results, and are omitted here for clarity of the presentation.

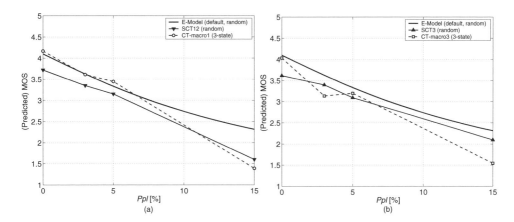

Figure 4.18: Conversation test results on integral speech quality under random packet loss and 3-state Markov model type loss. Abscissa: Packet loss percentage *Ppl*; ordinate: Quality (MOS). The depicted test results reflect those test conditions, where packet loss was the only degradation. For comparison, the predictions obtained with the E-model are depicted, too (using the random loss formula of Equation (4.5)). (a) Conversation tests on combinations of packet loss and circuit noise. (b) Conversation tests on packet loss and talker echo.

loss and noise (SCT12 and CT-macro1), and figure 4.18(b) the results for the tests on packet loss and talker echo (the mean opinion scores derived from the test ratings were linearly transformed according to the procedure described in Appendix D.3, Equation (D.3)). The comparison indicates a stronger impact of the macroscopic loss on speech quality:

- The tests on packet loss and noise show a stronger decrease of integral quality in case of 3-state Markov loss than in case of random loss.

- For high loss percentages (*Ppl* = 15%), the tests on packet loss and talker echo indicate considerably lower quality ratings for 3-state loss than for random loss.

- In both tests on 3-state loss, the results for high percentages of 3-state loss (15%, i.e. 30% within bursts) are by almost 1 point MOS lower than predicted by the E-model for the default of *BurstR* = 1 (random loss).

The observations are limited by the fact that they were made across different tests and not across different conditions of one test. However, the conclusion that the employed 3-state packet loss degrades integral quality more strongly than random loss is supported by the test results obtained for combined degradations (see Sections 4.4.2 and 4.4.3): In case of combined degradations, the degradation due to 3-state Markov loss is clearly the dominant one; when *random* loss is combined with other degradations, both packet loss and the other degradation type play a role for quality.

That quality is degraded more strongly by 3-state Markov than by random loss is confirmed also by listening to the (recorded) test conversations: During bursts, the subjects had considerable difficulties of understanding, and hence had to frequently ask their interlocutor for repetitions.

To investigate, whether the integral quality under 3-state Markov loss can be more accurately estimated with another approach than that employed in the current E-model (random loss, Equation (4.5)), three different modeling approaches are compared:

(1) In analogy to the procedure described in Section 4.2.1.1, integral quality is predicted based on *time-averaging* of estimated instantaneous quality. Instantaneous quality, in turn, is estimated using the 2-state impairment formula proposed as enhancement of the E-model in Section 4.1.2, Equation (4.14).

(2) In analogy to the procedure described in Section 4.2.1.2, integral quality is estimated based on the *average loss rate*, also using the 2-state prediction formula (see Section 4.1.2, Equation (4.14)).

(3) A completely different approach is taken up from Rosenbluth, who suggests the usage of an actual weighting of different degradation events, according to their importance for integral quality (ITU–T Delayed Contribution D.064, 1998). This approach is different from the *implicit* weighting by time-averaging of instantaneous quality (Gros, 2001; Gros and Chateau, 2001).

Estimating Integral Quality by time-averaging (1)

Using time-averaging of instantaneous estimates according to Gros and Chateau (2001), integral quality is predicted

(a) for a generalized profile corresponding to the 3-state model settings;

(b) based on estimates of instantaneous quality for the individual traces employed in the test, and averaging of the individual quality estimates over all traces applied for one test condition;

(c) according to the parametric approach proposed by Clark (2001), which can be viewed as a closed form expression of (a).

(a): The generalized profile used under (a) is described by the temporal borders t_i of adjacent segments of a given, microscopic behavior, with

$$t_i/s \in [0, 9, 27, 45, 63, 81, 99, 108].$$

These borders reflect the 3-state model parameter settings leading to average burst and gap durations of 18 s. The temporal borders separate alternating periods of a fixed 2-state loss behavior, according to the scheme $[0X0X0X0]$, that is, alternations between 0% and X% packet loss percentage ($X \in [6, 10, 30]$ to yield $Ppl \in [3\%, 5\%, 15\%]$). Starting from the three profiles obtained with these settings for X, the integral quality is determined as the time-average over the estimated instantaneous quality. The instantaneous quality associated with each profile is estimated using the 2-state extension of the E-model to predict the target quality levels at the temporal borders t_i (Equation (4.14)), and assuming an exponential decay for the stabilization after changes from burst to gap or gap to burst (i.e. changes in loss percentage from 0% to X%; see Section 4.2.1.1).

(b): In the process of trace creation, two files have been stored per trace: the actual trace, that is, the packet list indicating to the network simulation (NetDisturb) which packets are to be lost or received, and the corresponding sequence of state numbers (here, an entry at index i identifies the state the packet i is in). All stored state lists are processed to determine temporal borders corresponding to changes between the good and the bad state and vice versa. In addition, the actual loss rate during burst periods is calculated. From the individual trace information (timing and loss), individual estimates of the course of instantaneous quality are derived, following the same procedure as is employed for the generalized trace under (a). Subsequently, the quality estimates obtained per trace are averaged over all the traces.

(c): The modeling approach used in the monitoring tool 'VQmon' proposed by Clark (2001) is based on the translation of a given packet trace into 4-state Markov model parameters (see Figure 3.7). The resulting transition probabilities are used to estimate the average sojourn times in the gap and the burst 'states' (g and b respectively). Also calculated are the average loss rates associated with the sojourn in these states. The loss behavior described by the Markov model can be viewed as a periodic repetition of burst and gap periods, similar to the generalized profile discussed in the preceding text.

The algorithm proposed by Clark for E-model-based integral quality estimation can be outlined as follows: Ieg and Ieb are the impairment values associated with the loss levels during the gap and burst periods. I_1 is the estimated instantaneous impairment level linked to a change from burst condition Ieb to gap condition Ieg. I_2 is the corresponding instantaneous impairment level linked to a change from gap condition Ieg to burst condition Ieb. With these parameters and the assumption of an exponential decay with different time constants τ_1 and τ_2 for quality improvements and degradations, I_1 and I_2 can be expressed as:

$$I_1 = Ieb - (Ieb - I_2) \cdot e^{-\frac{b}{\tau_1}} \tag{4.23}$$

$$I_2 = Ieg + (I_1 - Ieg) \cdot e^{-\frac{g}{\tau_2}}. \tag{4.24}$$

Combining these equations gives an expression for I_2 independent of I_1:

$$I_2 = Ieg \cdot \left(1 - e^{-\frac{g}{\tau_2}}\right) + Ieb \cdot (1 - e^{-\frac{b}{\tau_1}}) \cdot e^{-\frac{g}{\tau_2}}. \tag{4.25}$$

The integration over a burst and a gap period leads to an expression for the average impairment associated with the profile:

$$\begin{aligned} Ie(av.) = \frac{1}{b+g} \cdot \Bigg[& Ieb \cdot b + Ieg \cdot g + \tau_1 \cdot (Ieb - I_2) \cdot \left(e^{-\frac{b}{\tau_1}} - 1\right) \\ & - \tau_2 \cdot \left(Ieb - (Ieb - I_2) \cdot e^{-\frac{b}{\tau_1}} - Ieg\right) \cdot \left(e^{-\frac{g}{\tau_2}} - 1\right) \Bigg]. \end{aligned} \tag{4.26}$$

Integral quality can be predicted using Equations (4.25) and (4.26) together with the E-model. Therefore, the baseline impairments Ieb during bursts and Ieg during gaps can be estimated with the random-loss formula Equation (4.5) (in case of microscopic random loss during bursts), or with the 2-state formula proposed in this book (Equation (4.14), for dependent loss during bursts). For details on the general approach see Clark (2001), or ITU–T Delayed Contribution D.105 (2003). In the analysis presented here, the 2-state formula was applied, and the exponential decay times found in Section 4.2.1.1 were used ($\tau_1 = 9$ s and $\tau_2 = 22$ s).

In Figure 4.19, the predictions derived with the approaches (a)–(c) are compared with the conversation test results. As can be seen from the left part of the figure, the predictions calculated by time-averaging based on the generalized loss profile (dashed curve; (a)) are almost identical to the time-averaging results for individual traces (dash-dotted curve, 'o'; (b)). This reflects the choice of a large number of traces per condition, so that averaging over per-trace quality estimates corresponds to estimating quality directly from an average profile. For lower loss percentages ($Ppl \leq 5\%$), the predictions according to (a) and (b) are in agreement with the test results. However, for an overall loss percentage of 15%, i.e. a loss percentage of 30% during bursts, the predictions are by almost 1 point MOS more optimistic than the mean ratings obtained from the test subjects. With this discrepancy, the time-averaging approaches (a) and (b) do not show any advantage over the predictions obtained using the random loss model (solid line, no marker).

The right part of Figure 4.19 depicts a comparison of the conversation test results and the predictions according to Equation (4.26), as proposed in approach (c) by Clark (2001); (dashed line). Also shown are the predictions obtained with the current E-model (random loss; solid line). It can be seen that the predictions delivered by approach (c) are slightly lower than the ones calculated using approaches (a) and (b) (depicted in the left picture). The dash-dotted, horizontal line in the right part of Figure 4.19 illustrates the worst-case

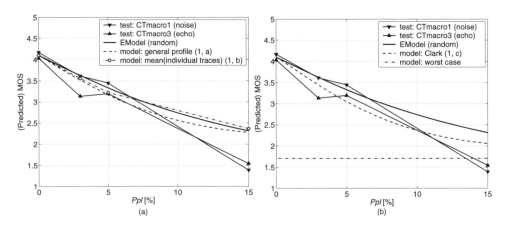

Figure 4.19: Integral quality ratings obtained in conversation tests on 3-state Markov model type loss, and the predictions derived with the 2-state enhancement of the E-model according to different procedures of time-averaging of estimated instantaneous quality (see text for more information). Abscissa: Packet loss percentage Ppl; ordinate: MOS. The depicted test results reflect the test conditions for which packet loss was the only degradation. Left part: Test results ('▼', and '▲'), and time-averaging predictions for generalized loss profiles ((a): dashed line), and for the individual traces used in the test ((b): dash-dotted line 'o'). The solid line shows the predictions derived with the random loss E-model version 4.1.1. Right part: Test results ('▼', and '▲'), and predictions obtained with the formula by Clark (dashed line (c): Clark, 2001). Also depicted are the predictions delivered by the current E-model (solid line). The dash-dotted line represents a worst-case estimation, which thus does not vary with the packet loss percentage (see text).

condition for the time-averaging approach: Assuming a generalized loss profile according to (a), the lowest quality is obtained for alternating periods of no loss ($R = 82.2$) and maximal loss ($R = 0$). The corresponding worst-case prediction is more optimistic than the test results for the highest packet loss rate of $Ppl = 15\%$. The hypothetical worst case prediction clearly indicates the limitations of the time-averaging approach: Due to the implicit weighting based on *fixed* exponential decay times, maximally bad periods are treated in a similar manner as weakly degraded periods. Instead, the test results show that subjects weight periods of very bad degradation more strongly with regard to integral quality than periods of lower degradation. Since packet loss during bursts was as high as 30%, with a considerably higher number of consecutively lost packets than for random loss, intelligibility during bursts was reduced.

Estimating Integral Quality from Average Loss Rate (2)

In analogy to the procedure described in Section 4.2.1.2, integral quality can be estimated from the average loss percentage.

Figure 4.20 depicts a comparison of the conversation test results on 3-state packet loss to predictions delivered by the 2-state enhancement of the E-model. The dashed curve

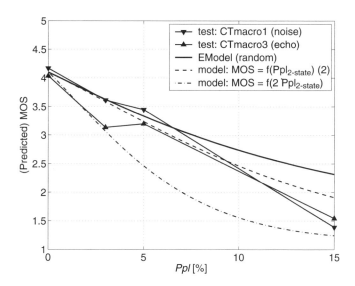

Figure 4.20: Integral quality ratings obtained in conversation tests on 3-state Markov model type loss, and the predictions derived with the 2-state enhancement of the E-model based on the average loss rate (see text for more information). Abscissa: Packet loss percentage *Ppl*; ordinate: MOS. The depicted test results reflect the test conditions for which packet loss was presented without additional degradations. Continuous lines ('▼', and '▲'): Test results. Dashed line: Prediction from 2-state E-model enhancement based on average loss rate *Ppl*. Dash-dotted line: Prediction from 2-state E-model enhancement based on average loss rate *during bursts* ($2 \cdot Ppl$). The solid line shows the predictions derived with the current version of the E-model (random loss approach; ITU–T Rec. G.107, 2005).

shows the predictions calculated from the average loss rate Ppl, and the dash-dotted curve the predictions using the loss rate during bursts ($= 2 \cdot Ppl$). As revealed from the figure, the 2-state model delivers relatively accurate predictions of integral quality. For the high loss percentage of $Ppl = 15\%$, however, a safer, i.e more pessimistic prediction is reached when the loss rate during bursts is employed. This finding emphasizes that – for integral quality – subjects ascribe greater importance to periods of strong degradation.

Estimating Integral Quality by Direct Segment-Weighting (3)

A different approach for predicting the integral quality related to a time-varying channel behavior has been suggested by Rosenbluth (ITU–T Delayed Contribution D.064, 1998). His modeling approach is based on individual weights given to passages of degraded quality. He collected listening test data on quality under periods of interruptions or of additive noise. Therefore, he employed short speech stimuli (approximately 8 s) that he concatenated to longer samples (60 s). On one hand, his test subjects had to make quality ratings of individual short passages degraded in different manners. On the other hand, the subjects had to rate the quality of the longer stimuli, which were composed by positioning the degraded short passages at different locations (i.e. beginning, middle and end of the long stimulus). From the quality ratings obtained for individual passages, and those obtained for combinations thereof, i.e. the long stimuli, Rosenbluth calculated the weights of the individual passages for integral quality.

The weights he determined are related to the intensity of the quality degradation of the passage, and to the relative position within the complete stimulus or connection (see Section 3.3.5.7, p. 76 ff.). The corresponding weighting of passage i is given by Equation

$$ W_i = \max\left[1, 1 + \left(0.038 + 1.3 \cdot L_i^{0.68}\right) \cdot (4.3 - MOS_i)^{\left\{0.96 + 0.61 \cdot L_i^{1.2}\right\}}\right], \qquad (4.27) $$

where W_i is the relative weight of passage i, MOS_i is the mean quality rating for the degradation period i (in MOS), and L_i is the location of the degradation period i within the complete stimulus or connection. L_i ranges from 0 to 1, with 0 being the beginning and 1 the end of the complete stimulus. In Rosenbluth's approach, the midpoint of the passage i is used to compute L_i. The integral quality of the complete stimulus or connection is calculated as

$$ MOS_E = \frac{\sum_i W_i \cdot MOS_i}{\sum_i W_i}, \qquad (4.28) $$

where MOS_E is the predicted integral MOS, and i denotes the indices of the individual degradation periods that make up the complete stimulus.

To compare this approach to the integral quality ratings obtained in the conversation tests, a corresponding generalized loss trace is employed. Since the conversation tests presented in this section only produced integral quality ratings for long passages, the weighting approach had to be adopted without accounting for the exact long and short stimuli durations employed by Rosenbluth. The positions in time of the burst and gap periods are described by the relative vector Ls, which contains the midpoints of the periods, using 108 s as the long stimulus duration.

$$Ls = [4.5, 13.5, 22.5, 31.5, 40.5, 49.5, 58.5, 67.5, 76.5, 85.5, 94.5, 103.5]/108.$$

The employed time periods were of 9 s duration, to lie close to the 8 s underlying Rosenbluth's model. The 9 s durations correspond to one half of the average burst and gap durations used in the conversation tests. Consequently, the associated packet loss levels were defined as

$$Ppls = [0\ X\ X\ 0\ 0\ X\ X\ 0\ 0\ X\ X\ 0], \quad X \in [0; 30].$$

For each burst period, the individual speech quality MOS_i was estimated using the E-model 2-state formula presented in Section 4.1.2, Equation (4.14). The resulting quality predictions are presented in Figure 4.21. As can be seen from the graph, the predictions are in good agreement with the test results, in particular for the high loss rate of $Ppl = 15\%$. This result indicates that for strong degradations, an individual weighting of periods of different quality levels is more appropriate for integral quality prediction than the time-averaging approaches discussed above: For 15% macroscopic packet loss, time-averaging had revealed predictions that are by more than 0.5 points MOS more optimistic than the predictions based on *explicit* weighting.

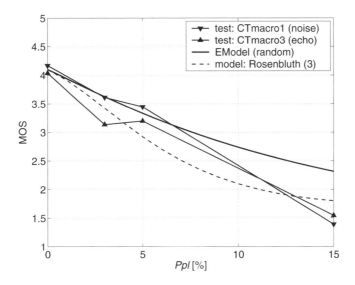

Figure 4.21: Integral quality ratings obtained in conversation tests on 3-state Markov model type loss, and the predictions derived with the 2-state enhancement of the E-model, employed in the framework of the weighting model proposed by Rosenbluth (ITU–T Delayed Contribution D.064, 1998). Abscissa: Packet loss percentage; ordinate: Mean opinion score (MOS). The depicted test results reflect the test conditions for which packet loss was the only degradation. Continuous lines ('▼', and '▲'): Test results. Slashed line: Prediction from Rosenbluth approach used with the 2-state E-model enhancement. The solid line shows the predictions derived with the current E-model version (random loss approach; ITU–T Rec. G.107, 2005).

Summary 3-State Loss

In this section, it was discussed how integral speech quality can be predicted from quality estimates for adjacent segments of dependent loss behavior. Three modeling approaches were compared:

(1) Time-averaging of instantaneous quality estimates.

(2) Quality prediction from average packet loss.

(3) Individual weighting of burst and gap periods based on their importance and relative positions within an entire stimulus.

In all cases, a good agreement with the conversation test results was found, as is summarized in Table 4.3.

The additional observations made for the different modeling approaches can be summarized as follows:

- The accuracy of speech quality predictions based on time-averaging of estimated instantaneous quality is limited, when individual periods of a connection are characterized by strong degradations that may affect intelligibility. In an extreme case, the connection may almost be interrupted completely during periods of bursts.

- For higher loss percentages, the importance of the periods of bad quality is increased. In this case, a worst-case estimate can be obtained from the 2-state loss model enhancement of the E-model, based on the average loss percentage related only to the bad periods.

- In case of high packet loss percentages, best prediction results seem to be obtained by individual weighting of degradation periods according to their importance (Rosenbluth, ITU–T Delayed Contribution D.064, 1998).

- The predictions derived with the 2-state E-model enhancement (Equation (4.14)) based on the average loss percentage show the highest correlation with the test results, and the lowest mean squared error (see Table 4.3). Moreover, this approach leads to acceptable quality estimates even in case of high packet loss percentages. Hence, it is

Table 4.3: Predictions derived with the different modeling approaches (1)–(3): Correlations between conversation test results averaged over the two conversation tests 'CT-macro1' and 'CT-macro3', and root mean squared error of the predictions (in MOS).

Model	Correlation with Test Results	Root mean Squared Error
Time-averaging, generalized trace	0.972	0.462
Time-averaging, individual traces	0.971	0.419
Time-averaging, burst and gap length (Clark)	0.969	0.328
Average loss percentage (random)	0.987	0.443
Average loss percentage (2-state)	0.990	0.250
Individual weighting (Rosenbluth)	0.969	0.261

proposed to use this simple model for *general* predictions of integral speech quality, as it is required for network planning.

4.3 Interactivity

The speech quality perceived by users under transmission delay depends on the conversation task to be carried out over the connection (see Sections 3.3.4 and 2.1.1.4 Kitawaki and Itoh, 1991). Here, the interactivity of the conversation between the users seems to play an important role. Whether speech quality under *packet loss* depends on the employed conversation scenario, has not yet been investigated.

Two of the test series on packet loss and additional degradations were carried out focussing on packet loss and transmission delay ('SCT2' and 'CT-macro2'). Besides introducing packet loss of two different types of distributions in these tests, different conversation scenarios were used, to study the following questions:

(1) Can the impact of the conversation scenario on the interactivity of the corresponding conversations be measured instrumentally by analyzing recordings of the conversations?

(2) Does speech quality under delay – that impacts the communication flow between conversation partners (see Section 3.3.4) – depend on the conversation scenario?

(3) Does speech quality under packet loss depend on the conversation scenario?

To study these questions, the interactive Short Conversation Test (iSCT) scenarios described in Section 2.1.1.4 were employed in a dedicated conversation test[8]. With the new iSCT scenarios, it was aimed at provoking more interactive conversations between the test subjects.

The iSCT scenarios were evaluated with regard to the impact on the communication behavior of the subjects. Therefore, Hammer *et al.* (2004a) developed a metric that allows the instrumental characterization of recorded telephone conversations. The metric is motivated by the state model of telephone conversations proposed by Brady (1965, 1968), and the state-based artificial conversational speech defined in ITU–T Rec. P.59 (1993). It employs a 4-state Markov process to classify phases of the conversations according to four conversation states:

A Only speaker 'A' talks.

B Only speaker 'B' talks.

M None of the two speakers talks (mutual silence).

D Both speakers talk at the same time (double talk).

The possible bidirectional transitions between states are shown in Figure 4.22. The transitions $D \leftrightarrow M$ and $A \leftrightarrow B$ are rarely encountered in real conversations and are thus not considered.

---

[8]This test series complements the other conversation tests described in Chapters 4 and 5 of this book, which were all carried out with the SCT scenarios (see also Section 2.1.1.4; Möller, 2000).

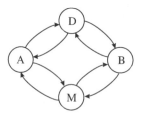

Figure 4.22: Illustration of the 4-state Markov model underlying the conversation analysis (Hammer *et al.*, 2004a).

The measures extracted based on this model, are the state probabilities for all four states, and the corresponding average sojourn times. As an additional measure, the *speaker alternation rate* (SAR) quantifies the number of alternations between the talkers A and B per minute. Such transitions comprise either alternations separated by mutual silence (A–M–B and B–M–A), or separated by phases of double talk (A–D–B and B–D–A). In the latter case, one talker is interrupted by the other.

A number of the recorded conversations from the test series on 3-state packet loss and additional impairments were analyzed according to this metric (from the tests 'CT-macro1'– 'CT-macro3', see Section 4.4.3). For the comparison of the two scenario types, only conversations held over clean, PCM A-law encoded connections were considered (Hammer *et al.*, 2004a). For the chosen sample scenarios, differences between iSCTs and SCTs could be observed in the state probabilities and sojourn times for mutual silence (state M), and the SARs. While the state probability and sojourn time for mutual silence are lower in case of the iSCT scenarios, the SAR was found to be higher in this case. These differences can be interpreted as an indication that the iSCT scenarios lead to more interactive conversations.

In spite of the higher interactivity of the iSCTs implied by the conversation analysis, no influence of the scenario type on the speech quality under transmission delay could be observed (see Figure 4.23(a)). Instead, transmission delays of up to 600 ms did not show any considerable impact on speech quality, regardless on the type of conversation scenario. Only in case of a delay as high as 1000 ms, quality is significantly decreased relative to a connection without delay. The test results are similar to findings reported in Möller (2000) and Karis (1991), who also found only a minor impact of delay on quality. The little quality impact of delay observed in the tests stands in contrast with how the E-model predicts it: Here, speech quality decreases rapidly for delays higher than 200 ms (Figure 4.23(a)).

Different factors may be responsible for the little effect of delay on speech quality in the conversation tests presented here:

(1) The packet loss impairment is much more obvious than that due to delay. Since the conditions were presented to the subjects in a randomized order, the degradation on the basis of which the subjects mainly made their judgments was packet loss.

(2) The perceptual effects caused by packet loss were understood as part of the system's quality, while the delay, although recognized by the subjects, was not attributed to the connection. Instead, it was either interpreted as a degradation not in the scope of our study, or as being related to a slow reaction of the interlocutor, for example, due to the conversation task.

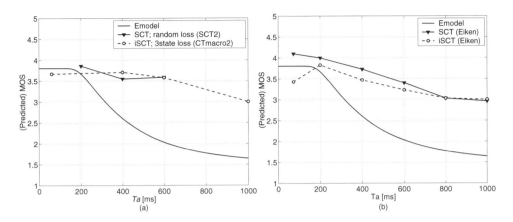

Figure 4.23: Results for 0% packet loss under transmission delay. (a) Author's own tests; 'SCT2' (SCTs, random loss), '▼'; 'CT-macro2' (iSCTs, 3-state loss), 'o'. (b) Conversation tests carried out by Eiken (2004). The corresponding predictions by the E-model are depicted for comparison (continuous line without markers).

(3) With regard to delay, the E-model does not predict the users' opinion. Instead, it predicts how delay affects the *efficiency* of a conversation held over a connection, which may not necessarily be reflected in the quality as *perceived* by the subjects.

(4) The cognitive load required by the conversation task – and in case of the iSCTs the additional instruction to the subjects to carry out the conversations as fast as possible – may have drawn the subjects' attention away from the characteristics of the line and from the quality rating task, and toward the conversation task.

Only few studies reported in the literature showed a strong impact of transmission delay on speech quality: For example, Kitawaki and Itoh (1991) compared the effect of delay on speech quality for different conversation tasks. The severe effect of delay they reported was found under conditions that are not representative for the average usage of a telephone link: Only for tasks like reading random numbers, a strong effect of one-way delay was found, using delays in the range [0, 2000 ms]. For less interactive conversation tasks, the effect of delay on quality was much less marked, but still much stronger than in the author's tests described here. Since the subjects recruited for the author's tests were naïve with respect to the test situation and task, the considerably higher impact of delay found by Kitawaki and Itoh (1991) may also be related to some training the subjects underwent.

Similar results, as obtained in the author's own conversation tests summarized in the preceding text, were obtained also by Eiken (2004). His conversation tests employed 12 conversations (17 test pairs), of which the first six were carried out using the SCT scenarios, and the second six using the iSCT scenarios. The test results can be summarized as follows (see Figure 4.23(b)):

- Delay showed only a very small impact on speech quality, although it was the only degradation investigated in the study.

- The results for the iSCT scenarios are slightly lower than those for the SCT scenarios. Since the iSCT scenarios were always presented after the SCT scenarios, order effects may play a role.

- From the analysis of the conversations, a number of observations were made:

  ○ Similar to the analysis described by Hammer *et al.* (2004a), a considerably higher number of speaker alternations was found for the iSCT scenarios than for the SCT scenarios (for one test pair on average 43 for the SCTs, and 81 for the iSCTs).

  ○ The mean durations of the conversations do not differ much between the conversation scenarios (150-190 s on average over all conversations).

  ○ The duration of the conversations increases with delay. The increase approximately corresponds to the overall delay introduced in each connection times the number of speaker alternations.

  ○ From the notes the subjects took during the conversations, it was revealed that considerably more mistakes were made in case of the iSCT scenarios than in case of the SCT scenarios. A tendency for less errors for longer delay could be observed for the iSCT scenarios.

This list supports the interpretation of an increased interactivity with the iSCT scenarios (speaker alternations), but also of a higher cognitive load (errors).

A comparison of the results obtained with the two scenario types for speech quality under packet loss in the author's own tests is depicted in Figure 4.24. To compare tests

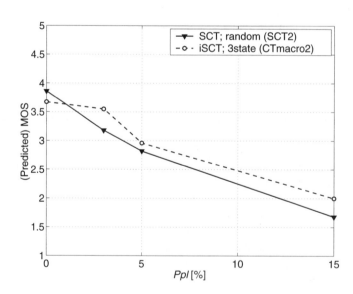

Figure 4.24: Comparison of the mean ratings of speech quality under packet loss obtained for the conversation tests on packet loss and delay (for the conditions with lowest transmission delay). 'SCT2' (SCTs, random loss): '▼'; 'CT-macro2' (iSCTs, 3-state loss): 'o'.

which were conducted under similar experimental conditions, the results from the two conversation tests on packet loss and delay were used here ('SCT2': SCTs, delay and random loss, see Section 4.4.2; 'CT-macro2': iSCTs, delay and 3-state loss, see Section 4.4.3). No remarkable difference between the two tests can be observed from this graph, although the test 'CT-macro2' involved not only more interactive conversation scenarios, but also 3-state packet loss. This highly time-varying loss type had been found to be more detrimental to speech quality than random loss (see Sections 4.2.2, and 4.4.3). Consequently, for 'CT-macro2', a stronger impact of packet loss was expected than for 'SCT2'.

The fact that speech quality under packet loss too, showed to be independent of the employed scenario type, and that it even lost its dependency on the packet loss *type*, substantiates the interpretation that interactivity may have been increased by the iSCT scenarios to the expense of a shift of the subjects' attention (see the preceding text, (4)).

4.4   Packet Loss and Combined Impairments

This section addresses the question of how integral quality is perceived when subjects are faced with combined degradations. In two series of conversation tests, this question is studied for combinations of different degradations (i.e. line noise, transmission delay and talker echo) with one of two types of packet loss:

(1) Random packet loss, as an example of macroscopically stationary packet loss ('SCT1'– 'SCT3'; Section 4.4.2).

(2) 3-state Markov packet loss, as an example of macroscopically nonstationary packet loss (CT-macro1–CT-macro3; Section 4.4.3).

The motivation for these studies was twofold:

- From a perceptual point of view, it is of interest to understand the combined impact of different types of degradations on speech quality perception. Since speech quality under packet loss has only recently become a topic of research, no experimental data are available on the quality perceived under packet loss and additional degradations. Typical analytical methods such as multidimensional scaling (MDS) used to explore the feature space related to quality are restricted to the listening situation. Hence, a better understanding of quality perception under combinations of packet loss with conversational degradations is sought.

- From the practical point of view of network planning, it is of interest to accurately estimate speech quality under combined degradations, as they may be encountered in complex network configurations. In the E-model, it is assumed that the different impairment factors quantifying the quality impact of certain degradations are 'additive'. The validity of this assumption for the impairment due to packet loss has to be investigated, before the proposed model enhancements can be applied to realistic network configurations.

In all the modeling approaches discussed in the present chapter, the equipment impairment factor $I_{e,eff}$ was used to express the quality impairment related to packet loss. In its current form, $I_{e,eff}$ is not dependent on any input parameter that it shares with another impairment factor of the E-model. Thus, the term 'additivity' as it is used in this context refers to the assumption that

(a) *Ie,eff* is *additive* to the other impairment factors, and, implicitly, that

(b) it is *separable* from other impairment factors, as it is calculated from a different set of input parameters.

From a practical point of view, the concept of *additivity* can help network designers to understand the nature of the influence particular network parameters have on the quality of their system. *Separability*, in turn, may help simplify the optimization of an envisaged network configuration, as the effect of different impairment classes can be treated separately. Separability also helps to easily enhance the prediction model to include new types of impairments, if these can be dealt with in separate impairment factors. In its current form, only the effective impairment factor *Ie,eff* is *separable* from other impairment factors. Instead of being motivated by the insight into the perceptual effects related to low bit-rate coding and packet loss, this is related to how these newer types of impairments were included into the E-model, namely, as an actual 'add-on' to the existing impairment factor framework.

4.4.1   Additivity and Multidimensional Feature Space

From a psychophysical point of view, the question is whether *additivity* and *separability* of impairments in some way relate to speech quality perception. As will be explained in the following, it is possible to establish a link between these two concepts and an *analytical* view on speech quality (Section 2.1):

One-way speech transmission quality was found to be of multidimensional nature (see Section 3.4). By MDS, a speech sample transmitted across a particular transmission link can be associated with a particular position in a multidimensional feature space. A given position in the feature space is related to the overall speech transmission quality of the respective link. Carroll distinguishes two categories of dimensions according to how quality or preference depends on them (Carroll, 1972): Dimensions that show a 'the more the better' type of behavior with regard to quality, and dimensions for which an 'ideal point' exists, at which quality is highest (for details on preference mapping see Section 2.1.2.4).

The dimensions of the category 'the more the better' reflect a 'vector model' of preference. In this model, the quality related to a particular point in the feature space corresponds to the projection of the point onto a subject's quality vector (Section 2.1.2.4, Figure 2.5). Due to the projection, the cosines of the angles between the quality vector and the dimensions reflect their importance for quality. Consequently, the dimensions are *linearly* related to the quality a certain subject perceives: Overall quality can be estimated from the linear combination of the different dimensions (Carroll, 1972; Mattila, 2001, pp. 208-48). The coefficients of the dimensions reflect their importance for quality, and depend on the orientation of the quality vector (see Section 2.1.2.4 for details).

If the dimensions show 'ideal points', the quality associated with a certain point in the feature space can be related to the squared Euclidian distance between the point and the ideal point (see Section 2.1.2.4, Figure 2.5: Unfolding model; Carroll, 1972). In the general case, quality can be expressed as a sum of the weighted squared dimensions, the weighted two-way interactions between dimensions and the weighted dimensions (elliptical point model or general unfolding model, see Section 3.4).

Obviously, both in the case of the vector model as well as in the case of the ideal point or unfolding model quality can be predicted from a sum of expressions involving the dimensions. Based on this observation, a link to *additivity* and *separability* can be established:

In the vector model, overall quality is expressed as a linear combination of the dimensions: *Additivity* of the components holds. Now, on the level of the dimensions, *separability* also holds, as no interaction between dimensions occurs. If certain dimensions depend only on disjoint quality elements, that is, instrumentally measurable network parameters that are not shared with other dimensions, *separability* also holds on the level of quality elements. This is what is implicitly assumed for Ie,eff in the E-model, although the impairment factors of the E-model have not initially been developed based on actual quality dimensions.

In case of the general unfolding model including dimensions with ideal points, *additivity* holds, too: Quality can be expressed as the sum of weighted dimensions, weighted squared terms of the dimensions, and weighted two-way interactions between the dimensions, which all act as additive components. In turn, already on the feature level *separability* into additive components does not hold, due to the interaction terms. Thus, it can be concluded that *separability* on the input parameter level, as it is assumed for Ie,eff in the E-model, cannot hold. This holds true even if all dimensions depend on distinct sets of the input parameters, since the additive components interact in case of this preference model.

Ie,eff in the E-model covers both the impairment due to packet loss and the impairment due to low bit-rate coding. For speech transmission quality of coded speech, Hall (2001) has shown that the feature space obtained in MDS experiments can be mapped onto quality ratings with reasonable accuracy by using a linear combination of the dimensions. In his study, the speech transmission quality of coded speech could obviously be predicted based on additive components. However, the only input parameter of the E-model Hall implicitly studied was the equipment impairment factor Ie expressing the impairment due to low bit-rate coding; *additivity* to other impairment factors cannot be validated with this study.

For time-varying distortions like packet or frame loss, particular quality dimensions are reported by Mattila (2001) as well as by Bernex and Barriac (2002). Mattila identified one dimension that is unique for degradations like frame loss ('smoothness of speech'; it corresponds to 'mute, clipping' found by Bernex and Barriac for packet loss; see Section 3.4). A component corresponding to an 'impairment factor' can be calculated from this dimension by determining an appropriate function for its mapping onto quality. Consequently, it can be said that in principle *additivity* to other impairments seems to be a valid assumption for the impairment due to packet or frame loss. In his work, Wältermann (2005) has provided a proof of this assumption, since he could show a high correlation between Ie,eff and the 'interruptedness' dimension he had found in his multidimensional quality analysis (ITU–T Delayed Contribution D.071, 2005).

For the dimension 'smoothness of speech', Mattila observed an interaction with the dimension 'bubbling', which showed to have a measurable impact on quality (see Sections 2.1.2.4, and 3.4; Mattila, 2002a). This dimension was interpreted to relate to fast fluctuations of the speech signal typical of signal processing like low-quality linear predictive coding (LPC). Such processing is currently handled in the same E-model impairment factor as packet loss (Ie,eff, see Equation (4.5)). In turn, a (however, less important) interaction for 'smoothness of speech' was found also with the dimension 'noisiness', and an interaction of the dimension 'bubbling' with the dimension 'dark-bright'. These dimensions are not explicitly handled in *Ie,eff*. Hence, some indication can be derived from the observed interactions that *separability* of *Ie,eff* may not be a valid assumption.

It has to be noted that the above considerations apply solely to a listening-only situation. No studies have been reported on the dimensions underlying conversational speech quality. No test methodology is available that allows the dimensionality of perceptual features encountered only in the conversational situation, such as those due to delay or echo, to reliably be evaluated. These features are not only related to the auditory perception of what is transmitted from the interlocutor, but to the whole process of conversation.

4.4.2   Microscopic Loss Behavior: Random Packet Loss

With the conversation tests on combined degradations described in this section we study whether additivity or even separability hold for combinations of random packet loss with other types of degradations. Conversation tests were chosen to include conversational degradations like talker echo and delay. From the test results, input parameters are identified that interact with packet loss in terms of their quality impact. Each test series was conducted with *random* packet loss of different loss percentages *Ppl* combined with a second parametric degradation:

SCT11 and SCT12: Different subscriber line-noise levels *Nfor* and packet loss percentages *Ppl*.

SCT2: Different mean absolute delay values *Ta* and packet loss percentages *Ppl*.

SCT3: Different levels of talker echo loudness rating (TELR) (with fixed echo delay *T*) and packet loss percentages *Ppl*.

The tests were run using the simulation system described in Appendix B. The short conversation test scenarios described in Section 2.1.1.4 were employed to stimulate the conversations. After each of the 14 test conversations, the test subjects were asked to make quality ratings on the MOS scale (Figure 2.1), and on the CR-10 scale (Borg, 1982; ITU–T Rec. P.833, 2001, Appendix I). More details on the test setup, the test procedure and the test subjects are provided in Appendix G.3. Note that only one or two reference connections were presented in the tests, to preserve as many conditions as possible for the actual test connections. In all tests, the G.729A was used as the main codec, and one clean connection with PCM A-law encoding (G.711) was presented as reference condition.

All observations described in the following text apply to both the ratings collected on the MOS scale and the ratings obtained on the CR-10 scale. For clarity of the presentation, only the MOS-data will be presented here[9].

4.4.2.1   Results and Discussion

For the conversation tests SCT11, SCT12 and SCT2, the results obtained on the MOS scale for the clean connection with logarithmic PCM (G.711) are in agreement with what

---

[9]Note also that some subjects seem to have encountered problems in the usage of the CR-10 scale, so that more outliers could be observed for this scale.

the E-model predicts for the corresponding input parameter settings. Consequently, the MOS-data presented in this section has not been transformed. In SCT3, however, the MOS scale was not entirely used by the subjects. Hence, a linear transformation of the SCT3 MOS-data was carried out for comparison, according to the procedure described in Appendix D.3, Equation (D.4). For the following comparisons of test results and E-model predictions, all results obtained on the MOS scale were transformed onto the R-scale using equations (E.34)–(E.36). This choice was made as additivity is assumed to hold on a ratio scale such as the R-scale, not on the MOS scale.

Due to the small number of reference conditions that can be presented in a conversation test, a normalization or transformation of our test results using a large number of anchoring conditions of known impairment was impossible. Consequently, a mismatch between test results and E-model predictions might stem from the choice of test conditions rather than from wrong predictions delivered by the model. Thus, an attempt was made to evaluate impairment *additivity* and *separability* in an additional way other than by directly comparing the test results to E-model predictions: For each SCT, the average judgments obtained for those test conditions, where only one individual impairment type was presented, were added to form overall ratings for combined impairments. In other words, in order to reflect the *separability* assumed by the E-model for Ie,eff, the impairment under packet loss $I(Ppl)$ and the additional impairment $I(X)$ (corresponding to the different SCTs, with X denoting noise, delay or talker echo) were added up to form a test-oriented predictor of quality under combined impairments. The corresponding expected, combined Transmission Rating Factor R_C was calculated according to Equation (4.29):

$$R_C(Ppl, Imp) = R_{729} - I(Ppl) - I(X) \qquad (4.29)$$
$$= R_{729}(X) + R_{729}(Ppl) - R_{729}$$

Here, R_{729} is the Rating Factor obtained for the connection with G.729A under loss-free conditions in the different tests (Table G.6; SCT1x: condition #3; SCT2/3: condition #2). This rating serves as the reference for the relative impairments $I(Ppl)$ and $I(X)$. $R_{729}(Ppl)$ stands for the Transmission Rating under the impairment due to packet loss, and $R_{729}(X)$ for the Transmission Rating under the additional type of impairment X varied in the different tests (i.e. due to noise, delay or talker echo). The only limitation in comparing the expected R_C to the R obtained directly from the MOS-ratings can be due to the transformation necessary for converting MOS to R. However, the results obtained on the CR-10 scale (which can linearly be transformed to the R-scale, see Möller, 2000, pp. 147–155) are very similar to the ones obtained on the MOS scale, so that the following discussion of transformed MOS results seems valid.

The standard deviations of MOS in the different tests are below 1 MOS for most conditions, and, on average, of 0.7–0.8 points MOS, which is within the range typical of conversation tests involving this scale (see Raake, 2004, for more details).

4.4.2.2 Packet Loss and Noise (SCT1x)

The test results for combinations of packet loss and noise are shown in Figure 4.25, in comparison to E-model predictions (conditions #4–#14, Table G.6, Appendix G.3).

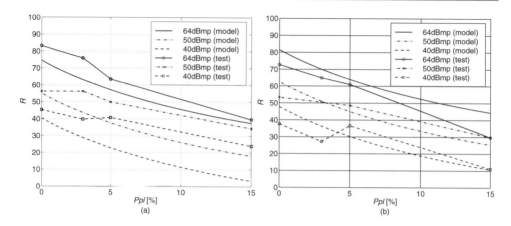

Figure 4.25: Test results from SCT11 and SCT12 (Raake, 2004). Lines with markers: Transmission rating factor R (ordinate) as transformed from MOS-ratings for G.729A under random packet loss of different loss percentages Ppl (abscissa) and different levels of Noise Floor $Nfor$ (curve parameter). Lines without markers: E-model predictions given for comparison. (a) $SLR = 13$ dB. (b) $SLR = 8$ dB.

The main observations can be summarized as follows:

- Figure 4.25 shows that the attenuation of the speech signal (send loudness rating, SLR) and the artifacts due to G.729A with and without packet loss on the line seem to interact in their influence on speech quality: A higher attenuation seems to make the degradations due to G.729A coding with and without packet loss less audible. For SCT11, the test series run with higher attenuation, the results are *more positive* than for the test series run with lower attenuation, SCT12. This stands in contrast to how the E-model predicts it. Here, higher attenuation than the optimum, which corresponds to the $8\,dB$ used in SCT12, leads to lower quality predictions.

- For conditions of additional noise, the decrease of quality over packet loss percentage is lower for SCT11 than for SCT12. The artifacts due to packet loss are partially masked by the additional noise. A similar effect is reported in Möller (2000) (pp. 167–172) for combinations of room noise and low bit-rate coding. Obviously, some of the artifacts due to packet loss are less perceivable the higher the noise level, and the lower the speech level. This masking of the packet loss impairment by noise is not covered by the E-model.

- *Separability* of the effective equipment impairment factor Ie,eff and the transmission rating due to the basic signal-to-noise ratio Ro, as it is assumed by the E-model, clearly does not hold.

In Figure 4.26, the test results are directly compared to the expected, combined Rating Factor R_C obtained using Equation (4.29).

Figure 4.26(a) shows the results for high attenuation (SCT11), and Figure 4.26(b) shows those obtained for low attenuation (SCT12). Obviously, the test curves for SCT11 do

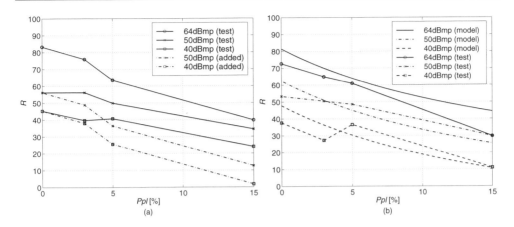

Figure 4.26: Test results from SCT11 and SCT12 (Raake, 2004). Solid lines: Transmission rating factor R as transformed from MOS-ratings for G.729A under random packet loss of different loss percentages Ppl (x-axis) and different levels of Noise Floor $Nfor$ (curve parameter). Slash-dotted lines: Combined transmission rating R_C obtained by adding the impairment due to individual impairment types. (a) $SLR = 13$ dB. (b) $SLR = 8$ dB.

not correspond to the expected R_C-curves (dash-dotted lines). *Separability* of the two impairment types noise and packet loss is not given. Instead, the higher attenuation of the line leads to a masking of the packet loss degradation by the line noise, in addition to the attenuation of the artifacts. The results for lower attenuation (SCT12) show a similar effect, which is, however, less expressed, as the packet loss artifacts are less attenuated and thus better audible.

These observations indicate that the decomposition of integral quality into a transmission rating due to the basic signal-to-noise ratio (Ro) and an independent effective equipment impairment factor Ie,eff is principally not permissible, since no separability is given. Instead, the noise level has to be included in Ie,eff, or the effect of packet loss in the transmission rating due to the basic signal-to-noise ratio Ro.

4.4.2.3 Packet Loss and Delay (SCT2)

A comparison of the test results and model predictions for combined impairments on the R-scale is provided in Figure 4.27 (see conditions #3–#14, Table G.6, Appendix G.3). In this surface plot, the delay Ta is plotted on the x-axis, and the packet loss percentage Ppl on the y-axis. The Transmission Rating Factor R is plotted on the z-axis. The upper, transparent surface displays the test results, the lower continuous surface the E-model predictions.

As can be seen from the graph, even delays of up to 600 ms had almost no influence on the quality perceived by the subjects, as already pointed out in Section 4.3. This stands in stark contrast to the model predictions, which imply a considerable quality impact of delay. For more insight into the relation between delay and speech quality see Sections 3.3.4 and 4.3.

In summary, it can be said that for the test series 'SCT2', involving nonhighly interactive conversations and naïve test subjects, speech quality under combinations of packet loss and

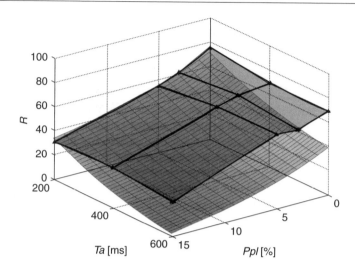

Figure 4.27: Test results from SCT2 (Raake, 2004): Transmission rating factor R as transformed from MOS-ratings for G.729A under random packet loss of different loss percentages Ppl and different overall delays Ta (upper surface, denoted by '▲'). E-model predictions are given for comparison (lower surface).

delay is not well predicted by the E-model. This is due to the fact that the effect of delay is not well predicted. Hence, *additivity* or *separability* cannot be validated, since for one of the two individual parameters no impairment could be measured in the quality test. In Section 4.4.3, the additivity of packet loss and delay impairment is discussed based on tests carried out with more interactive conversation scenarios.

4.4.2.4 Packet Loss and Talker Echo (SCT3)

In Figure 4.28, the results of the conversation test on combinations of random packet loss and talker echo are compared with predictions from the E-model. Figure 4.28 (a) shows the actual test results, and (b) the linearly transformed test data (after applying the normalization according to Equation (D.4)).

The following observations can be made for the nontransformed data shown in Figure 4.28(a):

- For this test series, the E-model predictions do not agree with the test results. In contrast to how it was expected, neither the results for talker echo as main degradation (G.729A coding, no packet loss) nor those for packet loss as main degradation (with a very low amount of $TELR = 50\,dB$, Table G.6) are well predicted by the E-model.

- The ratings are in general more positive than predicted by the model, especially for low MOS-values (i.e. low R-values).

As the effect of talker echo was found to be predicted reasonably well by the E-model in previous studies (Möller, 2000, pp. 164–166), the transformation onto the impairment scale

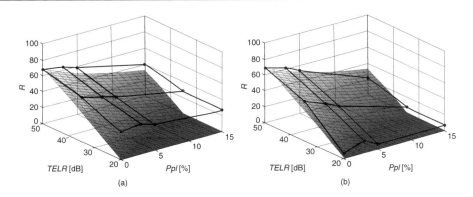

Figure 4.28: Test results for SCT3 (Raake, 2004): Transmission rating factor R as calculated from test results (MOS-ratings) for G.729A under random packet loss of different loss percentages Ppl, combined with different levels of $TELR$, mean one-way delay $T = 100$ ms. Test results: Upper surface, denoted by '▲'. E-model predictions are given for comparison (lower, continuous surface). (a) Test results transformed to R-scale. (b) Normalization of the MOS-results to cover the whole relevant MOS-range from 1 to 4.4.

for $TELR$ can be assumed to work well. The transformation of packet loss alone onto the impairment scale was found to work well in the present work, see Sections 4.1.1.

The more positive results found in SCT3 are thought to be related to two possible reasons:

- Firstly, a large number of very bad connections were presented in the test (i.e. five connections with a predicted MOS of 1). As subjects tend to judge different categories in one category rating test equally often, they did not ascribe a rating of MOS = 1 to all five out of 14 conditions (see e.g. Möller, 2000, pp. 116–117).

- Secondly, it could be observed that different subjects gave different weights to the two degradation types resulting in – on average – higher quality ratings for these bad connections.

A more detailed analysis of the test results revealed that the subjects can be clustered into three groups: One group which mainly based their quality judgements on the impairment due to packet loss (Ppl), another group of subjects judging mainly based on ($TELR$), and a third group of subjects affected by both Ppl and $TELR$ or the interaction $Ppl * TELR$. This result was obtained using an ANOVA for MOS on a per-subject basis. To increase the number of considered observations per-factor level, the settings for Ppl and $TELR$ were grouped into two levels each (Ppl: $0 - 3\% \equiv$ level 0, $5 - 15\% \equiv$ level 1; $TELR$: $50-65\,dB \equiv$ level0, $20-40\,dB \equiv$ level 1). The resulting variables were used as fixed factors in the ANOVA. The results of the ANOVA were further processed: For each subject, the values of the weighted sum of squares (MST) for each factor were normalized by the sum of the MST values obtained for the two factors Ppl and $TELR$ and the interaction (see Table 4.4). This normalization was carried out because the analysis was aimed at identifying the main sources of variation in the rating behavior of different subjects, however, ignoring the amplitude of the variation for better intersubject comparability. The three groups of

Table 4.4: Results of ANOVA on a per-subject basis for MOS with two fixed factors *Ppl-level* {0,1} and *TELR-level* {0,1}.

Group	Main Factor	Subject #	Ppl	TELR	Ppl*TELR
			\multicolumn{3}{c}{MOS Variance Due to}		
1	Ppl	9	0.96^a	0.02	0.02
		12	0.89^b	0.10	0.01
		19	0.88^a	0.02	0.10
2	TELR	7	0.00	1.00^a	0.00
		20	0.01	0.98^a	0.01
		14	0.03	0.95^a	0.03
		18	0.03	0.95^a	0.03
		11	0.01	0.93^a	0.06
		5	0.14^b	0.85^a	0.00
		4	0.16	0.84^b	0.00
		15	0.20	0.80^c	0.00
		21	0.03	0.71^c	0.26
		1	0.24	0.53^c	0.24
3	Both Ppl and TELR and/or Ppl*TELR	13	0.53^a	0.11	0.36^a
		8	0.58^b	0.21	0.21
		6	0.55^a	0.31^b	0.14^c
		2	0.59^c	0.30	0.11
		10	0.35^b	0.62^b	0.04
		16	0.40^c	0.58^b	0.02
		3	0.58^a	0.40^b	0.02
		17	0.50^c	0.50^c	0.00

a: $p < 0.01$.
b: $p < 0.05$.
c: $p < 0.1$. See text for details.

subjects were identified in a hierarchical cluster analysis on the normalized ANOVA results. It can be observed from Table 4.4 that three out of the 21 subjects belong to the group mainly affected by *Ppl*, 10 subjects to the group affected mostly by *TELR*, and eight subjects to the group effected by both or by *Ppl* ∗ *TELR*. Thus, it seems to depend on the focus of the individual subject, if packet loss, talker echo or both dictate the perceived quality.

To normalize for the scale-range effects, the test data were linearly transformed on the MOS scale to cover the whole scale-range. Therefore, the connection with G.711 (neither echo nor packet loss) and that with G.729A without packet loss but with the largest amount of echo, were chosen as extreme points of 'known' impairment (conditions #1 and #4, Table G.6, Appendix G.3; 'known' impairment because the impairment under talker echo was shown to be well predicted by the E-model (Möller, 2000, pp. 164-166)). The transformation was carried out using the procedure described in Appendix D.3, Equation (D.4), by using the clean connection with G.711, and the worst connection with talker echo but no packet loss as extreme points (corresponding to $MOS = 4.41$, and $MOS = 1.0$, respectively).

In Figure 4.28(a), values derived from both the actual test results (a) and the linearly transformed data are depicted (b). From Figure 4.28(a), no direct conclusions on impairment additivity or separability can be drawn. The linearly transformed ratings depicted in Figure 4.28 (b), however, support the assumption of *additivity* and *separability* of the echo and the packet-loss impairments. Here, the surface plots representing the test results and the model predictions are in relatively good agreement. As can be seen from Figure 4.28(b) though, the transformed results for quality under a high percentage of packet loss (15%) and without additional echo (rear corner of the mesh-plot representing the test results, 'o') lie below the E-model predictions, while the transformed results for the conditions with talker echo and no packet loss are consistently higher than the predictions (front right border of the mesh-plot). Obviously, the E-model gives more weight to the impairment due to talker echo than the test subjects in our tests did, while the opposite holds for packet loss.

Figure 4.29 depicts a comparison of the untransformed average test results to the theoretical, combined Transmission Rating factor R_C derived using Equation (4.29). At first sight, the figure implies that *additivity* and *separability* hold, since the two surface plots lie very close together (apart from the worst condition with $Ppl = 15\%$). However, it has to be kept in mind that different subjects obviously gave different weight to the two degradation types. Only if each subject weighs the impairments similarly as the others, something like *additivity* and *separability* can hold. Furthermore, an interaction between the two impairment types can be observed: The mean ratings are considerably higher than expected from tests where only one of the two impairments was presented (Section 4.1.1 for packet loss, and Möller, 2000 for echo). This observation is thought to be related to the different attention the individual subjects paid to the two types of degradation. Seemingly, the subjects

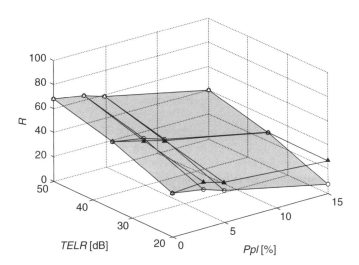

Figure 4.29: Test results for SCT3 (Raake, 2004): Transmission rating factor R as calculated from test results (MOS-ratings) for G.729A under random packet loss of different loss percentages *Ppl*, combined with different levels of *TELR*, at a mean one-way delay $T = 100$ ms. Test results: Upper surface, denoted by '▲'. The surface identified by 'o' presents the combined transmission rating R_C (see text for details).

could not concentrate well on both degradation types, on average leading to more positive quality ratings. This stands in contrast to the case of packet loss and noise (SCT1x), where a masking of one impairment by the other was observed. Such a masking typically takes place on lower levels of the perception process. In contrast, the interaction between echo and packet loss is thought to occur at higher cognitive processing levels.

Summary

In a series of four short conversation tests, the 'additivity' of the impairment due to random packet loss on G.729A coded speech with the impairments resulting from other types of degradations – as it is assumed by the E-model – was investigated. It was found that the perceptual effects due to packet loss, line noise and the attenuation of the line are not independent, and that the corresponding impairment factors cannot be regarded as separable. The additivity of the impairments due to packet loss and delay could not properly be evaluated, as delay did not show any major influence on the quality ratings in the tests. The impairment stemming from talker echo can be considered as additive to and separable from the impairment due to packet loss, if only a general prediction of speech quality is sought. Regardless of the limited additivity, the E-model predictions lie on the safe side for all impairment combinations investigated in the test series, that is, they are more pessimistic than the test results.

4.4.3   Macroscopic Loss Behavior: 3-State Markov Loss

In the previous section, random, i.e. *stationary* macroscopic packet loss and combined degradations have been discussed. The focus will now be on the question whether speech quality under combinations of nonstationary, macroscopic packet loss with other degradations is perceived in a similar manner.

Instead of random packet loss, 3-state Markov type loss was introduced into the connections of this test series. The exact loss conditions have already been discussed in Section 4.2.2. More details on the test setup, procedure and subjects can be found in Appendix G.4.

The particular 3-state packet loss is characterized by alternating periods of dependent (2-state) loss and no loss, with an average duration of these burst and gap periods of 18 s. For the dependent loss, an average loss of two consecutive packets has been chosen. With these settings, the 3-state loss used here perceptually differs in two ways from the random type loss employed in the conversation tests SCT1–SCT3:

- It leads to a clearly time-varying impairment.

- It has a higher impact on intelligibility, because

   o more consecutive loss events occur, and

   o an entire call is segmented into gap and burst periods of on average equal duration, so that the overall loss rate for the call translates into twice this rate during bursts.

So far, no studies have been reported in the literature, which analytically explore the perception of quality features that themselves vary in time. Hence, it is not clear at this point, whether strong packet loss variations are generally associated with a quality dimension, for example, corresponding to the 'smoothness' dimension reported by Mattila (2001). Then, the variation could simply be perceived as the variation of the feature magnitude, without necessarily forming an additional, new feature (in that case corresponding to something like 'slow variation'). Until now, it is not known, how variations of speech quality feature magnitudes are reflected in the differentiation between stimuli, for example, in MDS tests.

Similar to the way it was done for random packet loss and additional degradations, the results are presented in comparison to both E-model predictions and expected values calculated according to Equation (4.29). To enable a direct comparison and thus identification of differences between the results obtained for random and for 3-state packet loss, the results of the conversation tests SCT12, SCT2 and SCT3 will briefly be re-presented here.

4.4.3.1 Packet Loss and Noise (CT-Macro1)

The results for the tests on packet loss and noise are depicted in Figure 4.30. Figure 4.30(a) shows the conversation test results for 3-state loss and noise ('CT-macro1'), and corresponding predictions obtained with the E-model, which has been modified according to the 2-state loss impairment formula, Equation (4.14). Figure 4.30(b) shows the results for random packet loss and noise, and the respective predictions obtained with the current version of the E-model ('SCT12'; see Section 4.4.2.1, and Appendix G.3).

As stated earlier in this chapter (see Figure 4.18), a closer analysis of the test results reveals that the impairment due to 3-state packet loss is more dominant than that due to random loss:

- Speech quality for 3-state packet loss and noise in Figure 4.30(a) shows far less differences for different noise levels than the test results for random packet loss

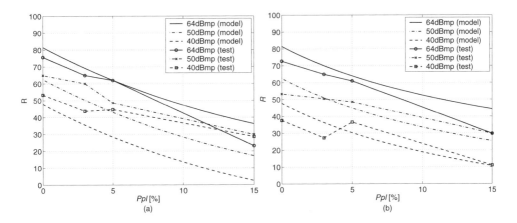

Figure 4.30: Test results transformed onto the R-scale for packet loss and noise floor. Abscissa: Packet loss percentage Ppl; ordinate: Mean quality ratings and E-model predictions (on the R-scale). Noise level $Nfor$ [dBmp]: curve parameter. (a) Burst loss; (b) random loss (see Section 4.4.2).

and noise in Figure 4.30(b). For 3-state loss, the impact of line noise on speech quality decreases with increasing packet loss percentages. At the highest packet loss percentage employed in the 3-state test ($Ppl = 15\%$), additional noise did not lead to any further speech quality degradation (Figure 4.30(a)). In contrast, at the highest *random* packet loss percentage ($Ppl = 15\%$), the highest noise level *did* lead to a further quality degradation, by as much as $\Delta R = 20$ (Figure 4.30(b)).

- An ANOVA was carried out for the conversation test results on random loss and noise ('SCT12') as well as for the test results on 3-state loss and noise ('CT-macro1'). In both cases, the loss percentage *Ppl* and the noise level *Nfor* were used as fixed factors. Note that for both tests, the ANOVA lead to almost identical results regardless of whether the MOS- or the CR-10-ratings were used for the analysis; thus, we only present the MOS-results here. The highly significant F-ratios ($p < 0.001$) obtained for both factors in the two ANOVAs indicate that the test subjects considered noise as the more important of two similarly relevant quality degradations in the test on random loss (see Appendix G.3, Table G.7: $F(Ppl)|_{random} = 42.765$ and $F(Nfor)|_{random} = 69.378$), and packet loss as the most important degradation in the test on 3-state loss (see Appendix G.4, Table G.11: $F(Ppl)|_{3-state} = 68.146$ and $F(Nfor)|_{3-state} = 19.898$).

- Due to the additivity/separability assumption, the E-model predicts the impacts of noise and packet loss to be independent of each other. This, however, is neither valid for random nor for 3-state packet loss.

- For both conversation tests, the ANOVAs revealed a small, though highly significant ($p \leq 0.001$) interaction between the speech quality impacts of packet loss percentage *Ppl* and line noise *Nfor* (see Appendix G.3, Table G.7: $F(Nfor * Ppl)|_{random} = 3.734$; Appendix G.4, Table G.11: $F(Nfor * Ppl)|_{3-state} = 4.321.$).

- Especially in case of additional noise, the E-model predictions are consistently more pessimistic than the test results, and hence lie on the safe side. However, the agreement between test results and model predictions is much lower than in case of random packet loss. Since the deviations are largest in case of high packet loss percentage and high noise level, they are little likely to stem from mis-predictions for individual impairments, but from the invalid assumption of separability and additivity.

The fact that separability does not hold for the degradations due to 3-state packet loss and line noise is underlined by Figure 4.31. Here, the MOS-results transformed onto the R-scale are compared with the *expected* values R_C, which were determined from the speech quality ratings for conditions of packet loss only, and conditions of line noise only (see Equation (4.29)).

The comparison of the test results ('▲') and the ratings to be expected when additivity and separability hold ('○') underlines the lack of separability for both random (b) and 3-state packet loss (a). In case of 3-state loss, the deviation between the test results and the surface representing separability increases with increasing loss percentage and increasing noise level. At lower loss percentages, additivity and separability are reasonable: The loss artifacts seem to stand out against the additional noise. For random packet loss and noise, already at lower noise levels and packet loss percentages separability is obviously not

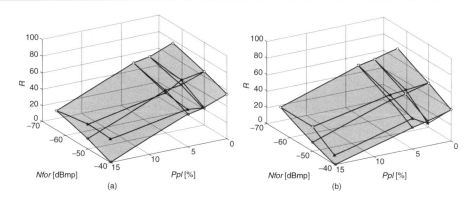

Figure 4.31: Transmission Rating Factor R obtained from test results (upper surface, '▲'; z-axis) and theoretical Combined Transmission Rating Factor R_C (lower surface, 'o'; z-axis). The two independent parameters are the packet loss percentage Ppl and the line noise level $Nfor$ [dBmp]. (a) Burst loss. (b) Random loss (see Section 4.4.2).

given. A possible explanation for this effect is the that the impairment due to random loss is partly masked by the line noise. In case of 3-state loss, the situation seems to be different. Seemingly, the attention of the subjects is drawn to this strongly time-varying type of packet loss, since

- the impairment during bursts is much more important than that due to random loss of the same overall loss percentage, and may impact intelligibility, and

- time-varying perceptual events are typically given higher relevance than stationary events, as auditory information processing is mainly based on dynamics (see Section 1.1.2).

4.4.3.2 Packet Loss and Delay (CT-macro2)

In the test series on packet loss and delay, the test subjects' quality ratings almost did not reflect the delay inserted in their connections, in contrast to how the E-model predicts it. The situation is depicted in Figure 4.32. Figure 4.32(a) shows the results obtained in the burst packet loss series, and Figure 4.32(b) shows the results obtained for random packet loss (see Section 4.4.2.3). As can be observed from the graphs, the little impact of delay on speech quality can be observed for both 3-state and random packet loss[10]. Possible reasons for the little effect of delay on speech quality were discussed in more detail in Section 4.3.

Kitawaki and Itoh (1990, 1991) reported a conversation-task-dependency of the delay impairment. The possibility that the E-model may overestimate the delay impairment for free conversations was also pointed out by (Möller, 2000, pp. 166-167). In the conversation test

---

[10]Note that due to the little effect of delay already observed in the random tests described in Section 4.4.2.3 ('SCT2'), the test conditions in the 3-state conversation test series described here were slightly changed. Instead of a minimum delay of 200 ms as in the random loss tests, the minimum delay in the burst test series was set to 60 ms, which corresponds to the minimum delay of the simulation system. Moreover, an additional connection with a delay of 1000 ms was included in the tests in order to study the effect of delay in case of 0% packet loss for delay values in the range [60 ms, 1000 ms].

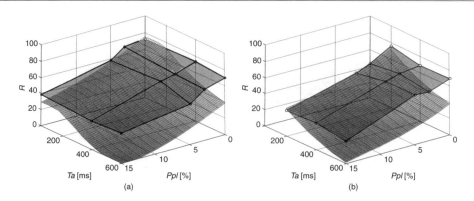

Figure 4.32: Transmission Rating Factor R obtained from the test results (upper surface, o) and E-model predictions (lower surface, no marker). The independent x- and y-variables of the graphs are the transmission delay Ta [ms], and the packet loss percentage Ppl [%]. (a) 3-state loss test. (b) Random loss test (see Section 4.4.2).

on 3-state packet loss and delay presented here, more interactive conversation scenarios than the SCT scenarios employed by Möller (2000) and in the tests described in Section 4.4.2 were used, described in more detail in Section 4.3. Nevertheless, the impact of delay on quality is found to be comparable to the random loss test series, where the standard SCT scenarios were used (see Section 4.4.2).

As can be seen in Figure 4.32, the effect of delay on speech quality is very similar to the random and the 3-state loss cases, in spite of the different conversation scenarios employed. The fact that in the burst test the effect of delay is as small as in the random case may be due to the stronger effect of burst packet loss: As the burst losses were strongly time-varying and thus unpredictable by the subjects, the subjects may have paid attention to this impairment type, instead of including delay in their ratings. This effect may have been counterbalanced by the more interactive iSCT scenarios, together leading to very similar results as for random loss. The subjects' attention was attracted to packet loss especially in case of high losses of 15% overall loss percentage, where intelligibility was considerably affected. This was confirmed also by informal listening to the recorded conversations. Hence, packet loss was a far more critical problem to a successful and fast finalization of the different conversation tasks than transmission delay.

For general application, it can be stated that the E-model predictions lie on the safe side, as they are more pessimistic than the mean ratings found in our tests. Since delay did not show any considerable impact on speech quality, the combination of delay and packet loss will not be discussed further at this point. In turn, a discussion of the role of the conversation scenario on the interactivity of a telephone conversation, and of the effect of interactivity on speech quality is provided in Section 4.3.

4.4.3.3 Packet Loss and Talker Echo (CT-macro3)

In Section 4.4.2, the combined effect of talker echo and random packet loss on speech quality was investigated based on a conversation test ('SCT3'). The test results were compared with predictions by the E-model. The test results were found to be much more optimistic

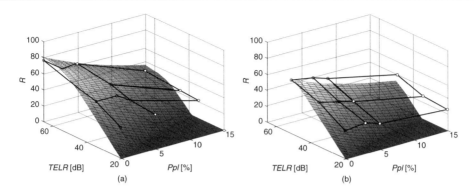

Figure 4.33: Conversation tests on packet loss and talker echo: Transmission Rating Factor *R* obtained from MOS test results (upper surface, 'o'), and E-model predictions (lower surface, no marker). Independent parameters: Talker echo loudness rating *TELR* [dB]; packet loss percentage *Ppl* [%]. (a) 3-state loss test ('CT-macro3'). (b) Random loss test ('SCT3'; see Section 4.4.2).

than the predictions delivered by the model. This was partly ascribed to a fairly large number of very bad connections used in the test, which were predicted to yield an MOS of 1. To reduce a similar tendency in the tests on 3-state packet loss and talker echo presented in this section, the settings for talker echo were changed as compared with the random test described in Section 4.4.2. With the modifications, only three instead of five conditions were expected to yield an MOS of 1 (for the exact test conditions see Appendix G.3, Table G.6, and Appendix G.4, Table G.9).

The test results are depicted in Figure 4.33. Figure 4.33(a) shows the results obtained for 3-state packet loss, and Figure 4.33(b) the results for random packet loss provided already in Section 4.4.2. In case of 3-state packet loss, the model predictions are too optimistic for high *TELR* (low amount of echo), and too pessimistic for low *TELR* (high amount of echo). The situation is different in case of random loss: Here, the model predictions are generally too pessimistic, that is, they lie on the safe side.

In case of 3-state packet loss the test results show a smaller dependency on talker echo than expected: Packet loss clearly was the dominating degradation type in this test run. This could be substantiated by an ANOVA using *TELR* and *Ppl* as fixed factors. In contrast to random packet loss, the 3-state packet loss percentage led to a much higher F-ratio than the talker echo loudness rating[11].

For random loss, it was found that the two degradation types packet loss and echo interact with each other: Different subjects showed a stronger dependency of the ratings on packet loss, others on talker echo, and a third group on combinations of the two (see Section 4.4.2.4). On average, however, something like impairment additivity was found to result from the different subjects' weighting of the two impairment types with regard to integral quality.

In case of 3-state loss, additivity of the impairments due to packet loss and talker echo is limited. This can be seen from Figure 4.34, where the Transmission Rating Factors *R*

---

[11] $F(Ppl)|_{random} = 16.3$ and $F(TELR)|_{random} = 36.9$: Appendix G.3, Table G.8; $F(Ppl)|_{3-state} = 68.7$ and $F(TELR)|_{3-state} = 22.1$: Appendix G.4, Table G.12; in all cases, $p < 0.001$.

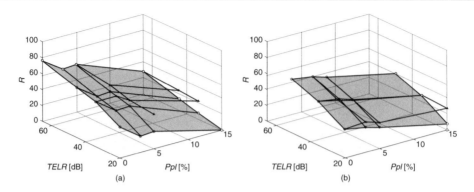

Figure 4.34: Conversation tests on packet loss and talker echo: Transmission Rating Factor R obtained from MOS test results (surface identified by '▲'), and theoretical Combined Transmission Rating Factor R_C (surface identified by '∘'). (a) 3-state loss test ('CT-macro3'). (b) Random loss test ('SCT3'; see Section 4.4.2).

obtained from the test and the theoretical, combined Transmission Rating factor R_C derived with Equation (4.29) are compared. Figure 4.34(a) shows the R-values obtained for burst packet loss and talker echo, and Figure 4.34(b) the results obtained in the test series on random packet loss and talker echo (see Section 4.4.2.4).

For random packet loss, the two surface plots lie very close together, see Figure 4.34(b). It can be argued, that adding individual impairments leads to the same impairment as that directly reflected by quality ratings for combined degradations. For 3-state packet loss, in turn, the two surfaces show a gap for all combinations of talker echo and packet loss. The impairment types cannot be considered as additive. For an increasing amount of talker echo (decreasing talker echo loudness rating *TELR*), the gap between the surfaces increases. A small interaction between echo and 3-state packet loss degradation could be confirmed in the above-mentioned ANOVA, too (see Appendix G.4, Table G.12: F-ratio $F = 2.880$, $p \leq 0.01$).

4.4.4   Model: Packet Loss and Combined Impairments

As discussed in the previous subsections, the separability the E-model assumes for packet loss and other degradation types does not hold neither in case of packet loss and noise, nor in case of packet loss and talker echo[12]. The lack of separability was confirmed both for random and 3-state packet loss, i.e. macroscopic packet loss. For the latter, the deviation from separability and additivity is even more expressed.

Two different causes seem to be responsible for the interaction of packet loss with the quality element 'noise' and the quality element 'talker echo':

- On the signal level, the degradations may directly influence each other.

  ∘ Additional line noise leads to a masking of some artifacts stemming from the packet loss degradation, so that the integral quality degradation is less expressed

---

[12]For packet loss and transmission delay, no conclusions could be drawn, since delay did not play any role for the MOS/CR-10 quality ratings.

than expected from the individual impairment types. Similarly, the speech intelligibility and quality may be improved when some of the information lost with a packet is restored based on the perceptual mechanisms of phonemic restoration (see Section 1.1.2.5).

- ○ Depending on the extent of double talk during a conversation, echo may mask some of the packet loss artifacts. In turn, packet loss may occur on the encoded echo signal, so that its effect may be diminished.

- In turn, on a purely perceptual level, the different quality features related to different degradation types may play a role.

  - ○ A strong impairment due to packet loss may distract the user from an additional noise or echo impairment. This is particularly important in case of strongly time-varying macroscopic packet loss (such as the 3-state loss discussed above), which seems to dominate the users' attention.

  - ○ The processes involved in the perception of the two feature types packet loss and talker echo can be considered as fundamentally different: Talker echo is a pure conversational degradation, while packet loss is a degradation perceived already in the listening situation. In case of medium feature magnitudes, i.e. medium degradation, different users may give considerably different weight to the combined degradation types in their integral quality judgements[13].

From the results of the conversation tests presented in Sections 4.4.2 and 4.4.3 it seems valid to say that under combined degradations users differ in their weighting of packet-loss-related quality features for integral quality. For all tests, analyses of variance carried out with packet loss percentage and the additional degradation (echo or line noise) as fixed, and the test subject as random factors revealed a statistically significant impact due to the subject. This effect was observed to be more prominent in case of packet loss and talker echo than in case of packet loss and line noise.

These observations can result in two different strategies for quality prediction under combinations of packet loss and other degradations:

(1) The separability and additivity assumed by the E-model is exploited for worst-case estimates, since it was shown that the E-model mostly *underestimates* quality for combined impairments (Sections 4.4.2 and 4.4.3). This approach takes individual, critical subjects into consideration, for whom separability and additivity are valid assumptions.

(2) For the prediction of the quality perceived on average by users, the interaction between packet loss impairment and other degradations is taken into consideration. This strategy helps to reduce the number of planning decisions for more costly network equipment than necessary, or too careful and thus inefficient trade-offs between different network configurations.

To conclude the present chapter, we now present an approach for considering the interaction between the impairments due to packet loss and due to line noise. For packet loss

---

[13]For example, as it could be observed during our conversation tests, some users were not affected by talker echo in their speaking style, while others spoke considerably slower or started to stutter.

and echo, the considerable dependence of the quality ratings on the subject, and the generally much more optimistic ratings obtained in the tests ('SCT3' make and 'CT-macro3') let it appear to be premature at this point to derive a corresponding E-model modification, especially in the light of the sparse test data available for this case. Similarly, more auditory test data are necessary to determine a stable model for the interaction between packet loss, line attenuation and line noise. However, the perceptual effects appear to be more obvious than in case of packet loss and talker echo.

The approach we chose is based on three assumptions resulting from the conversation test results (Sections 4.4.2 and 4.4.3):

(a) For higher line attenuation than the E-model default of $OLR = 10$ dB, the artifacts duc to packet loss are less perceivable. In turn, for lower line attenuation, the artifacts become even more disturbing.

(b) Additional noise may mask some of the artifacts. This effect is even more obvious, when the speech signal is strongly attenuated by the channel.

(c) In case of strongly time-varying packet loss, such as the 3-state Markov model type loss used in the conversation tests 'CT-macro1–3', the users' impression of quality is dominated by the packet loss impairment, and additional degradations on the line (here noise) do not contribute to quality.

These assumptions have been transformed into a corresponding extension of the E-model.

To take assumption (a)-(c) into consideration, we propose to use an interaction term, for example, by modifying the basic formula of the E-model, Equation (2.5), to

$$R = Ro - Is - Id - Ie,eff + A + E_{Ppl \circledast Nfor}, \qquad (4.30)$$

where

$$E_{Ppl \circledast Nfor} = Ie,eff \cdot \left(1 - \left(\frac{Ro}{95}\right)^{c_1} \cdot e^{-(OLR/dB - 10)/c_2}\right) + \frac{(95 - Ro) \cdot (Ie,eff)}{(95 - Ro + Ie,eff)} \cdot S_1 \cdot S_2$$

$$(4.31)$$

The 'enhancement' factor $E_{Ppl \circledast Nfor}$ represents the quality improvement due to the interactions between packet loss, line noise and line attenuation. Here, the term $(Ro/95)^{c_1}$ accounts for the masking of packet loss artifacts due to noise, according to assumption (a). The exponent c_1 is a constant to be determined, for example, by curve fitting. The term

$$e^{-(OLR/dB - 10)/c_2}$$

reflects the impact of the line attenuation on the audibility of the artifacts caused by packet loss. It corresponds to the assumption (b), with c_2 as a constant that quantifies the line level dependency. We have to view this latter expression as a first approach, since only two different settings for the attenuation in send direction SLR have been used in our conversation tests. So far, no other auditory tests have been reported in the literature on the effect of loudness rating on speech quality under coding or packet loss: In listening tests, the speech level used for playback is typically adjusted to a fixed value.

In the third, additive term, $(95 - Ro)$ is a measure of the 'impairment' related to the transmission rating for the basic signal-to-noise ratio Ro. The term $(Ie,eff - Ie)$ reflects the fact that we want to cover the dominance of strongly time-varying impairments due to packet loss. The two terms

$$S_1 = \frac{1}{1 + \left(\dfrac{\mu_{13} \cdot T_p}{c_3}\right)^{-c_4}} \qquad (4.32)$$

and

$$S_2 = \frac{1}{1 + \left(\dfrac{Pb}{c_5 \cdot (Ppl + 0.01)}\right)^{-c_6}} \qquad (4.33)$$

are logistic functions serving as 'switches': They ensure that this part of the interaction factor is only activated when the packet loss is strongly varying in time. In more detail, the two switches reflect two different aspects:

- The first term considers how far apart in time different loss events are spread. The distance of burst periods is calculated for the parameters of a 4-state Markov model (see Section 3.3.5.1, Figure 3.7, and Equation (3.18)). As discussed in Section 4.2.2, we can calculate the 4-state model parameters μ_{13}, μ_{31} and Pb (i.e. the loss percentage during 'bad' periods) in a similar manner for the 3-state loss model used in the conversation tests 'CT-macro1'–'CT-macro3' (see Section 4.4.3). The factor μ_{13} is the average number of steps, i.e. packets necessary for the loss model to transit from the 'good' to the 'bad' state. The time T_p is the packet size employed for the connection. Two additional model parameters have to be used (c_3 and c_4), which are a measure of the time-variability threshold (c_3), and the slope of the logistic function (c_4).

- The second term is the relation between the loss percentage during 'bad' periods Pb and the overall loss percentage Ppl. It is one of several possible measures for the variation of packet loss in terms of loss percentages. The term 0.01 is used to avoid that the denominator turns zero, so that the formula can be applied also in case that $Ppl = 0$. Two further constants are employed, c_5 and c_6. The constant c_5 quantifies the threshold for the relation between the loss percentages (Pb/Ppl) at which the logistic function becomes relevant, and the constant c_6 is a measure of the function's slope.

With these modifications, we can considerably improve the agreement between the test results and the model predictions, as is illustrated in Figure 4.35. The settings for the constants c_1 to c_6 employed for this comparison are shown in Table 4.5.

With this choice of parameters, the correlation between the test results and the predictions can be improved from $\rho = 0.896$ to $\rho = 0.946$, and the root mean squared error of the prediction from $RMSE = 10.34$ to $RMSE = 6.68$ (expressed on the R-scale). Moreover, the modification does not introduce new cases for which speech quality is considerably overestimated by the model. Nevertheless, the tendency of quality underestimation could be reduced by this modification. Although the E-model lies on the safe side with its predictions for the combined distortions investigated in this section, such a reduction of cases of underestimation is necessary to make more cost-efficient planning decisions.

The suggested interaction approach or further elaborations of it have to be complemented and validated by additional auditory test data, before it can be transformed into a new E-model. As we have stated earlier, the two conversation test series presented in this

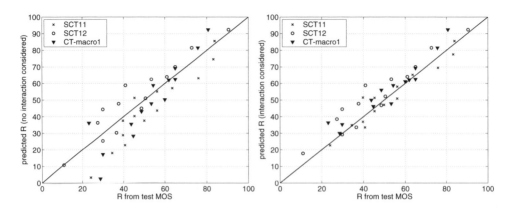

Figure 4.35: Correlation between model predictions (ordinate) and conversation test results (abscissa). (a) Test results and predictions by the 2-state enhancement of the E-model (according to ITU–T Rec. G.107, using Equation (4.14)). (b) Test results and predictions taking interactions between packet loss and line noise into consideration (according to Equations (4.30)–(4.33)).

Table 4.5: Settings for the constants c_1 to c_6 employed in Equations (4.31) through (4.33) to obtain the quality predictions depicted in Figure 4.35(b).

Constant	c_1	c_2	c_3	c_4	c_5	c_6
Value	0.3	16	12	10	1.8	10

section are the only ones reported in the literature that are documented well enough to serve as basis for parametric modeling. Furthermore, it has to be noted that the developed expressions are limited to the case that packet loss does not affect the transmitted (circuit) noise, that is the noise that is introduced after a possible IP-trunk. At this point, there is no auditory test data available to extend the approach to a more general case, where packet loss may also effect noise transmission.

4.5 Time-Varying Distortion: Summary

Based on the author's auditory tests and data from the literature, the speech quality related to different types of loss distributions was investigated. A variety of achievements have been made:

- A parametric model has been developed to predict speech quality under random packet loss, which has been adopted by the ITU–T in an update of the E-model (ITU–T Rec. G.107, 2005). Based on further auditory tests, the model has been refined to also include packet loss distributions that show short-term dependencies between loss events (*microscopic* loss behavior: Equation (4.14)).

- It was shown that *general* speech quality predictions can validly be made based on the average loss percentage, when the model presented in this book is used, and a mean short-term loss correlation is taken into consideration. In most cases where the predictions do not agree with the speech quality ratings collected in auditory tests, the model underestimates quality.

- Based on a series of conversation tests, it has been shown that predictions with this approach are too optimistic only in those cases, where packet loss shows strong variations, and leads to periods where the intelligibility between the conversation partners is considerably affected. Different models proposed in the literature have been investigated that implicitly or explicitly take a weighting of periods of high and low degradation into consideration.

- In order to investigate the impact of the interactivity of a conversation on speech quality under packet loss, particular conversation scenarios have been developed. In conversation tests, however, neither an impact of the scenarios on speech quality under packet loss, nor on speech quality under delay could be observed. A number of possible reasons for this finding have been discussed.

- From conversation tests on combinations of packet loss with other, better known degradations such as noise, delay and talker echo it could be concluded that the related impairments are not additive. Instead, they interact with each other in their impact on speech quality. It was discussed that the combined impact on quality is less severe than expected from the individual impairments. Consequently, the parametric model proposed in this book underestimates speech quality under packet loss and additional degradations.

- A first approach has been proposed to reflect the interaction between packet loss and circuit noise in the E-model.

Most results described in this chapter were obtained for one speech codec, i.e. the G.729A. While general observations on speech quality under packet loss will be valid for other codecs too, the exact modeling approaches suggested in this chapter have to be validated and possibly adapted to make them applicable to other coding algorithms.

5

Wideband Speech, Linear and Non Linear Distortion: Quality Features and Modeling

The second overall quality element addressed in this book is linear distortion. Linear distortion is of particular relevance for Voice over IP (VoIP), as it is directly linked to two major characteristics: (a) The flexibility of employing different user interfaces, and (b) the potential of wideband (WB) transmission (Chapter 3).

In VoIP, the transmission bandwidth may be extended to WB, for example employing one of the low bit-rate WB speech codecs that recently became available (Section 3.3.2.2). From a fundamental as well as an application-oriented point of view, it is desirable to dispose of a quantitative measure of the quality improvement that can be achieved with WB transmission. For network planning, it becomes necessary to develop a WB network planning tool similar to the narrowband (NB) E-model, which allows predictions of speech quality to be made, before the network has been setup. Up to now little knowledge is available on how exactly such a model would look.

The traditional NB telephone bandpass (BP) filter and the corresponding user interfaces employed at the send and receive sides introduce a linear distortion of the initially WB input speech signal (Sections 3.3.13 and 3.3.14). For the user at receive side, NB telephone speech typically appeals to a certain mode of 'telephone speech perception' (Section 3.7). Thus, relative to the evoked reference inherent to the user, a clean NB telephone channel is typically not perceived as being distorted. This situation may change, when WB speech has been introduced to a larger market. An important factor for the appreciation of WB transmission is the employed user interface: Only when high and low frequency components are appropriately coupled to the listener's ear, the advantage of WB speech may be perceived. In contrast to NB telephony, where the handset-telephone is still the most widely employed[1] user interface, no corresponding 'wideband user interface' has been defined.

---

[1]Both wireline and cordless phones are referred to here.

Speech Quality of VoIP: Assessment and Prediction   Alexander Raake
© 2006 John Wiley & Sons, Ltd

The user interface has two implications for speech quality:

1. It introduces a linear distortion of the transmitted speech signal.

2. Due to both acoustic and nonacoustic properties, it is associated with a particular expectation by the user.

The linear distortion introduced by a user interface is linked to specific quality features. These features need to be specifically taken into consideration, when the quality related to a particular network configuration is to be predicted. Of special importance in this context is the combined effect of the introduced linear distortion with additional impairments known already from NB networks, such as talker echo. Current network planning models, like the E-model or the SUBMOD-model (see Section 2.2.2), consider linear distortion only in an insufficient way (Möller, 2000, pp. 176–183). So far, the available auditory test data does not enable the extension of the current models or the development of new ones.

This chapter contributes to quantifying the impact of *linear* distortion on the speech quality of WB and NB speech. A set of research questions is discussed:

(a) How can we quantify the WB improvement over NB speech? (Section 5.1)

(b) How can the impairment due to linear distortion like BP-filtering be quantified? (Section 5.2)

(c) Which factors related to the quality assessment of WB speech may impact its desired nature (i.e. the expectation of the user)? (Section 5.4)

5.1 Wideband Speech: Improvement Over Narrowband

The advantage of WB over NB telephony can be expressed on different scales. A few tests have been reported in the literature that express the WB improvement on the mean opinion score (MOS) scale (ITU–T Contribution COM 12–11, 1993; Krebber, 1995; Pascal, 1988). In these tests, the advantage was found to be around 1.3–1.5 points MOS (see Section 3.3.13).

As mentioned earlier in this section, clean NB connections are typically not perceived as being degraded, when no WB conditions are presented in the same test. If WB conditions are presented in addition to NB conditions, the latter are rated lower than in case of only NB conditions. Hence, the measured advantage in MOS is a rather speculative one, since it is based on a compression of the MOS scale. A way forward to cope with this problem is to compare tests with NB conditions to tests with mixed NB and WB conditions. The NB conditions presented in both tests may then serve to interpolate the actual improvement by expressing it relative to the ratings on the NB-scale.

We take a similar approach in this book, which, in large parts, is concerned with the expression of speech quality on the E-model R-scale (see Equation (2.5), p. 45; ITU–T Rec. G.107, 2005). Since the R-scale can be assumed to have ratio-properties, it may be more appropriate to express the WB improvement on this scale rather than on, for example, the MOS scale. Moreover, an extension of the R-scale is needed, if the E-model is to be extended to WB speech (or alternatively a WB E-model is to be setup). The R-scale currently ranges from 0–100, where the upper end reflects the best possible NB quality. An extension of this range allows a direct quantification of the improvement due to WB relative to NB channels.

The procedure of comparing NB- and mixed NB/WB tests can be questioned by the fact that most users may not be able to rely on a stable reference of WB telephone speech, in contrast to the ones evoked for NB telephone speech, or for WB radio speech (see Section 5.4). To account for this problem, we have conducted a listening-only test (LOT) on speech transmission quality with a particular WB handset: The 'Hi-Fi phone' described in Section B.5.3 had been found to be suitable for the assessment of WB speech transmission, since it (a) enables appropriate coupling of all frequency components to the listener's ear, (b) can be assumed to more readily be associated with a telephone context, and thus (c) may be more flexibly dissociated from the well established references evoked when using headphones or 'normal' handset telephones (see Section 5.4.1).

The test ('LOT Hi-Fi phone') was initially carried out to assess the impact of BP-filtering on the quality of WB speech. The exact test conditions and more details on the test results are presented in Section 5.2. In two conditions, this test, carried out with both WB and NB connections (Table 5.2) is comparable to a purely NB listening test described in Möller and Jekosch (1998). In the respective test (LOT 'VR8'), different conditions of high-pass, low-pass and BP-filtering were presented, as well as conditions with different levels of signal-correlated noise (modulated noise reference unit, MNRU; ITU–T Rec. P.810, 1996). In both of the tests, a NB connection with IRSmod-filtering was used (Intermediate Reference System, ITU–T Rec. P.48, 1989; ITU–T Rec. P.830, 1996). Moreover, in the listening test 'LOT VR8', a connection with a band-limitation of 500–2500 Hz has been employed, and in the author's listening test 'LOT Hi-Fi phone' a condition with a 400–2400 Hz BP filter. To enable a better comparison of the latter two conditions, the test results of the author's tests have been interpolated to obtain an estimated rating for 500–2500 Hz. More details on the interpolation can be found in Section 5.2. The obtained test results for the different conditions are shown in Table 5.1.

From the comparison of the NB and NB/WB tests, the R-scale extension necessary to account for an optimum WB connection is determined based on a linear or nonlinear

Table 5.1: Complementary test conditions of the speech quality listening test on bandpass-filtered speech (Section 5.2) and of the listening test 'VR8' reported in Möller and Jekosch (1998). The corresponding test results are used for the estimation of the R-scale extension in case of WB. The value printed in italic has been obtained by an interpolation of the MOS-results as it is described in more detail in Section 5.2.

Bandpass [Hz]	Filter Shape	Test Results [MOS]	
		'LOT VR8' (NB)	'LOT Hi-Fi phone' (NB/WB)
200–7000	Flat	–	4.16
300–3400	IRSmod	4.14	3.21
400–2400	Flat	–	2.28
500–2500	Flat	3.02	(2.26)

extrapolation toward the optimum WB connection[2]. The general extrapolation procedure can be outlined as follows:

(a) The test results are either employed directly, or at first normalized, according to one of the procedures described in Appendix D.

(b) Subsequently, the test results are transformed to the E-model R-scale using the NB transformation MOS $\rightarrow$ R given by Equation (E.34). Since this transformation was determined for the NB case, it may be bound to a loss of information.

(c) The R-ratings for the NB case are plotted versus the R-ratings obtained for the NB/WB case. The relation between the two sets of R-values is later used for extrapolation.

(d) The maximum transmission rating factor for WB, $R_{O_{NB/WB}}$, is derived from the rating for the WB case by linear or nonlinear extrapolation (see Figures 5.1 and 5.3).

As an example for the application of this procedure, the WB extension of the R-scale has been determined for the test ratings listed in Table 5.1. The corresponding extrapolation procedure is depicted in Figures 5.1 and 5.2. While Figure 5.1 shows the test R-values obtained by direct transformation of the MOS ratings, Figure 5.2 depicts the R-values resulting after prior normalization of the MOS ratings. Here, the equivalent-Q method has been applied for normalization of the NB results ('LOT VR8'). Since no actual reference conditions were employed for the NB/WB test ('LOT Hi-Fi phone'), a linear normalization was carried out in this case. Therefore, the method described in Appendix D, Section D.3

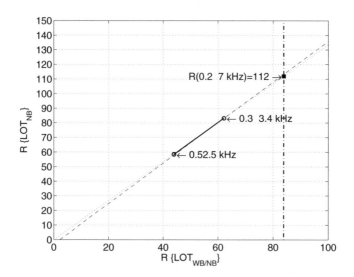

Figure 5.1: Linear extrapolation of narrowband results to estimate the R-scale extension for wideband (see text for details). No normalization of test results. NB-test: 'LOT VR8'; NB/WB test: 'LOT Hi-Fi phone'.

---

[2]In case of the listening test described in Section 5.2, the highest mean quality ratings (MOS and speech sound) were obtained for the BP 200–7000 Hz, instead of the 50–7000 Hz one.

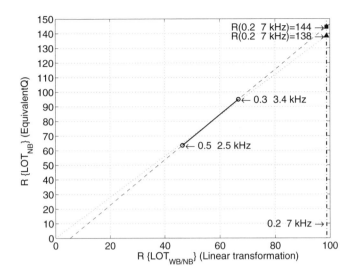

Figure 5.2: Linear extrapolation of narrowband results to estimate the *R*-scale extension for wideband (see text for details). Prior normalization of NB-results according to equivalent-Q-method (see Appendix D, Section D.1), and linear transformation of NB/WB data (see Section D.3). NB-test: 'LOT VR8'; NB/WB-test: 'LOT Hi-Fi phone'.

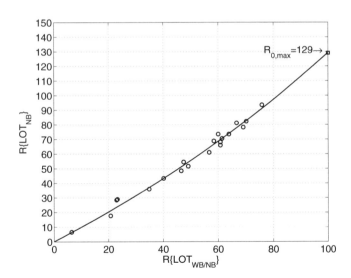

Figure 5.3: Exponential extrapolation of NB results to estimate the *R*-scale extension for WB (see text for details). Underlying MOS-results ('∘') for NB- and mixed NB/WB tests from Barriac *et al.* (2004).

was employed in such a way, that an MOS of 4.49 was achieved for the best connection in the test (instead of the MOS of 4.16 listed in Table 5.1). As can be seen from the graphs, the initial extrapolation lines show a slight offset, which has been corrected for by connecting the point $(0, 0)$ with the midpoint of the line between the two NB data-points.

With the nontransformed data, a more conservative estimate is derived ($R_{WB} \cong 112$), as opposed to the relatively high estimate obtained after prior transformation ($R_{WB} \cong 138$)[3]. However, the linear extrapolation procedure obviously has two weak points:

1. The *narrowband* transformation $MOS \rightarrow R$ is employed. Apart from possibly being inapplicable, a problem related to this transformation rule occurs for large MOS ratings, since the transformation is not defined for values larger than $MOS = 4.5$.

2. A linear extrapolation is used, which may not reflect the actual relation between the ratings.

The problem of linear extrapolation (2) can be avoided when more data-points for NB connections are available that are common for both the considered NB- and the mixed WB/NB-tests.

Recently, a corresponding series of two listening tests has been reported by Barriac *et al.* (2004). In one test, only 22 'NB'-conditions were used, including the following:

- Four standard codecs in single and tandem operation.

- Three WB codecs at different bit-rates, the output being down-sampled to 8 kHz.

- A clear channel reference down-sampled to 8 kHz.

The second test was conducted with mixed WB and NB conditions, comprising:

- The same 22 conditions as in the NB test.

- An additional set of three WB codecs operating at different bit-rates.

- A clean channel reference sampled at 16 kHz.

From the list of conditions it becomes clear that the particular NB case with almost 4000 Hz bandwidth was already improved as compared with the traditional 300–3400 Hz BP.

In each of the two tests, 24 listeners judged the quality of the processed speech samples on the MOS scale. When the R-values obtained from the MOS ratings[4] for the mixed NB/WB test are plotted over those obtained from the MOS ratings for the NB test, a curvilinear relation is revealed. A good approximation of this relation can be achieved by curve fitting based on Equation (5.1),

$$R\{LOT_{NB}\} = a \cdot \left(\exp \left(\frac{R\{LOT_{NB/WB}\}}{b} \right) - 1 \right), \tag{5.1}$$

with $a = 169.38$ and $b = 176.32$. If this relation is employed instead of the linear extrapolation of step (d), the R-extension is estimated as $R_{0NB/WB} \cong 129$ (see Figure 5.3).

---

[3]Similar estimates are obtained when the same procedure is applied to the test results reported by Krebber (1995).

[4]By transformation using Equation (E.34).

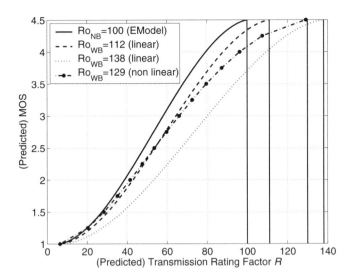

Figure 5.4: Possible transformation rules between NB/WB R-estimates and corresponding MOS ratings.

Based on the different estimates for the R-scale extension, a direct transformation rule between MOS and $Ro_{NB/WB}$ can be defined, corresponding to the one used in the current E-model version for NB (Equation (E.33)). Depending on whether a linear or nonlinear extrapolation has to be used, different transformation rules may result (Figure 5.4). To determine one valid estimation of the relation between MOS ratings obtained in mixed WB/NB tests and corresponding WB R-values, a larger amount of auditory test data should be available.

5.1.1 Summary

From the extrapolation of the available results of NB and mixed WB/NB tests, different estimates have been derived for the R-scale extension necessary to capture the quality of WB speech. A more conservative (low) R-value estimate lies around a value of 110, while the most progressive (positive) estimate is of $R \cong 138$. A good compromise seems to be the intermediate value of $R \cong 129$. Further auditory tests on mixed NB/WB and purely NB connections have to be carried out to substantiate the proposed procedure. Other, more analytical approaches can be envisaged that rely on a paired comparison of different WB- and NB conditions with a WB reference, where subjects rate the degradation with regard to this reference employing a ratio scale, such as the CR-10 scale (Borg, 1982; ITU–T Rec. P.833, 2001, Appendix I).

5.2 Bandpass-Filtered Speech

This section deals with the question, how the quality related to WB speech transmission, and the quality impact of linear distortions of WB can be quantified, and be related to a parametric description of the channel characteristics. As a starting point, the linear distortion due to BP-filtering is analyzed, which may be extended to more complex types of linear distortions in future work.

5.2.1 Listening Test

A listening test on quality under bandwidth limitations was carried out for this purpose. To provide appropriate acoustic conditions, and realize a situation recognized by the subjects as telephone-typical, the WB handset telephone 'Hi-Fi phone' described in Section B.5.3 was employed in this test (see Section 5.4.1 for considerations on the suitability of this user interface for WB speech assessment).

5.2.1.1 Test Setup and Procedure

The employed filter-conditions are listed in Table 5.2.

As source material, a shortened version of the phonemically balanced Eurom text material was employed (in German; Gibbon, 1992). In its shortened version, it consists of meaningful passages of approximately 8–10 s duration composed of a small number of interconnected sentences. Each of the 40 available text passages was read aloud by six non-professional speakers in a natural way (three female, three male). The recordings were made in an anechoic environment using a condenser microphone Type AKG C414 B-TL with spherical directivity. A distance of 30 cm between the talker's mouth and the microphone was chosen as a compromise between the near-field (telephone) case and larger distances as in hands-free telephone applications. The recorded samples were further processed with the line-simulation system described in Appendix B.

The presentation of the samples as well as the collection of ratings was controlled using a computer program. The subjects were asked to make ratings of integral quality on the MOS scale (Figure 2.1), and of speech sound quality on the 11-point scale depicted in Figure 2.3. Twenty-five listeners participated in the test, who were, to their own account

Table 5.2: Bandpass filters used in the test. All filters have a flat frequency response in the passband. Where indicated, an *IRSmod* filtered version was used in addition (Intermediate Reference System, frequency response according to Annex D of ITU–T Rec. P.830, 1996).

f_{up} [Hz]	f_{low} [Hz]					
	50	100	200	300	400	600
2000	X			X		X
2400					X/IRS	
3400	X		X	X/IRS	X	X
4000			X			
5000	X	X		X		
7000	X		X	X	X	X

normally hearing (14 female, 11 male). Their age ranged from 20 to 59 years, with an average of 29 years.

5.2.1.2 Test Results

The mean ratings obtained from the subjects on the MOS scale are shown in Figure 5.5. In this graph, the lower and upper cut-off frequencies of the employed BP-filters serve as independent variables (f_{low} and f_{up}, respectively; both in Hz).

In contrast with what could be expected intuitively, the condition with the BP 200–7000 Hz was considered as the best one in this test (both for the MOS and on the 11-point sound quality scale), and not the WB condition (50–7000 Hz). As pointed out in Section 3.3.13, there is generally no agreement between test results in this respect: In some tests, a preference for the full WB was observed (e.g. Krebber, 1995; Voran, 1997), while others have reported a preference for some bandwidth restrictions of lower frequency components (e.g. Moore and Tan, 2003). From the information provided in the literature, it cannot be concluded whether this observation is caused by the specific recording procedure, or whether it may be due to the particular listening device (see Section 3.3.13). If viewed from an ecological perspective, our finding may be explained with the fact that the perception of lower frequency components is less plausible for one-ear listening than for two-ear listening (head-shadowing for higher frequency components versus diffraction around the head for lower ones).

To obtain a broader picture of the speech transmission quality as a function of the lower and upper cut-off frequencies of BP filters, the results have been interpolated using

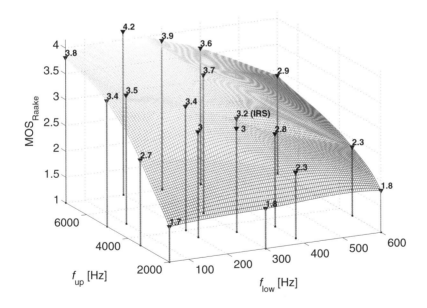

Figure 5.5: Mean speech transmission quality ratings (MOS) obtained in the author's test on BP-filtered speech. Independent variables: Lower and upper cut-off frequencies (f_{low}, and f_{up}, respectively [Hz]). The depicted surface plot is based on an interpolation of the test results (see text for details).

MATLAB's function 'griddata.m' (see Figure 5.5). It is based on a Delaunay triangulation of the MOS data points. It interconnects data points to their neighbors in such a way that no data points are enclosed by the circle connecting the three data points forming one triangle. A cubic interpolation procedure was chosen. The triangulation and interpolation algorithms employed by the respective MATLAB functions are based on Barber *et al.* (1996). Note that the small waves visible in Figure 5.5 show artifacts stemming from the interpolation carried out on a surface that was subsampled in some parts.

5.2.2 Parametric Impairment Model

A parametric model was established predicting the *impairment* values related to the BP-filtering conditions used in the listening test described in the preceding text. Although it was challenged by some authors, whether *impairment* is an appropriate measure in the context of speech sound quality (e.g. Möller, 2000), it has been employed here reflecting the notion of coloration. As it is used in this book, the term coloration implies an undesirable deviation from a reference timbre, that is an impairment.

The impairment values related to the different BP-filters were determined as follows:

1. The MOS ratings were transformed onto a WB R-scale, using one of the transformation rules discussed in Section 5.1 (see Figure 5.4).

2. The impairment values were derived from the R-scale data as the difference between the best R-value obtained from the test ratings and the R-value for the considered BP.

Depending on the extrapolation procedure employed for the transformation MOS $\rightarrow$ R, and the prior normalization of the MOS test results (see Section 5.1), different impairment values may be obtained. However, the model described in this section is not principally affected by different impairment estimates. Instead, only the parameters of the model have to be adapted. Due to the relative independence of the exact impairment calculation, a particular example may be used to illustrate the model.

In Figure 5.5, quality was displayed over the lower and upper cutoff frequencies of the BP-filters. However, the quality impact of the filters can be better captured, when the *impairment* is plotted over the bandwidth expressed on the Bark-scale (see Section 1.1.2)

$$z_{bw} = z(f_u) - z(f_l), \tag{5.2}$$

and over the center frequency expressed by

$$f_c = \sqrt{f_l \cdot f_u}. \tag{5.3}$$

The conversion to the Bark scale is carried out using Equation (1.1). A corresponding surface plot is depicted in Figure 5.6. From this plot it becomes apparent that certain contours of equal impairment exist. Corresponding 'iso-impairment contours' for this test plotted over center frequency and bandwidth in Bark are shown in Figure 5.7 (following the idea of iso-preference curves described by Munson and Karlin, 1962).

It can be observed from this plot, that to the left and right of a particular center frequency value the iso-impairment contours have a linear shape. The center frequency can be viewed as the ideal point of the contour, since it is associated with the lowest bandwidth that still yields the same level of impairment. Interestingly, the optimum center frequency is not

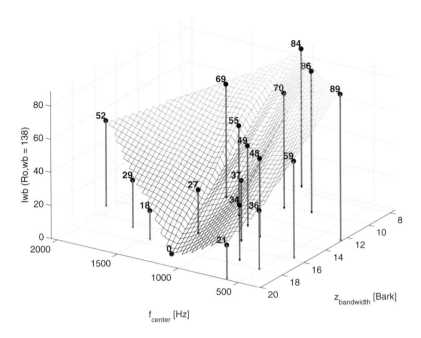

Figure 5.6: Impairment values plotted over bandwidth in Bark and center frequency in Hz. For this example, the R-scale extension using linear extrapolation and prior test data normalization has been employed, so that $Ro_{NB/WB} = 138$ (see Section 5.1).

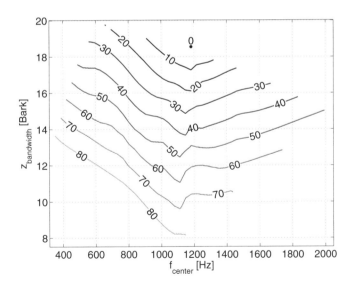

Figure 5.7: 'Iso-impairment contours' for the impairment due to bandpass filtering, plotted over bandwidth in Bark and center frequency in Hz.

constant: With decreasing impairment level, the optimum center frequency slowly increases. Another observation can be made: The iso-impairment contours, which represent increments of an impairment of 10 points on the R-scale in this case, are equally spaced along the bandwidth axis in Bark. This indicates a linear relationship between bandwidth in Bark and the level of impairment.

These findings are illustrated by a different presentation of the impairment data. In Figure 5.8(a), the impairment is plotted over the center frequency, using the bandwidth as curve-parameter. In the representation shown in Figure 5.8(b), the roles of bandwidth and center frequency have been reversed. Figure 5.8(b) clearly shows the linear dependency of the impairment on the bandwidth in Bark. In turn, Figure 5.8(a) substantiates the existence of an optimum center frequency, which seems to be bandwidth-dependent to some extent.

From these representations, a simple parametric model can be derived to predict the impairment due to bandwidth restrictions I_{bw}:

$$
I_{bw} = a_1 \cdot \left| \frac{f_c}{\text{Hz}} - a_6 \cdot \left(\frac{z_{bw}}{\text{Bark}} + a_5 \right) \right|
$$

$$
- a_2 \cdot \left(\frac{f_c}{\text{Hz}} - a_6 \cdot \left(\frac{z_{bw}}{\text{Bark}} + a_5 \right) \right) \tag{5.4}
$$

$$
- a_3 \cdot \frac{z_{bw}}{\text{Bark}} + a_4
$$

By curve fitting of the impairment data from the test using Equation (5.4), the values a_1 to a_6 are determined as listed in Table 5.3.

In Figure 5.9, two examples are shown underlining the good performance of the model. The accuracy does not depend on whether or not the test ratings have been normalized prior to calculating the impairment values. It can be seen from the correlations and root mean squared error values listed in Table 5.3, that the model performance is not affected by the transition to a more complex extrapolation procedure.

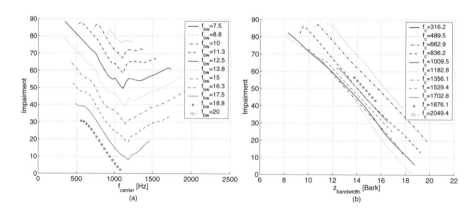

Figure 5.8: (a) Impairment plotted over center frequency [Hz] using the bandwidth [Bark] as curve parameter. (b) Impairment over bandwidth [Bark], with the center frequency as the curve parameter.

Table 5.3: Model parameters determined by curve fitting of the impairment values derived from the listening tests (see Section 5.2.1 and Barriac *et al.* (2004)). The last two columns show the correlation with the test results, and the root mean squared error of the prediction. The four cases represent: $Ro_{NB/WB} = 112 \equiv$ direct listening test data, linear extrapolation. $Ro_{NB/WB} = 138 \equiv$ linearly normalized test data, linear extrapolation. $Ro_{NB/WB} = 129$ (raw) $\equiv$ direct listening test data, exponential extrapolation based on test data by Barriac *et al.* (2004).

$Ro_{NB/WB}$	a_1	a_2	a_3	a_4	a_5	a_6	Corr.	RMSE
112 (raw)	$2.5 \cdot 10^{-2}$	$6.9 \cdot 10^{-3}$	5.5	98.9	132.8	7.7	0.993	2.01
138 (linear)	$3.8 \cdot 10^{-2}$	$7.7 \cdot 10^{-3}$	8.3	152.6	90.2	11.1	0.992	3.25
129 (raw)	$3.5 \cdot 10^{-2}$	$6.7 \cdot 10^{-3}$	7.4	129.2	101.8	9.9	0.992	3.39

A linear dependency of the logarithm of naturalness ratings on the logarithmic transmission bandwidth has been reported by Lawson and Chial (1982). The authors have departed from the idea of Stevens (1957) that the power law can serve as a fundamental descriptor of psychophysical phenomena. Their listening test employed a magnitude estimation technique for naturalness rating. The data obtained in our test and the model presented in this section cannot directly be linked to their observations. In our case, a quasi-linear relation between the bandwidth in Bark and the impairment can be observed, and not with the logarithm of the impairment. However, it seems questionable, whether their observation of an exponential relationship actually agrees with their own data, which appear to deviate from the linear shape assumed for the logarithmic data display.

5.2.2.1 Applicability to Other Filter Shapes

So far, the BP impairment model given by Equation (5.4) has been discussed for flat filters. To make it applicable to other filter shapes, an approach based on the notion of the 'equivalent rectangular bandwidth' (ERB) is chosen. Instead of using this approach in the linear frequency and amplitude domains as it is typically done for technical applications, the Bark scale was employed for the bandwidth, and a log-amplitude representation. Since a 50 dB stopband-attenuation was chosen for the filters, 50 dB have been added to the frequency response of the filter, to exclude irrelevant contributions to the ERB-estimate. The *ERB* [Bark] is calculated as

$$ERB = \frac{\text{area}(curve)}{\max(curve)}. \tag{5.5}$$

For determining the bandwidth in Bark, we set $z_{bw} \equiv ERB$. The center-frequency z'_c [Bark] is obtained as the center of gravity of the surface. We then calculate the equivalent lower and upper cut-off frequencies z_l and z_u (still in Bark) as

$$z_l = z'_c - \frac{ERB}{2} \tag{5.6}$$

$$z_u = z'_c + \frac{ERB}{2}. \tag{5.7}$$

We obtain the lower and upper frequency borders f_l and f_u [Hz] necessary to calculate the center frequency f_c in Hertz by Bark-to-Hertz conversion according to Equation (1.2). The procedure is illustrated in Figure 5.10.

As the only example that could be evaluated with this approach based on the available test data, the concatenation of the NB channel bandpass (ITU–T Rec. G.712, 1992) and the frequency response typical of handset usage in send direction was employed (IRSmod ITU–T Rec. P.830, 1996). The corresponding test result and prediction are shown in Figure 5.9 ('■'). It can be observed from the Figure that the quality prediction is in very

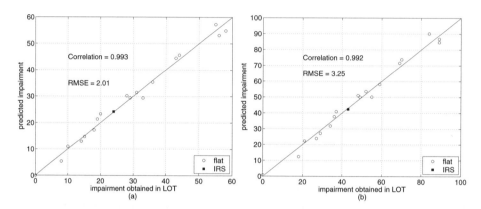

Figure 5.9: Predicted impairment (ordinate) vs. impairment derived from auditory test data (abscissa). (a) $Ro_{NB/WB} = 112$. (b) $Ro_{NB/WB} = 138$.

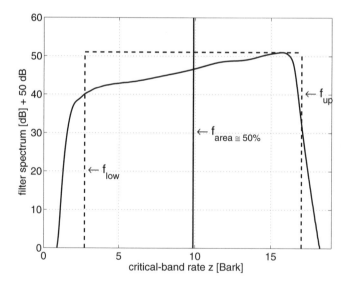

Figure 5.10: Perception-motivated 'Equivalent Rectangular Bandwidth' (see text) for the example of an IRSmod channel at send side (intermediate reference system; ITU–T Rec. P.830, 1996).

good agreement with the test data. Before this approach can be applied more generally, however, a larger number of test cases have to be considered.

5.3 Wideband Codecs

The E-model handles the impairment introduced by low bit-rate coding by using Equipment Impairment Factors (Ie; see Section 2.2.2.2). For most NB codecs, Ies are tabulated in ITU–T Rec. G.113 Appendix I (2002), which are typically determined in auditory listening tests.

Following the same approach for WB, we have determined impairment factors for three WB speech codecs operating at different bit-rates. Therefore, we started from the maximum $Ro_{NB/WB}$ value for the clean WB channel, and used results of five NB/WB tests carried out by the 3rd Generation Partnership Project quantify the quality impairment introduced by the codecs (for details see ITU–T Delayed Contribution D.029, 2005). These three codecs then serve as references for anchoring the impairment due to other WB codecs, which have been assessed in pure WB tests. In the example summarized here, a linear relation between NB and NB/WB-results is assumed, and the conservative WB-advantage estimate of $Ro_{NB/WB} = 112$. The procedure for the other WB extensions described in Section 5.1 follows the same principle:

1. MOS values obtained in the respective test are transformed to R values in the NB range [0;100], using the transformation given in ITU–T Rec. G.107 (2005).

2. The derived R values are expanded to the WB range (here [0;112]) by linear scaling (i.e. multiplying the initial R-values by 1.12), or by an exponential approach as described in Section 5.1.

3. For each test condition, we determine a WB impairment factor Ie_{WB} by subtracting the R-value obtained for the test condition under consideration (R(test cond.)) from either:

   (i) a clean WB channel:

   $$Ie_{WB} = R(\text{clean WB}) - R(\text{test cond.}), \tag{5.8}$$

   (ii) a standard NB channel (including G.711 coding and a default circuit noise level), and then adding the difference between the clear WB (here 112) and the standard NB channel (here 93.2), which corresponds to 18.8 in our example:

   $$Ie_{WB} = 93.2 - R(\text{test cond.}) + 18.8. \tag{5.9}$$

Some WB impairment values are depicted in Figure 5.11. It has to be noted that the derivation procedure results in a fixed relationship between NB (Ie) and WB (Ie_{WB}) impairment factors; thus, the new Ie_{WB} values rely on the same additivity principle that is used for NB channels. We have listed some examples of impairment values in Table 5.4. It can be observed from the table that the advantage of WB over NB can almost fully be exploited when a codec like the AMR-WB is used, operating at bit-rates starting from 19.85 kbit/s (ITU–T Rec. G.722.2, 2002). In this case, for our example of $Ro_{NB/WB} = 112$, the impairment is as small as $Ie_{WB} = 2$.

According to the additivity assumption of the E-model, the overall impairment linked to a series of two codecs (codec-tandem) is the sum of the individual impairments. In case of

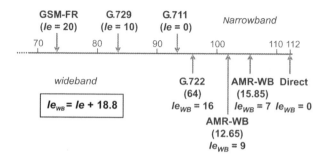

Figure 5.11: Illustration of the anchoring of WB impairment factors on the R-scale (example for the scale extension to 112).

Table 5.4: Examples of WB impairment factors, assuming $Ro_{NB/WB} = 112$.

Codec	Bit-rate	Ie_{WB}	# Tests
G.722	64.0	16	5
G.722	48.0	24	4
G.722.1	24.0	12	3
AMR-WB	6.6	29	4
AMR-WB	8.85	18	4
AMR-WB	12.85	9	5
AMR-WB	14.25	7	4
AMR-WB	15.85	7	5
AMR-WB	18.25	5	4
AMR-WB	19.85	2	4
AMR-WB	23.05	2	3
AMR-WB	23.85	5	3

twin tandems, the additivity seems to be acceptable, with some exceptions (see Table 5.5). However, the derivation of impairment factors for different mixed WB codec-tandems reveals an order effect (see Table 5.6). For example, for the tandem AMR-WB(12.65) * G.722(48) we obtained $Ie_{WB,tand} = 41$, while the reverse order, i.e. G.722(48) * AMR-WB(12.65), lead to $Ie_{WB,tand} = 29$.

Note that in the meantime, the extension of the R-scale to 129 has been adopted by the ITU-T in an upcoming Appendix II to the E-model (ITU-T Rec. G.107, 2005). Corresponding WB impairment factors for this extension value (both for NB and WB codecs) can be found in the actual version of Appendix II to ITU-T Rec. G.113 (see also Section 6.3).

5.4 Desired Nature

In Section 3.7, different factors have been identified that impact the desired nature of telephone speech. In this work, the desired nature or expectation of the user is addressed primarily with regard to WB speech: Due to the lack of a long-term experience with the

Table 5.5: Verification of impairment additivity in case of codec twin tandems. Note that the codecs listed in column #1 were applied in tandem operation, i.e. twice. Column #5: Impairment obtained from auditory tests; column #6: Impairment obtained from adding individual impairments (for details see ITU–T Delayed Contribution D.029, 2005).

Codec Tandem	MOS	R [0;100]	R [0;112]	Ie_{WB} Tandem	$\sum Ie_{WB}$ Individual
Direct	4.15	83.5	107.7	0	0
G.722(64)	3.57	69.4	89.5	18	32
G.722(48)	3.77	73.8	95.2	13	48
G.722.1(24)	3.02	58.5	75.5	32	24
AMR-WB(6.6)	2.27	44.1	56.9	51	58
AMR-WB(8.5)	3.05	59	76.1	32	36
AMR-WB(12.65)	3.7	72.2	93.1	15	18
AMR-WB(14.25)	3.88	76.4	98.6	9	14
AMR-WB(15.85)	3.93	77.6	100.1	8	14
AMR-WB(18.25)	4.06	81	104.5	3	10
AMR-WB(19.85)	4.09	81.8	105.5	2	4
AMR-WB(23.05)	4.14	83.2	107.3	0	4
AMR-WB(23.85)	3.91	77.1	99.5	8	10

Table 5.6: Verification of impairment additivity in case of mixed codec-tandems. Column #5: Impairment obtained from auditory tests; column #6: Impairment obtained from adding individual impairments (for details see ITU–T Delayed Contribution D.029, 2005).

Codec Tandem	MOS	R [0;100]	R [0;112]	Ie_{WB} Tandem	$\sum Ie_{WB}$ Individual
Direct	4.38	91.8	118.4	0	0
AMR-WB(12.65)*G.722(64)	3.41	66.1	85.3	33	25
AMR-WB(12.65)*G.722(48)	3.09	59.8	77.1	41	33
AMR-WB(12.65)*G.722.1(24)	3.56	69.2	89.3	29	21
G.722(48)*AMR-WB(12.65)	3.56	69.2	89.3	29	33
AMR-WB(15.85)*G.722(64)	3.6	70.1	90.4	28	23
AMR-WB(15.85)*G.722(48)	3.16	61.2	78.9	39	31
AMR-WB(15.85)*G.722.1(24)	3.69	72	92.9	26	19
G.722(48)*AMR-WB(15.85)	3.57	69.4	89.5	29	31

related *improvement* over the quality of traditional wireline telephony, users cannot yet have established any corresponding mode of perception. If presented with WB speech today, the perceived features may evoke reference to other 'schemas' of speech perception, like that of radio speech. On the other hand, when WB telephony has been available for some time, the associated changes to the desired features may lead to a depreciation of traditional *narrowband* telephone speech (see Section 3.7).

Jekosch (2000, 2005b) has explained the problem based on Piaget's work in the following way: When a user is confronted with an uncommon signal, that is a signal so far unknown to him, a disequilibrium may result between the cognitively still unprocessed perceptual event(s) and the expected or learned *schema*[5]. This disequilibrium has to be resolved by the user, which can be done following either of two adaptation mechanisms:

- In the process of *assimilation*, the perceived is integrated into to the already existing schema, that is recoded.

- In the process of *accommodation*, not the mental representation of the perceptual event, but the existing schema is modified, and adapted to the percept. The schema remains flexible until the iterative adaptation process, which involves cognition, enables the integration of the percept into the schema. Ultimately, this process may result in the creation of new schemata: Due to learning (long-term experience, see above), a stable reference (i.e. schema) will be established.

The present work implicitly addresses the *appeal* to certain schemata, and not their *formation*: Based on two examples, the impact of nonacoustic quality elements on the quality perception of WB and NB speech is investigated. The examples are motivated by the decisions that had to be taken during the design of the listening test for the assessment of the quality related to WB speech transmission, and its impairment due to bandwidth restrictions (see Sections 5.1 and 5.2). The two questions studied in this section are as follows:

(a) To what extent does speech transmission quality depend on the nonacoustic properties of the employed user interface?

(b) Does the content of transmitted speech influence its perceived sound quality?

Question (a) was raised during the selection of an appropriate user interface to be employed in the listening tests. Question (b) was brought up when the text material to be used in the tests had to be chosen.

5.4.1 User Interfaces

A listening test was performed to investigate the psychological role of different user interfaces for the assessment of speech transmission quality as a function of the channel bandwidth.

Most studies reported in the literature on the speech transmission quality of linearly distorted speech were carried out to assess the influence of the channel frequency response (Gabrielsson *et al.*, 1985; Gleiss, 1970; Moore and Tan, 2003; Pascal and Boyer, 1990; Voran, 1997). The effect of the applied user interface on the *desired nature* of the perceived speech has not been studied. When both the handset and the headphones were used for presentation in the test, and the terminal was taken into consideration, the main emphasis was on the quality difference between WB (50–7000 Hz) and NB (300–3400 Hz) transmission.

In the test presented here, electrostatic headphones and specially prepared handset telephones with ideal-typical acoustical properties were employed as user interfaces. With this choice, three different aspects of speech sound quality perception can be studied:

---

[5]Such schemata can be considered as not cognitively reflected, similar to the preprocessed percept.

(a) The psychological difference between an 'ideal' headphone (STAX Lambda Pro, monotic presentation, see Section B.5) and an 'ideal' handset-telephone ('Hi-Fi phone' built
   from handset-telephone and STAX earphone, see Section B.5.3).

(b) The subjective effect of the acoustical difference between an 'ideal' and a nonideal
   handset due to imperfect acoustic coupling to the ear ('Hi-Fi phone' vs. WB handset
   'B5', see Section B.5.3).

(c) Comparison of monotic and diotic presentation using the ideal headphone.

5.4.1.1 Test Setup

The speech files for the corresponding listening-only test were recorded with six different
speakers (3 f, 3 m) in an anechoic chamber. The microphone was placed 30 cm in front
of the speakers mouth reference point (MRP), as a compromise between telephone- and
hands-free-terminal-typical acoustical conditions. A shortened version of the Eurom sentence material (Gibbon, 1992) was read aloud by the speakers in a natural way. The recorded
samples were further processed using the real-time telephone-line simulation established for
this book (see Appendix B). Five different BP filter settings were employed, as listed in
Table 5.7. A list of the four user interfaces applied in the test is provided in Table 5.8.

Each filter-condition was applied to different sentences uttered by the six speakers.
The active speech level of the samples was adjusted to -26 dB rel. ovl. (i.e. relative
to the overload of the digital system) to yield telephone-typical conditions. The transmission characteristics of the WB handset 'B5' were adjusted to the Stax-characteristics,

Table 5.7: Bandpass filters employed in the listening test on the
impact of the user interface on speech transmission quality.

Lower Limit f_l [Hz]	Upper Limit f_u [Hz]	Bandwidth [Hz]	Filter Characteristics
50	7000	6950	Flat
100	5000	4900	Flat
300	3400	3100	Flat
300	3400	3100	IRS
400	2400	2000	Flat

Table 5.8: User interfaces employed
in the test, and abbreviations used for
identification.

User Interface	Abbreviation
Modified handset	'B5'
Hi-Fi phone	'hifi'
Headphone (monotic)	'm.'
Headphone (diotic)	'd.'

which were regarded as reference, using a corresponding correction filter *RLR1 /RLR2* (see Figure B.1). Although spectral correction of the transmission characteristics of the phone 'B5' was applied, imperfect coupling to the listener's ear may occur – especially of lower frequencies – so that the handset 'B5' may not provide appropriate WB capabilities. For the diotic headphone presentation, the speech level was lowered by 6 dB to account for the effect of binaural loudness (Zwicker and Fastl, 1999, pp. 311–313).

It has to be noted that no additional anchoring conditions were used during the test, since only relative differences between the user interfaces were sought. Although presentation of additional 'bad' reference connections allows better comparison to other tests, the resulting quality ratings tend to decrease the ability to distinguish between connections of relatively high quality, as they were presented in this test.

The listening-only test (LOT) was divided into two sessions of 60 sample–terminal combinations each. During the tests, the speech files and the corresponding terminal equipment were randomly selected. The test was controlled using a computer program that indicated to the subjects, which user interface they were to use for the next listening sample. With the help of a switch operated by the program, the sample was then played out over the corresponding user interface.

Two different absolute category-rating scales were employed for the sound and quality ratings collected from the subjects. Quality ratings were given on the ITU-T recommended 5-point ACR-scale (MOS scale, see Figure 2.1), and speech sound quality was rated on the unidimensional continuous 11-point scale shown in Figure 2.3.

Seventeen naïve listeners with normal hearing abilities (2 female, 15 male) participated in the test. They were mostly recruited from the university's student body, and their age range was between 21 and 37 years.

5.4.1.2 Results

The results of the LOT show that an interdependence exists between the applied BP filter, the user interface and the quality ratings collected from the subjects. In an analysis of variance (ANOVA) of the MOS-ratings using the interface, the BP filter and the speaker as fixed factors, highly significant effects due to the BP and due to the interface were found ($p < 0.001$). The F-ratio, however, was found to be by more than a factor 10 higher for the bandwidth factor than for the interface factor (F(interface) $\cong$ 17.4, and F(bandpass) $\cong$ 178.3). The interaction between the BP and the user interface was found to be a highly significant, however, less important factor ($F \cong 4.9$, $p < 0.001$). It can be seen from Figure 5.12, that the results for integral speech transmission quality and speech sound quality are very similar. Since linear distortion was the only quality element introduced in the test, this observation is not surprising. Due to the similarity between the ratings, only the MOS-results are discussed in the following text.

The overall quality MOS for the flat filters indicate a slight preference for the Hi-Fi phone in case of NB presentation (Figure 5.12(a) and Figure 5.13(a)). However, only the difference between the Hi-Fi phone and the handset 'B5' were found to be statistically significant (on a 95% confidence level). For higher bandwidths, the diotic headphone was rated better than the Hi-Fi phone. This stands in contrast to the results of a mirror-experiment carried out by Berger at T-Berkom, Berlin (unpublished). In that experiment, the handset 'B5', the Hi-Fi phone and the diotic headphone presentation using a Stax Lambda Pro headphone, were employed, too. However, instead of providing the test subjects with the

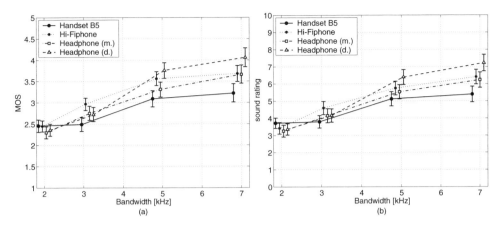

Figure 5.12: Mean speech transmission quality ratings (MOS, left), and mean speech sound quality ratings (sound, right). Abscissa: Bandwidth; curve parameter: User interfaces. The error-bars indicate the 95% confidence intervals.

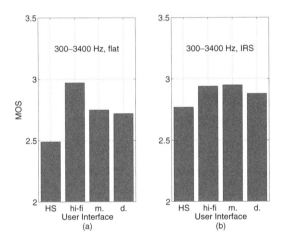

Figure 5.13: Mean speech transmission quality ratings for the NB case (300–3400 Hz). (a) Flat frequency response; (b) IRS-shape. Abscissa: Employed user interfaces.

actual user interfaces, this test was focused entirely on the acoustic contribution of the user interfaces to quality: The speech samples were recorded using a dummy head, and presented to the subjects over one pair of headphones. Here, the diotic presentation was *always* preferred to that using the (monotic) Hi-Fi phone. This points to a general preference of listening diotically to the speech samples presented in this test series.

Between the results for the NB and the WB conditions of the author's tests, which are depicted in Figure 5.12, a remarkable increase of quality by more than 1.5 MOS points can be observed for diotic headphone presentation. The comparison of the monotic headphone and the Hi-Fi phone reveals a small preference for the Hi-Fi phone throughout the whole bandwidth range, but decreasing for higher transmission bandwidth. The *acoustical* properties of the headphone (monotic) and the Hi-Fi phone can be regarded as identical. Thus, it can be concluded that the quality difference between monotic headphone presentation and Hi-Fi phone presentation is related to the listeners' expectation toward the terminal equipment. Here, two influencing factors can be pointed out:

- Monaural presentation does not correspond to a common headphone usage, from which the listeners expect diotic presentation. The Hi-Fi phone is assumed to be perceived as handset-telephone, and for this reason – especially for lower bandwidths – the perceived nature corresponds more to the expected nature, resulting in higher quality (see Section 3.7). On the other hand, the unusual design of the Hi-Fi phone will possibly have altered listener's expectation and – for higher bandwidths – make it appear to yield similar quality as monotic headphone presentation.

- A comparison of diotic and monotic headphone presentation shows clear preference for the diotic, especially for increasing transmission bandwidth. In case of WB transmission, the diotic presentation is rated 0.4 points MOS better, than the Hi-Fi phone and monotic headphone, presentations. This appears to be in line with the high expectation toward the acoustic properties of headphones as well as the advantage of listening diotically to high-quality signals.

- The handset is rated worse than all other user interfaces. Especially for WB transmission, the quality rating for the handset is significantly lower (about 1.0 MOS points) as compared to diotic headphone presentation. The relatively poor performance of the handset for higher bandwidths can have several reasons, for example:
  ○ The transmission of lower frequency components might cause distortions due to mechanic resonance of the handset.
  ○ The handset does not couple low frequency components equally to all the listeners' ears, due to differences in application pressure and handset position.
  ○ Wideband speech does not match the specific expectation toward classical handsets.

Another, remarkable observation can be made from the results depicted in Figure 5.13(b): When IRS characteristics were chosen for the NB presentation (300–3400 Hz), the IRS-shape equalized out the differences between the user interfaces. Coming back to the appeal to perceptual schemata, this can be interpreted as follows:

- The user interface may evoke the usage of a particular schema. Some evidence from the test presented here can be derived from the increase of quality for diotic headphone presentation between NB and WB transmission.

- The interaction between the transmission bandwidth and the user interface reflected in the quality ratings indicate that for the flat frequency responses no fixed schema has been established which would correspond to the test situation. In turn, the acoustic properties in case of IRS-filtering seem to evoke such a fixed schema, so that the differences between the user interfaces are neutralized.

From the test results, no doubts arouse with regard to the choice of the Hi-Fi phone as the user interface for the more detailed listening tests on bandwidth restrictions (see

Section 5.2). This phone achieved the highest rating in case of the flat NB filter (300–3400 Hz), which indicates that it can be associated with a telephone context by users.

5.4.2   Content

This section addresses the question of whether users judge the listening quality of particular speech samples independent of the content, i.e. of what is said. On one hand, the answer to this question is of interest for the choice of an appropriate text material used for listening tests. On the other hand, the question relates to a more fundamental issue, namely, whether the assessment of the *form* of a speech sample can be separated from the interpretation subjects make of it, that is from its content and *function* (see considerations on semiotics in Section 1.1).

Speech is *interpreted* as a sign, both by the speaker and by the listener. Since an acoustic speech signal or written text turns into a sign only due to the interpretation, the sign is *different* for the speaker and the listener (see Section 1.1; Jekosch, 2000, 2005b).

For speech quality listening tests, unprocessed source-speech material is selected that is processed with the system under test. During quality assessment, different factors related to the speech material may come into play:

(a) The phonemic properties of the material.

(b) The intonation and articulation of the speaker, i.e. *how* she/he has uttered the material.

(c) The features of the speaker's voice.

(d) The impact of the transmission system on the acoustic signal.

(e) The content, i.e. what is said.

In the past, different types of text material were used in various listening tests for speech quality assessment, for example:

- Phoneme-specific material to study the influence of specific signal-processing on phonetic properties of speech (Huggins and Nickerson, 1985).

- A limited number of short, meaningful sentences to study the multidimensionality of speech quality in a telecommunications context; the attention of listeners is drawn to the measurement object and away from the specific speech material (see Sections 2.1.2 and 3.4).

- A greater number of meaningful sentences to investigate the overall quality of speech, reflecting the variety of utterances typical for real-life telephone conversations (e.g. ITU–T Rec. P.800, 1996)

- Syntactically correct, but semantically unpredictable sentences (SUS), initially developed to study the intelligibility of speech-synthesis systems (Benoît, 1990; Benoît *et al.*, 1996). For example, SUS-material has also been used to study the quality of higher quality speech-processing devices such as low bit-rate codecs (Bappert and Blauert, 1994).

- Different other text material at syllable, word and sentence level, were primarily developed to investigate the intelligibility of speech-synthesis systems (see Section 1.1.2.4; for an overview see Jekosch, 2000, 2005b)

However, with respect to the quality of natural speech, the content or predictability of the text material has only been considered in terms of its influence on *intelligibility* (e.g. Stickney and Assmann, 2001). In these studies, it was confirmed that sentences providing little semantic contextual information are less intelligible than sentences with larger amounts of context information (see Section 1.1.2.4).

The influence of content in tests related to speech *quality* can be avoided, if only a small number of meaningful sentences are employed (e.g. the same sentence for each condition, as in many analytical tests, see Section 2.1.2). This stands in contrast to utilitarian listening-only quality tests typically carried out in the context of telephony, where a more general evaluation of a system is sought (ITU–T Rec. P.800, 1996). Here, a larger number of, more or less, independent sentences are used, in order to reduce training effects, and to represent a real-life telephone situation.

Based on the above considerations, the results of two auditory listening tests are discussed in the following (Raake, 2002). The question to be addressed is, whether subjects are able to judge the (acoustic) form of a speech sample independent of the content.

5.4.2.1 Source Material

To this aim, a small database of four different types of French text material read by six native speakers of French (four female, two male) was established. The text material consists of:

- Semantically unpredictable sentences (SUS; Benoît, 1990; Gibbon *et al.*, 1997)

- Everyday sentences likely to occur, for example, in a telephone conversation. The text material is taken from the EUROM-database, see Gibbon *et al.* (1997). The French version of the sentences has been taken from the Institut de la Communication Parlée (ICP) in Grenoble, France. The sentences were shortened to show a length similar to that of the SUS-material.

- Short passages taken from the book 'Le Petit Prince' by Antoine de Saint-Exupéry.

- More complex sentences of philosophical content, namely, short excerpts from the collection of thoughts – 'Pensées' – by Blaise Pascal.

The untrained speakers were instructed to read the material aloud in a natural way. The speech material was digitally recorded in an anechoic environment at 32 kHz sampling-rate, with the microphone placed 30 cm in front of the speaker's mouth. The active speech level of the recorded samples was adjusted to -26 dB (rel. ovl. of the digital system) according to the procedure described in ITU–T Rec. P.56 (1993).

5.4.2.2 Speech Processing

The recorded samples were BP-filtered using three different transmission bandwidths (Table 5.9; the 150–3550 Hz condition corresponds to the telephone channel filter defined in ITU-T Rec. G.712).

Table 5.9: Bandpass filters applied in the listening tests. All filters have a flat frequency response in the passband. The filter for the 150–3550 Hz passband is an implementation of the narrowband channel bandpass (ITU–T Rec. G.712, 1992).

Lower Limit f_l [Hz]	Upper Limit f_u [Hz]	Bandwidth Δf [Hz]
350	2550	2200
150	3550	3300
50	7000	6950

For each text type, a different sentence was used so that every sentence was unknown to the listeners, yielding $6 \times 4 \times 3 = 72$ short speech samples (six speakers, four text types, three BP-filters). With the BP-filters, the speech samples were linearly distorted to represent different transmission conditions. The BP-filters have flat spectral characteristics in their transmission band, and were adjusted to yield equal loudness ratings according to ITU–T Rec. P.79 (1999), using the WB loudness ratings given in ITU–T Rec. P.79, Annex G (2001).

5.4.2.3 Test Setup and Subjects

For sample presentation, the specifically built WB handset, the 'Hi-Fi phone', was used (see Section B.5). This way, an appropriate coupling of the WB frequency components could be assured, at the same time attempting to provide a 'normal' telephone situation[6].

In two series of auditory tests, the set of speech files was presented in randomized order to two different groups of listeners:

1. French native speakers.

2. German listeners without knowledge of French.

With this choice, one group of listeners was able to understand the presented speech, while the other group was more restricted to formal aspects of the samples.

The subjects were asked to judge the sound quality of the transmitted speech on the 11-point, one-dimensional rating scale depicted in Figure 2.3 (IEC Publ. 268-13, date unknown). The instructions for the listeners were a crucial point in the experiment, as only a one-dimensional rating aiming at the speech sound quality was asked from the subjects. In order to reflect the application to telecommunications, the subjects were instructed to ignore the characteristics of the voice itself as far as possible, and to concentrate mainly on the effect due to the transmission. The instructions handed to the test subjects in written form read as follows (English equivalent):

---

[6]Evoking a corresponding perception schema (see Section 5.4.1).

[...] The aim is to judge the sound quality of speech transmitted across telephone lines of different transmission quality.[...] What impression do you have of the sound quality of the transmitted voice[7]?

The presentation of the samples as well as the collection of the ratings were controlled with a computer program. The subjects had the option of listening to the samples as often as they wanted, before their rating was taken into consideration by the system.

Twenty native French listeners, naïve with respect to the test task and conditions, participated in the first test series (10 female, 10 male). This first test series was carried out at the Laboratoire de Mécanique et d'Acoustique, CNRS, Marseille, France, and the test subjects were recruited from different labs of this institution. The age range of the subjects was from 22 to 58 years. The attention of the subjects was not specifically drawn to the content or sense of the speech material. However, it was pointed out to them that meaningless sentences also would be presented. This way, it was tried to avoid that the semantically peculiar SUS-sentences would be rated worse only because of their unexpected occurrence.

Fifteen listeners, who by their own account did not know the French language, took part in the second listening experiment carried out at IKA in Bochum, Germany (three female, 12 male). As they were not able to understand the presented speech samples, their judgments referred mainly to formal aspects, while they were still able to identify the signal as speech. Hence, it can be assumed that they were able to rely on analogies to the German language (or others). The age range of the German listeners was from 23 to 45 years. The test subjects, scientific and nonscientific members and students of the IKA, were naïve listeners with respect to the test.

5.4.2.4 Test Results

The results show that the sense of the text material obviously played a role in the listeners' judgments: The French listeners rated the SUS-sentence material significantly lower than the Eurom-sentences for all three bandwidths, as is depicted in Figure 5.14(a). The speech sound quality of the two literary text types were rated in between that of the SUS and that of the Eurom materials. An analysis of variance using the speaker, the text genre and the BP filter as fixed factors revealed that text type and bandwidth are highly significant factors for the variance of the sound ratings ($p < 0.001$, see Table G.13). The interactions between text type and speaker, and between bandwidth and speaker were found to be significant, too ($p < 0.05$).

It is interesting to note that no interaction effect between the text genre and the transmission bandwidth could be observed. Obviously, a particular transmission bandwidth is not associated with an expected content that impacts quality perception. Instead, the quality difference between the different source materials equally applies to all bandwidths employed in the test.

In the judgments of the German listeners, no significant difference between the four test sets could be found (Figure 5.14[b]), except for the WB case. The analysis of variance

---

[7]French instructions: '[...] Le but est de juger de la qualité sonore de la parole transmise à travers des lignes téléphoniques de différentes qualités de transmission [...] Quelle impression aviez-vous de la qualité sonore de la voix transmise?' German instructions: '[...] Ziel ist, die Klangqualität der über Telefonleitungen unterschiedlicher Übertragungsqualität übertragenen Sprache zu beurteilen. [...] Welchen Eindruck haben Sie von der Klangqualität der übertragenen Stimme?'

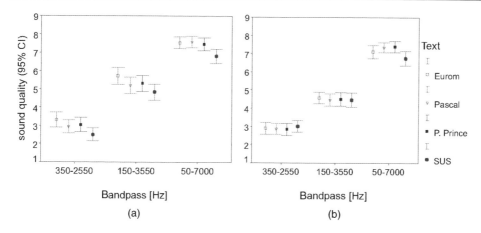

Figure 5.14: Impact of the employed text material on speech sound quality impaired by bandwidth limitations. (a) French listeners; (b) German listeners. Also indicated in the graphs are the 95% confidence levels for the mean ratings (line-endpoints).

of the ratings revealed no significant impact of the text genre, but a significant interaction between the genre and the speaker ($p < 0.05$, see Table G.14).

In case of WB transmission, the SUS-material was rated slightly lower than the Eurom-material by the non-French listeners. However, the excerpts from 'Le Petit Prince' and Pascal were rated significantly better than the SUSs. An analysis of the dependency of the judgments on the speaker showed that the SUS-material uttered by one specific (female) speaker ('D') was rated worse by the non-French listeners than the samples of all other speakers, and to some extent also by the French listeners, see Figures 5.15 and 5.16.

If this speaker is discarded in the data analysis, the average ratings of the non-French listeners no longer show any quality differences between the four text types. The average ratings of the French listeners, on the other hand, are only very little effected when this

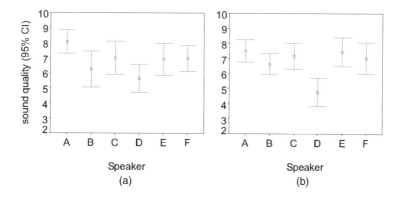

Figure 5.15: Speaker-dependency: Mean ratings for SUS-text under WB transmission (50–7000 Hz) as a function of the speaker. (a) French listeners; (b) German listeners. Also indicated in the graphs are the 95% confidence levels for the mean ratings (line-endpoints).

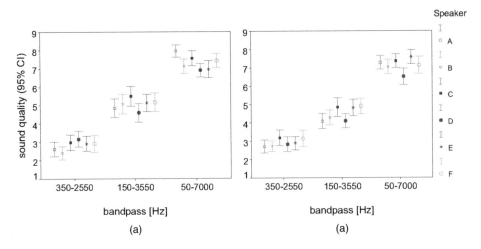

Figure 5.16: Speaker-dependency: Ratings averaged over text-genres as a function of the bandwidth and the speaker. (a) French listeners; (b) German listeners. Also indicated in the graphs are the 95% confidence levels for the mean ratings (line-endpoints).

particular speaker is omitted. The difference between SUS- and Eurom- sentence material is still statistically significant. This implies that the speaker has read the SUS material in a way that especially for non-French listeners was unusual or conspicuous: If the content of the speech sample is not understood, the particular form cannot be anticipated. This could be confirmed by informal listening to the corresponding sample, which is read in a monotonous way at higher pitch. The speaker has obviously expressed the senselessness of the SUS-sentences by a corresponding form, which is rated more negatively by the listener.

As stated in the preceding text, for WB the excerpts from 'Le Petit Prince' and Pascal were rated higher by the non-French listeners than the other two text types. This can probably be explained with the fact that this material is read in a literary way by the speakers and may correspond more to the expectation of how French should sound like (if the language is not understood). The difference between text types found in the ratings of the French listeners can have various reasons:

- The listeners are effected by the sense of the speech material. In this case, the nonacoustic features of the SUS-speech do not correspond to the desired ones, so that lower ratings are given[8]. In turn, the speech material is rated more positively, if the sense or content corresponds to the expectation that is raised by the telephone-typical situation.

- Another reason for lower ratings in case of SUS could be that other speakers than speaker 'D' expressed the senselessness of the SUS-material with a particular form, which concerns features that were not accessible for the non-French listeners.

---

[8]Whether the peculiarity of the employed SUS-sentences is related to an *affective* devaluation i.e. lower judgment of speech sound quality – as it was found e.g. by Västfjäll (2003) in the context of sound quality research – cannot be revealed from this test.

- If lower intelligibility or higher listening effort played a role in the more critical rating of the SUS-material by the French listeners, this effect is expected to be lower for WB than for the other two bandwidths. However, the offset between the mean ratings for SUS and the other text-types is bandwidth-independent (see Figure 5.14. Nevertheless, it cannot be excluded at this point that the observed effect is caused by a higher listening effort associated with the SUS-material.

In the WB case, the ratings averaged over the text types are found to be speaker-dependent, and the two groups of listeners show different preferences (see Figure 5.16). The French listeners seem to have preferred speaker 'A', a young female speaker with higher pitch and very clear articulation. The non-French listeners, on the other hand, had a preference for speaker 'E', a male, older speaker of lower pitch, who read the text material as if it was a fairy tale. This speaker, in turn, was rated worst by the French listeners. Informal listening revealed that his articulation was less accurate than that of most other speakers, an effect that cannot be perceived when the language is not understood.

For the French listeners, the speech material mainly introduced a constant shift on the sound quality scale. This has consequences for the selection of text material for specific tests: If the aim of a test is the subjective comparison of different sound-related channel characteristics, in principal, both low predictability sentences (SUS) and everyday sentences (Eurom) can be used. Other, more specific text types such as the literary 'Le Petit Prince' and the excerpts from Pascal, may introduce a conflict with the telephone-related expectation.

5.4.3 Summary

In a listening test on the quality of BP-filtered speech, an influence due to the employed user interface was found. Since a difference in the quality impact was observed for user interfaces, which had been determined before to be acoustically identical, the related quality difference is thought to be because of the desired nature of the perceived speech samples. The test results support the assumption that the desired nature is impacted by acoustic and nonacoustic properties of the channel: In the test, only the spectral shape typical of telephone channels lead to an equalization of the quality of all employed user interfaces.

In two further listening tests on the speech sound quality of BP-filtered speech, an influence of the employed text material on the quality ratings could be observed: If the sense of a speech sample is understood, it may impact the assessment of the form of the sample. A dependency of speech sound quality on the interaction between the text genre and the bandwidth could not be confirmed by the test. Instead, the test indicates that a particular bandwidth was not associated with a particular expected content. Thus, quality rank orders between different transmission conditions do not seem to be affected by the content of the underlying text material. However, the described tests show that the sense of the underlying text material has to be taken into consideration, when the *absolute* measurement of speech sound quality is aimed at. In speech-quality research, the separability of the acoustic signal and the speech sign as a whole is often assumed. The listening tests discussed here indicate that this assumption has to be reconsidered depending on the actual measurement task.

6

From Elements to Features: Extensions of the E-model

In previous chapters, several extensions of the E-model to effects typical of Voice over Internet Protocol VoIP have been described. In this chapter, these extensions will be summarized briefly, differentiating between the ones that have already made their way into the current E-model version, and those that have not or not yet been included in the standard (ITU–T Rec. G.107, 2005).

6.1 E-model: Packet Loss

In Chapter 4, different approaches for predicting general, integral speech quality under arbitrary packet loss distributions have been evaluated. Therefore, the perception-oriented description of loss distributions in terms of their macroscopic loss behavior, that is characterized by subsequent periods of microscopic packet loss (see Section 3.3.5) has been employed. According to this definition, microscopic loss behavior considers those segments of 'constant' behavior that do not yield time-varying quality, but are connected to the quality dimension of 'continuity' or 'interruptedness' (Section 3.4). In turn, the macroscopic behavior is defined as the concatenation of such passages of potentially differing microscopic behavior, which leads to an instantaneously perceived quality that varies over time.

To cover the speech quality impact of the microscopic loss behavior alone, a formula recently included in an update of the E-model has been proposed, see Equation (6.1), which has been developed in detail in Section 4.1.2.3, Equation (4.14). With this new 'Effective Equipment Impairment Factor' Ie,eff, the basic formula of the E-model reads as shown in Equation (6.2).

$$Ie,eff = Ie + (95 - Ie)\frac{Ppl}{\dfrac{Ppl}{BurstR} + Bpl} \tag{6.1}$$

$$R = Ro - Is - Id - Ie,eff + A. \tag{6.2}$$

Speech Quality of VoIP: Assessment and Prediction   Alexander Raake
© 2006 John Wiley & Sons, Ltd

Here, three new parameters have been introduced:

- *Ppl* [%] is the mean packet loss percentage during the considered segment;

- *Bpl* is the Packet Loss Robustness Factor that quantifies the codec and packet loss concealment (PLC) robustness against packet loss. Tabulated values for the Equipment Impairment Factor *Ie* already known from the initial E-model, ITU–T Rec. G.107 (2000), as well as for *Bpl* are listed in ITU–T Rec. G.113 Appendix I (2002).

- *BurstR* is the Burst Ratio, which was first proposed in ITU–T Delayed Contribution, D.020 (2001), and is now patented as US Patent 6,931,017 (2005). Equation (6.3) provides the definition of the Burst Ratio, which measures the relative occurrence of consecutive packet loss under dependent (bursty) loss as compared with random, i.e. independent loss.

$$BurstR = \frac{\text{average No. consecutively lost packets}}{\text{average No. consecutively lost packets for random loss}}. \tag{6.3}$$

Here,

BurstR = 1 identifies random, i.e. independent loss;

BurstR < 1 applies to sparse loss, i.e. showing a lower tendency for consecutive packet loss than for independent loss;

BurstR > 1 identifies burst loss, i.e. showing a higher tendency for consecutive packet loss than for independent loss.

For random loss with *BurstR* = 1, Equation (6.1) takes the form of the initial random loss impairment model introduced in Raake (2004):

$$Ie,eff = Ie + (95 - Ie)\frac{Ppl}{Ppl + Bpl}. \tag{6.4}$$

In Chapter 4, Section 4.2, the E-model predictions obtained for average loss parameter settings have been compared with more complex prediction approaches, for example, predicting integral quality based on estimated instantaneous quality profiles. In this analysis it was found that general predictions for network-planning purposes can be achieved already with the rather simple approach as given in Equation (6.1). In case of erroneous predictions, speech quality is rather under- than overestimated, making the model safe to avoid unsatisfied users. Only in case of severe impairments due to periods of interchanging quality levels and excessive degradation during bad sections (e.g. affecting intelligibility), Equation (6.2) overestimates the speech quality perceived by actual users.

6.2   E-model: Additivity

In Chapter 4, a variety of conversation test results on narrowband (NB) speech quality were discussed in the light of the E-model assumption that different types of degradations are additive on the ('psychological') *R*-scale. The main observation made from the available test data is a considerable interaction between the impairment caused by packet loss, and the stationary impairment caused by circuit noise. It has to be noted that in the tests, circuit noise was not affected by the packet loss, and a different type of interaction can be expected when packet loss affects circuit or background noise transmission (Section 3.3.11).

As a first proposal of how such interactions could be included in the E-model algorithm, an interaction factor is suggested, between the effective equipment impairment factor (Ie,eff) and the factor quantifying speech quality due to the basic signal-to-noise ratio (Ro). It covers several aspects:

(a) The masking of the impairment due to coding (with and without packet loss, i.e. Ie,eff) by line noise.

(b) The attenuation of the impairment due to coding (with and without packet loss; Ie,eff) by higher loudness ratings.

(c) The dominance of the impairment due to packet loss in case of severe losses, which may, for example, affect intelligibility.

The interaction factor is formed of three terms, and each reflects one of the aspects (a)–(c) (see Section 4.4.4 for details). No new input parameters to the E-model are required with this extension. With the interaction terms, the basic formula of the E-model has been rewritten as:

$$R = Ro - Is - Id - Ie,eff + A + E_{Ppl\circledast Nfor}, \tag{6.5}$$

where

$$E_{Ppl\circledast Nfor} = Ie,eff \cdot \left(1 - \left(\frac{Ro}{95}\right)^{c_1} \cdot e^{-(OLR/dB-10)/c_2}\right) + \frac{(95 - Ro) \cdot (Ie,eff)}{(95 - Ro + Ie,eff)} \cdot S_1 \cdot S_2, \tag{6.6}$$

and

$$S_1 = \frac{1}{1 + \left(\dfrac{\mu_{13} \cdot T_p}{c_3}\right)^{-c_4}}, \tag{6.7}$$

$$S_2 = \frac{1}{1 + \left(\dfrac{Pb}{c_5 \cdot (Ppl + 0.01)}\right)^{-c_6}}. \tag{6.8}$$

The above set of equations corresponds to Equations (4.30) through (4.33), Section 4.4.4, where details on the role of the different terms have also been provided. The constants c_1 through c_6 are curve-fitting parameters. A good fit of the test data was obtained using the settings summarized in Table 6.1. This approach serves a first suggestion to ITU-T's SG12

Table 6.1: Proposed settings for the constants c_1 to c_6, Equations (6.6) through (6.8).

Constant	c_1	c_2	c_3	c_4	c_5	c_6
Value	0.3	16	12	10	1.8	10

on how cases of deviation from the additivity-property can be handled in a quantitative manner. So far, it is limited by the few available test data on combined quality elements, as well as by the fact that no impact of packet loss on noise transmission was covered by the data.

6.3 E-model: Wideband, Linear and Non-Linear Distortion

In Section 5.1, an extension of the E-model R-scale has been proposed, to express speech-quality predictions for wideband (WB) networks on the same scale as applied to NB (ITU–T Rec. G.107, 2005). This extension was quantified to lie in the range $Ro_{NB/WB} \in [112, 138]$. A good compromise founded on a basis of solid auditory tests seems to be a value of

$$Ro_{NB/WB} = 129, \tag{6.9}$$

which implies an improvement of about 29% over the NB case. In the meantime, this WB extension value has been adopted by the ITU-T in a new Appendix II to ITU-T Rec. G.107.

Starting from the WB extension, a parametric model has been presented, quantifying the impairment Ibw due to linear frequency distortion (Section 5.2). With this additional bandwidth impairment factor, the (WB-extended) basic formula of the E-model reads:

$$R = Ro_{NB/WB} - Is - Id - Ie,eff - Ibw + A. \tag{6.10}$$

The bandwidth impairment factor Ibw is defined as follows:

$$
\begin{aligned}
Ibw = a_1 \cdot & \left| \frac{f_c}{\text{Hz}} - a_6 \cdot \left(\frac{z_{bw}}{\text{Bark}} + a_5 \right) \right| \\
& - a_2 \cdot \left(\frac{f_c}{\text{Hz}} - a_6 \cdot \left(\frac{z_{bw}}{\text{Bark}} + a_5 \right) \right) \\
& - a_3 \cdot \frac{z_{bw}}{\text{Bark}} + a_4
\end{aligned}
\tag{6.11}
$$

Here, $a_i, i \in [1, 6]$ are curve-fitting parameters that depend on the chosen R-scale WB extension (values for the extension of $Ro_{NB/WB=129}$ are given in Table 6.2).

The parametric impairment model is based on two new input-parameters to the E-model, the transmission bandwidth z_{bw} (in Bark, see Section 1.1.2.1), and the center frequency f_c [Hz]. In case of (almost) rectangular filter-shapes, these two parameters can be obtained from the lower and upper cut-off frequencies, according to Equations (6.12) and (6.13).

Table 6.2: Model parameters determined by curve fitting for $Ro_{NB/WB} = 129$.

a_1	a_2	a_3	a_4	a_5	a_6
$3.5 \cdot 10^{-2}$	$6.7 \cdot 10^{-3}$	7.4	129.2	101.8	9.9

- z_{bw} is the bandwidth (in Bark):

$$z_{bw} = z(f_u) - z(f_l); \tag{6.12}$$

- f_c is the center frequency (in Hz):

$$f_c = \sqrt{f_l \cdot f_u}. \tag{6.13}$$

In case of more complex filter-shapes, we assume that these parameters can be estimated from the amplitude spectrum. As a first approach, it is proposed to use the 'equivalent rectangular bandwidth' (ERB). ERB is calculated in terms of amplitude in dB and frequency expressed on the Barks scale, as opposed to the ERB used in technical contexts (here, amplitude is typically expressed on a linear scale, and frequency in Hertz). In this first approach, ERB is calculated from the graphical representation of the amplitude spectrum in dB, scaled in such a way that the maximum is set to 50 dB (see Section 5.2.2.1, Figure 5.10). This choice was made to exclude lower parts not contributing to the passband. The ERB is then calculated as

$$ERB = \frac{\text{area}(curve)}{\max(curve)}, \tag{6.14}$$

considering only those parts of the curve, that lie above 0 dB. The center-frequency in Bark z_c' is calculated as the center of gravity of the surface. From these intermediate parameters, the desired parameters f_c and z_{bw} are obtained as

$$z_{bw} = ERB \tag{6.15}$$

$$f_c = \sqrt{T_{Bark \to Hz} \left(z_c' - z_{bw}/2 \right) \cdot T_{Bark \to Hz} \left(z_c' + z_{bw}/2 \right)}, \tag{6.16}$$

Table 6.3: Provisional settings for WB Equipment Impairment Factors *Ie,wb* of WB codecs on the WB *R*-scale [0, 129] (ITU–T Rec. G.113, Appendix IV, 2006).

Codec type	Reference	Operating rate kbps	*Ie,wb* value
ADPCM	G.722	64	13
		56	20
		48	31
Modified Lapped Transform Coding	G.722.1	32	13
		24	19
CELP	G.722.2	23.85	8
		23.05	1
		19.85	3
		18.25	5
		15.85	7
		14.25	10
		12.65	13
		8.85	26
		6.6	41

Table 6.4: Provisional settings for WB Equipment Impairment Factors Ie,wb of NB codecs on the WB R-scale [0, 129] (ITU–T Rec. G.113, Appendix IV, 2006).

Codec type	Reference	Operating rate kbit/s	Ie,wb value
PCM	G.711	64	36
ADPCM	G.726, G.727	40	38
	G.721(1988), G.726, G.727	32	43
	G.726, G.727	24	61
		16	86
LD-CELP	G.728	16	43
		12.8	56
CS-ACELP	G.729	8	46
	G.729-A + VAD	8	47
VSELP	IS-54	8	56
ACELP	IS-641	7.4	46
QCELP	IS-96a	8	57
RCELP	IS-127	8	42
VSELP	Japanese PDC	6.7	60
RPE-LTP	GSM 06.10, Full-rate	13	56
VSELP	GSM 06.20, Half-rate	5.6	59
ACELP	GSM 06.60, Enhanced Full Rate	12.2	41
ACELP	G.723.1	5.3	55
MP-MLQ	G.723.1	6.3	51

where $T_{Bark \to Hz}(z)$ is the Bark-to-Hz-scale transformation as approximated by Equation (1.2), Section 1.1.2.1.

The bandwidth impairment model proposed here is currently being submitted to ITU-T SG12 for the discussions on extending the E-model toward WB transmission and linear frequency distortion.

Based on the extension of the R-scale to cover WB transmission, WB Equipment Impairment Factors (Ie,wb) have been derived in Section 5.3. As stated earlier, the extension value of 129 has recently been adopted by the ITU-T (ITU–T Rec. G.107, Appendix II, 2006). Following a procedure similar to the one discussed in Section 5.3, WB Equipment Impairment Factors (Ie,wb) have been determined recently for different WB as well as NB codecs (see Möller et al., accepted for publication, 2006). Corresponding values have been proposed to SG12 of ITU-T and can now be found in ITU–T Rec. G.113, Appendix IV (2006). For completion of this chapter, these Ie,wb-values are shown in Tables 6.3 (WB codecs) and 6.4 (NB codecs).

7

Summary and Conclusions

The range of quality that may be perceived with Voice over Internet Protocol (VoIP) networks is a wider one than that known from traditional telephone networks: The quality degradation due to packet loss may be considerable, but similarly high is the potential improvement due to wideband (WB) transmission.

With this book, an attempt has been made to bridge the gap between the VoIP-network world on one hand, and speech-quality perception by the users on the other hand. To this aim, the book first provides an overview of speech and speech quality perception (Chapter 1), of the methods that can be employed for speech quality assessment (Chapter 2), and of the effects on speech quality that can be encountered with VoIP networks (Chapter 3).

The quality of speech transmitted across a VoIP network is not only of multidimensional, but may also be of time-varying nature. Speech, in turn, owes its main functionality to its time-varying nature. Consequently, the presence of time-varying degradations on a speech signal may considerably impact the perception process. In the context of time-varying degradations, the notion of different timescales has proven to be a useful tool for describing the related perceptual effects (see Chapter 3): *Microscopic* refers to the short-term loss behavior. Depending on the type of microscopic loss, it may impact the performance of loss concealment strategies employed by a decoder, but also be responsible for the degradation of fundamental properties of the transmitted speech signal. For example, when the durations of degraded passages reach the average duration of phonemes, speech intelligibility may be reduced (Watson, 2001). An exact definition of the timescale for the microscopic loss behavior is possible only by differentiating it from the *macroscopic* loss behavior: Macroscopic behavior is related to the variation of perceived quality. In the instantaneous rating of speech quality under packet loss, for example, using a slider, variations of the loss rate are reflected only with some 'latency' by the subjects (Gros and Chateau, 2001). This latency seems to be an appropriate measure of the macroscopic loss timescale, since quality-variations that occur faster than this latency do not lead to time-varying quality judgments (a lower bound of this latency seems to lie between five and eight seconds). The two terms can best be understood when the possible rating strategy of subjects is taken into consideration (Gros and Chateau, 2001): Instead of evaluating lower level auditory properties such as loudness or signal-to-noise level, in case of packet loss, subjects seem to integrate over the number of impairment events perceived per unit time. The microscopic behavior then determines the perceptual strength of the impairment events, while the macroscopic behavior determines the temporal variation between different levels of microscopic behavior.

This classification has been employed to refer to different types of packet loss, both for loss simulation and for quality modeling (see Chapter 4). Starting from the simplest case of microscopic loss, namely, random loss, auditory test results were reviewed to define an appropriate parameter-based model of speech quality or quality impairment under packet loss. A first approach was described that quantifies the impairment of speech quality under random loss. The model was further refined to be applicable to dependent packet loss that can be described by a 2-state Markov chain. Apart from the (overall) packet loss percentage *Ppl*, the model uses only two additional parameters: One can be interpreted as the robustness of the codec and the loss concealment against packet loss, and the other as a factor that quantifies the tendency of the network to show consecutive packet loss (relative to the random case). The advantage of this approach is that the random formula can be regarded as a special case of this more general model for quality under dependent loss. In a series of conversation and listening tests, the model proved to be applicable even to the more general prediction of speech quality under macroscopic, i.e. strongly time-varying loss. Therefore, the overall loss rate of the connection has to be calculated, as well as the correlation of losses during phases of packet loss. In spite of its too optimistic predictions for severe levels of macroscopic packet loss, its overall performance was slightly better than that of the more complex investigated models, which carry out an implicit or explicit weighting of impairment events. Because of its simple appliccability to network planning, our model has recently been adopted by the ITU-T in an update of the E-model (ITU–T Rec. G.107, 2005).

In conversation tests on macroscopic, i.e. strongly time-varying packet loss and thus strongly time-varying speech quality, it was revealed that quality was affected considerably during periods of high loss. This effect is assumed to be caused by a significantly reduced speech intelligibility during such periods. When additional degradations were presented, packet loss clearly was the dominating degradation type. In case identical tests were carried out with random instead of strongly time-varying loss, the impact of packet loss was less dominant. At this point, another property of a distinctly macroscopic loss behavior is revealed: Since it is linked to strong quality variations, it seems to attract the entire attention of the test subjects (or users). In turn, in all observed cases the quality related to conditions impaired by combined degradations was higher than to be expected from from the ratings obtained when the degradations were presented in isolation. The underlying interactions between the degradations are thought to be due to masking in case of packet loss and noise, and due to higher-level perceptual processes in case of packet loss and talker echo.

The observations can be incorporated into the framework of quality as a multidimensional property: In his multidimensional analysis of the speech transmission quality in mobile communications, Mattila (2002a) found interactions between several quality dimensions. In turn, both Mattila (2001) and Hall (2001) have reported that the multidimensional feature spaces they had found, can *linearly* be mapped to speech-quality ratings with high precision. The latter result implies perceptual independence between the dimensions with regard to their role for quality. In a later study, Mattila (2002b) has reported interactions of the observed dimensions. In case of the conversation tests reported in this book, such an interdependence is also observed.

The conversation tests can also be interpreted in terms of the additivity assumed by the E-model: It claims that impairments that are appropriately transformed onto a perceptual quality scale are additive. On a low level of the algorithm, several of the E-model input parameters interact: They are interrelated to derive intermediate values and ultimately impairment factors. For example, the loudness rating in send direction not only considers the actual loudness, but also impacts the model settings for the echo attenuation. So far, the E-model handles the impairment under packet loss as completely independent of all other input-parameters to the model. This book shows that additivity may well hold, that is, that the overall impairment can possibly be partitioned into individual impairment factors. However, for packet loss the E-model also makes the assumption of separability, that is, that the parameters contributing to the different additive components do not interact with each other, which clearly does not hold.

Almost all model predictions that deviate from the test results obtained in this book *underestimate* quality. Hence, from the two possible errors that can be made during network planning, the more problematic *overestimation* of quality seems to be excluded to a large extent.

The quality of telephone speech can considerably be improved over that known from the traditional narrowband (NB) case, when WB transmission is introduced (Chapter 5). In terms of the mean opinion score (MOS) (see ITU–T Rec. P.800, 1996) that ranges from 1-5, the improvement may be higher than 1.5 points. If expressed on the E-model transmission rating factor scale, the *R*-scale (ITU–T Rec. G.107, 2005), an advantage of up to approximately 40 points, with a more realistic score of approximately 29 points ($\equiv$ 29%), may be obtained (relative to a maximum of 100 for the NB version of the model). This latter value has now been adopted by the ITU-T in ITU–T Rec. G.107, Appendix II (2006). Wideband transmission is typically achieved using WB speech coders. In this book, it is shown that these codecs may deliver almost the entire WB quality advantage to the user, and this already at transmission rates as low as around 20 kbps (as opposed to e.g. 64 kbps for the log. PCM employed in ITU–T Rec. G.711, 1988). Whether the improvement related to WB transmission is perceived and adequately recognized by users depends on the employed user interface. Both the acoustical and the nonacoustical properties of the user interface play a role for perceived quality. Relative to the acoustical properties, the nonacoustical ones are of much less importance. From the tests described in this book it becomes apparent, that the impact of the frequency response not only affects the perceived features of the transmission, but seems to impact also the desired features: Differences between user interfaces could be observed for quality of bandpass (BP) filters deviating from the traditional overall telephone frequency response, and disappeared when this traditional 300–3400 Hz bandwidth and frequency shape were employed. To date, no WB reference system has been defined that compares with its NB counterpart IRS (Intermediate Reference System, see ITU–T Rec. G.712, 1992; ITU–T Rec. P.830, 1996), neither in terms of the represented systems, nor in terms of the perceptual (*tele*-communication) experience on the side of the user.

For listening tests on speech-transmission quality under bandwidth restrictions, we developed a WB handset ('hi-fi phone') that on one hand enables appropriate coupling of all presented frequency components to the user's ear, and on the other hand seems to be perceived as being a *telephone* interface. The latter was investigated in a listening test. In this test, acoustically similar devices were used for comparison. It was found that the particular WB handset was rated best for the NB condition, and lower than diotic headphone presentation for the WB condition. The test indicates that diotic headphone representation is bound to a higher expectation, leading to more critical judgments when the acoustic conditions are bad (low bandwidth). In turn, the subjects judged the quality under diotic listening as higher than under monotic listening – both with the same headphone used in diotic listening, and with the hi-fi phone.

A detailed listening test on the quality perception under bandwidth restrictions provided evidence that for a given center frequency of the BP the bandwidth measured in Bark is linearly related to perceived speech quality. In other words, different bands seem to contribute to a similar extent to speech quality. The test further indicates a direct dependency of bandwidth impairment on the center frequency chosen for the BP. The optimum center frequency was found to be bandwidth-dependent. A parametric impairment prediction model was developed reflecting this behavior. A first way of applying the model to transmission spectra of arbitrary shape has been shown. Therefore, the same parameters of bandwidth (Bark) and center frequency (Hz) initially employed for the impairment by BP filters of flat spectra in the passband has been used. According to the employed method, a perception-based equivalent rectangular bandwidth (ERB) is calculated for the BP filter under consideration. Instead of making calculations in the linear amplitude and frequency domains, as it is the case for technical ERB-calculation, the Bark-scale and a log-amplitude representation were employed. Thus, the simplified method vaguely resembles the way in which excitation levels are obtained in auditory research (Zwicker and Fastl, 1999, pp. 164–171).

Finally, the extensions of the E-model we propose in this book have been briefly summarized, to serve as a quick overview to readers interested in planning VoIP networks.

The main achievements of this book are as follows:

- A broad overview has been provided on the quality elements and quality features related to VoIP.

- A large number of listening and conversation tests conducted by the author are described, which aim at the quality degradation due to time varying distortion on one hand, and at the quality impact of bandwidth extensions or reductions on the other hand.

- A parametric model of speech quality under random and dependent packet loss has been developed. Its predictions are shown to be in very good agreement with auditory test results. We have compared the modeling approach to alternative algorithms described in the literature, which it outperforms for the purpose of general speech-quality predictions for network planning. The model is now included in the E-model (ITU–T Rec. G.107, 2005).

- In conversation tests, speech-quality perception under combined 'stationary' and time-varying distortions has been investigated. It was revealed that different degradations interact with each other in their impact on quality. For a selected combination of degradations, the interaction was implemented in an improved version of the E-model.

- The quality improvement to be achieved with WB transmission has been quantified.

- Based on the estimate of the expected improvement, the degradation due to bandwidth limitation was investigated. A parametric model has been set up that permits quality estimates based on the magnitude frequency response of the transmission system. Moreover, a first proposal was made for quantifying the impairment due to WB speech coding.

8

Outlook

The auditory tests carried out in the course of this work were all utilitarian type of tests, i.e. tests focussing on integral quality judgment. Future research has to focus on analytical tests, in order to better understand the relation between the multidimensional nature of speech quality on one hand, and its time-varying behavior on the other. In this respect, it will be of particular interest to investigate, how single features, and not quality, are perceived instantaneously. This way, the features that determine the instantaneous quality rating of different types of degradations can be revealed. The knowledge of the instantaneous perception of individual features and of quality can be exploited for speech quality prediction in case of time-varying distortions. Note that time-varying network characteristics will be of even higher importance, when 4th generation (4G) mobile networks will evolve: To be always connected, the user is handed over between different types of networks of possibly differing quality features.

The perceived nature of transmitted speech has attracted the main interest in past speech-quality research. However, it has been shown that the desired nature or composition is just as important for the appreciation and – ultimately – acceptance of a service. Future research conducted into speech quality perception has to take the role of the desired nature into consideration. Examples of questions that need to be addressed are: How can the desired nature be measured? What are the quality elements that impact the desired nature, and how can they be controlled? What is the desired nature of a particular service or sound? How can the role of the desired nature for speech quality perception be quantified to serve for quality modeling?

With audio and multimedia over internet protocol (IP), stereo or even multiple audio streams are conceivable at low cost. This opens the door for advanced speech communication services involving virtual reality technology like spatial audio rendering in multiparty connections. Such services will be a leap forward in the paradigm shift away from narrowband telephony and reach far beyond the currently discussed wideband speech. Spatial rendering of multiple interlocutors in a virtual speech chat room application or conference scenario aids the auditory system to exploit its remarkable capability of speech source segregation in multitalker situations, which is typically referred to as the Cocktail Party effect (e.g. Bronkhorst, 2000). While a large amount of information is available on the

Speech Quality of VoIP: Assessment and Prediction   Alexander Raake
© 2006 John Wiley & Sons, Ltd

effect of different acoustic conditions on speech intelligibility, only little is known on user expectation and speech-quality perception related to such services, especially when speech is transmitted across voice over internet protocol (VoIP) links.

An important factor in this respect is the role of additional modalities: The convergence of voice and data networks enables multimedia services that may coexist with current voice services. The user may switch back and forth between different perception modes, one of which is 'telephone' speech. The perceived and desired nature related to all perception modes will impact each other in the integral quality associated with the entire service, but also with individual service options. A first example of multimodal communication services is videoconferencing. It currently seems to be (re-) born with the new possibilities offered by voice and media over IP software and services.

Another topic of relevance for future research is related to the effect of *affect* on speech quality in telecommunications. Up to now, ways to assess speech quality as it results in the unguided context typical of the everyday usage of telephone services have not gained considerable attention. However, affect may have to be distinguished from the analytic description of speech quality obtained for example from multidimensional scaling experiments: Affective reactions and the resulting quality may not necessarily rely on an 'analytic', dimension-based internal representation of the percept, as it is argued for example by Västfjäll (2003) for aircraft sound quality. This aspect may be of special importance for speech quality in networks like VoIP, for example, when it comes to individual, bad and strongly time-varying degradations, which may be associated with a high degree of annoyance. Future research has to focus on alternative quality assessment methods that leave behind the guided (and thus conscious) tests currently employed, and on ways to quantitatively relate network properties to affect and performance metrics.

A

Aspects of a Parametric Description of Time-Varying Distortion

A.1 4-state Markov Chain: Sojourn in Bad and Good State

An N-state Markov chain is defined by a probability vector $(p_1, \ldots, p_N)$, and a stochastic matrix $\mathcal{P} = (p_{ij})$. The probability vector contains the initial probabilities of being in state $1, \ldots, N$, and the matrix $\mathcal{P}$ expresses the transition probabilities between the states i and j (e.g. Behrens, 2000, p. 19). The transition probability matrix for the 4-state Markov chain depicted in Figure 3.7 is given by Equation (A.1)[1]:

$$\mathcal{P} = \begin{pmatrix} p_{11} & 0 & p_{13} & p_{14} \\ 0 & p_{22} & p_{23} & 0 \\ p_{31} & p_{32} & p_{33} & 0 \\ p_{41} & 0 & 0 & p_{44} \end{pmatrix}. \tag{A.1}$$

Following the matrix notation, the probabilities that the states $1, \ldots, N$ are occupied after k steps can be calculated as

$$\left(p_1^{(k)}, \ldots, p_N^{(k)} \right) = (p_1, \ldots, p_N)\, \mathcal{P}^k \tag{A.2}$$

$$= (p_1, \ldots, p_N) \begin{pmatrix} p_{11} & p_{12} & \cdots & p_{1N} \\ p_{21} & p_{22} & \cdots & p_{2N} \\ \vdots & \vdots & & \vdots \\ p_{N1} & p_{N2} & \cdots & p_{NN} \end{pmatrix}^k.$$

---

[1] A method for estimating the transition probabilities from captured network packet traces was proposed by Clark (2001).

Speech Quality of VoIP: Assessment and Prediction   Alexander Raake
© 2006 John Wiley & Sons, Ltd

If the Markov chain is ergodic[2], the power to k of the transmission matrix, $\mathcal{P}^k$, converges to a matrix $\mathcal{W}$ with increasing k. $\mathcal{W}$ is composed of N identical row vectors $\pi^T = (P_1, P_2, \ldots, P_N)$ (e.g. Behrens, 2000, pp. 50f.):

$$\lim_{k \to \infty} \mathcal{P}^k = \mathcal{W} = \begin{pmatrix} P_1 & P_2 & \cdots & P_N \\ P_1 & P_2 & \cdots & P_N \\ \vdots & \vdots & & \vdots \\ P_1 & P_2 & \cdots & P_N \end{pmatrix} \tag{A.3}$$

The row vectors π^T represent the *equilibrium distribution* of the Markov chain: The elements P_i correspond to the overall probabilities for occupying states i, $i \in \{1, 2, \ldots, N\}$, here referred to as state probabilities. For a given transition probability matrix P, the equilibrium distribution π^T can be calculated by solving the Eigenvalue problem according to Equation (A.4):

$$\pi^T \cdot \mathcal{P} = \pi^T, \text{ with} \tag{A.4}$$

$$\sum_{i=1}^{N} P_i = 1. \tag{A.5}$$

In case of the 4-state Markov model shown in Figure 3.7, the resulting state probabilities P_i are given by Equations (A.6)–(A.9):

$$P_1 = \frac{p_{23} p_{31} p_{41}}{D} \tag{A.6}$$

$$P_2 = \frac{p_{13} p_{32} p_{41}}{D} \tag{A.7}$$

$$P_3 = \frac{p_{13} p_{23} p_{41}}{D} \tag{A.8}$$

$$P_4 = \frac{p_{14} p_{23} p_{31}}{D}, \tag{A.9}$$

with

$$D = p_{23} p_{31} p_{41} + p_{13} p_{32} p_{41} + p_{13} p_{23} p_{41} + p_{14} p_{23} p_{31}. \tag{A.10}$$

In this model, packets are lost both in states '3' and '4'. Hence, the overall loss rate can be obtained by adding the corresponding state probabilities, see Equation (A.11):

$$p_{pl} = P_3 + P_4 = \frac{p_{13} p_{23} p_{41} + p_{14} p_{23} p_{31}}{p_{13} p_{23} p_{41} + p_{13} p_{32} p_{41} + p_{23} p_{31} p_{41} + p_{23} p_{31} p_{14}} \tag{A.11}$$

The equilibrium distribution is directly related to the expectation of the number of steps necessary to get from state i to state i, that is to return to the same state.

---

[2] An ergodic Markov chain is irreducible and aperiodic. An irreducible chain is one in which all states communicate, i.e. of which each state is occupied after some time, regardless of the starting point. A chain is aperiodic if no period (i.e. common divisor) greater than 1 exists in the different numbers of steps after which a starting point of the chain may be reached again, regardless of the starting point. For example, if a particular state is reached again after any of $0, 2, 4, 6, \ldots$ steps, its period is 2, and the chain is not aperiodic (see e.g. Behrens, 2000, pp. 23–52).

(Behrens, 2000, pp. 54f):

$$\mu_{ii} = \frac{1}{P_i} \tag{A.12}$$

In a more general fashion, the expectation of the number of steps necessary for a transition between states i and j to occur can be expressed as (Behrens, 2000, pp. 55ff):

$$\mu_{ij} = \frac{(\delta_{ij} - Z_{ij} + Z_{jj})}{P_j}. \tag{A.13}$$

Here, δ_{ij} is the Kronecker-symbol, with

$$\delta_{ij} = \left\{ \begin{array}{ll} 1 & if \quad i = j, \quad i, j \in \mathbb{N} \\ 0 & if \quad i \neq j. \end{array} \right. \tag{A.14}$$

Z_{ij} and Z_{jj} refer to elements of the Matrix

$$\mathcal{Z} = (Id - (\mathcal{P} - \mathcal{W}))^{-1}, \tag{A.15}$$

with $\mathcal{P}$ and $\mathcal{W}$ according to Equations (A.1), and (A.3), respectively. Id is the identity matrix, corresponding to δ_{ij}. Obviously, setting $i = j$ in Equation (A.13) leads to $\mu_{jj} = \frac{1}{P_j}$, in accordance with Equation (A.12).

Now, for the 4-state Markov model of Figure 3.7, with $i, j = 1, 2, 3, 4$ in Equations (A.13) to (A.15), the average numbers of steps the chain stays in the 'good' and the 'bad' state can be calculated by solving Equation (A.13) for μ_{13} (sojourn in 'good' state) and μ_{31} (sojourn in 'bad' state). The resulting average numbers of steps are shown in Equations (3.18) and (3.19)

A.2 Impact of Other Quality Elements on Packet Loss

A.2.1 Forward Error Correction

As stated previously, examples for realizing forward error correction (FEC) are the usage of parity check codes (IETF RFC 2733, 1999) or Reed-Solomon codes (Jiang and Schulzrinne, 2002; Rizzo, 1997).

According to such methods, a number k of packets is summarized to an FEC-block (i.e. handling the large information units of packets similar to bits as in the 'classical' application of linear block codes). $(n - k)$ redundant FEC data packets are added to the FEC block, leading to an overall of n data units per FEC block. If at least k data units of a particular block were successfully received, up to $n - k$ losses can be recovered. A block code according to this description is typically referred to as (n, k) code. In IETF RFC 2733 (1999), some methods for generating and adding redundant data are illustrated, which allow single or even multiple consecutive losses to be recovered. The latter, however, is achieved at the expense of considerable additional delay. For example, two packet sequences $A\ B$ and $C\ D$ may be added with FEC data based on a $(3, 2)$ Reed-Solomon Block code using the bit-wise XOR operation $\otimes$: $FEC(A, B) = A \otimes B$. Similarly, $FEC(C, D) = C \otimes D$, and

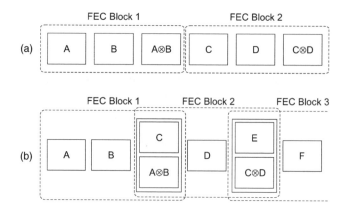

Figure A.1: Examples of FEC implementations, (3, 2) Reed-Solomon code. (a): Additional FEC packets are sent (from IETF RFC 2733, 1999). (b): Piggybacking of FEC-data of one FEC-block onto packets of the next FEC block (example from Jiang and Schulzrinne, 2002).

the resulting data stream could be realized as two subsequent FEC blocks, as shown in Figure A.1(a).

If packet A is lost, it can be recovered as

$$A = B \otimes FEC(A, B) = B \otimes (A \otimes B).$$

The amount of additional data to be sent with FEC can be reduced, when the redundant packets are combined with the actual data packets, as less header data is necessary in this case. An example is the *piggybacking* of packets of one FEC block onto data packets of the next FEC block (Figure A.1(a). See Jiang and Schulzrinne, 2002, for more details).

Some information is available from the literature on the translation of network packet loss into residual packet loss due to FEC:

- A formula for converting Bernoulli, that is random packet loss into the effective loss in case of (n, k) type FEC without piggybacking is derived in Rosenberg *et al.* (2000).

- The loss distribution resulting from 2-state Markov loss with (n, k) type FEC and no piggybacking is calculated in Frossard (2001).

- For $(3, 2)$ type FEC with piggybacking, Jiang and Schulzrinne (2002) provide a formula for calculating the overall loss probability p_{FEC} under 2-state Markov loss. It reveals that FEC is increasingly sensitive toward packet loss with an increasing number of consecutively lost packets $\mu_{10} = 1/q$.

A.2.2   Jitter and Jitter Buffer

In the Internet, jitter itself was observed to consist of *spike*-like events superimposed on a delay varying less dynamically around its mean (Bolot, 1993; Jiang and Schulzrinne, 2000; Markopoulou *et al.*, 2002). Delay spikes are typically characterized by a more or less abrupt step in delay and subsequent (almost linear) decay back to a delay value near the mean delay (decay after a small number of packets, see Bolot, 1993; Markopoulou et al., 2002). In delay measurements in Internet backbone networks, Markopoulou et al. (2002) also observed periodic spike clusters, or interchanging plateaus of frequent and sparse clusters (e.g. depending on the provider of the network connection). While different models exist to describe the behavior of packet loss in Internet protocol (IP) networks (section 3.3.5), delay jitter models are not readily available. Hence, to investigate the effect of jitter buffers on packet loss and transmission delay, many authors have relied on packet traces collected from real network traffic (e.g. Rosenberg et al., 2000; Sun and Ifeachor, 2003).

Recently, some attempts were made to parametrically quantify the jitter distributions collected from packet traces. Already in the mid-90s, Internet traffic was observed to be *self-similar* (e.g. Crovella and Bestavros, 1996, 1997), that is to show similar properties regardless on the timescale it is regarded on. In other words, similarities exist between short-term dependencies and long-term dependencies of traffic events. Due to this property, delay variations in packet traces are sometimes modeled using Pareto distributions, which are considered suitable for modeling self-similar traffic (Ding and Goubran, 2003). Crovella and Bestavros (1996), however, found that the degree of self-similarity depends on the applied transport protocol: If the connection oriented transmission control protocol (TCP) is used, self-similarity can indeed be observed. In turn, when connectionless protocols like user datagram protocol (UDP) are applied, as it is the case for Voice over Internet Protocol (VoIP), the degree of self-similarity is considerably lower. This observation is complemented by a recent finding reported by Sun and Ifeachor (2004), who showed that the delay variations of real network traces can much better be modeled with a Weibull than with a Pareto distribution.

Different proposals were made to quantify the relation between jitter and the jitter-buffer implementations on one hand, and the resulting packet loss and overall transmission delay on the other hand (e.g. Cole and Rosenbluth, 2001; Jiang and Schulzrinne, 2000; Sun and Ifeachor, 2003, 2004). In general, the impact of the jitter buffer can be characterized as follows: If a fixed jitter buffer is applied, the introduced delay is constant. In case of individual packet delays exceeding the fixed buffer threshold, a considerable amount of packets may get lost. In turn, excessive loss can be avoided by adaptive jitter buffer implementations, which adjust the buffer size to the congestion situation of the network. In this case, small amounts of packet loss are achieved at the expense of potentially considerable transmission delay, for example, during periods of network congestion. Actual jitter buffer implementations may work entirely adaptive and try to achieve an optimal trade-off between packet loss and delay (e.g. Ramjee et al., 1994; Rosenberg et al., 2000).

To discuss this issue in more detail, the nomenclature provided in Ramjee et al. (1994) can be applied, as depicted in Figure A.2 (see Table A.1 for the definitions of the parameters). Here, it is assumed that the jitter buffer chooses a certain playout delay D_i for a given talkspurt i, with $D_i = \hat{d}_i + \mu \cdot \hat{v}_i$. Here, $\hat{d}_i$ is the mean delay, and $\mu \cdot \hat{v}_i$ is a certain safety margin related to the expected delay variation $\hat{v}_i$. Both values are typically estimated by the jitter buffer based on the jitter monitored on previously received packets.

Then, the probability that a packet j is received on time and thus played out is given by (Rosenberg et al., 2000):

$$p_R = (1 - p_n) \cdot P \left[n_j < D_i \right]. \tag{A.16}$$

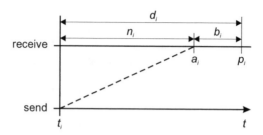

Figure A.2: Delay and jitter parameters (from Ramjee *et al.*, 1994). For the parameter definitions see Table A.1.

Table A.1: Variables used to describe VoIP network delay and its variation, as well as the delay introduced by the jitter buffer (see Figure A.2).

Variable	Description
t_i	Send time
a_i	Arrival time
n_i	Network delay
b_i	Jitter buffer delay
d_i	Playout delay

Here, p_n refers to the network packet loss probability, which is assumed to be equal and independent for all packets in this case (random loss). For the network delay as the random variable under consideration, the function

$$F(\hat{d}_i) = P\left[n_j < D_i\right] \tag{A.17}$$

is the delay cumulative distribution function (CDF). Following Equation (A.16), the total probability that a packet is lost can be calculated as

$$p_{pl} = 1 - p_R = 1 - \left\{(1 - p_n) \cdot P\left[n_j < D_i\right]\right\}. \tag{A.18}$$

To minimize the total loss probability, an adaptive jitter buffer will attempt to optimize D_i so as to render

$$P\left[n_j < D_i\right] = 1, \quad \text{with } D_i \text{ minimal.} \tag{A.19}$$

For the case of static jitter buffers, an estimate of the upper bound of buffer-induced packet loss can be found in Cole and Rosenbluth (2001). Typical adaptive jitter buffer implementations seek to adjust D_i to the current network situation, with different dynamic properties. Depending on the implementation, these may range from more conservative slow adaptation to approaches which accurately track delay spikes, switching back to less timely adjustment when the spike period is over (Ramjee *et al.*, 1994, for a performance analysis see Sun and Ifeachor, 2003).

An analytic description of packet loss due to jitter buffering was proposed by Sun and Ifeachor (2004), based on the finding that the jitter observed in real packet traces can be modeled using a Weibull distribution. In this particular case, the CDF is described by

$$F(D_i) = P\left[n_j < D_i\right] = 1 - e^{-((D_i-\mu)/\alpha)^\gamma}, \tag{A.20}$$

with μ chosen as the minimum of the network delay. α and γ are the so-called scale and shape parameters, respectively, which were determined by Sun and Ifeachor using maximum likelihood fitting of actual trace data. According to this approach, the overall packet loss can be determined as (again assuming random-type network loss):

$$p_{pl} = p_n - (1 - p_n) \cdot e^{-((D_i-\mu)/\alpha)^\gamma} \tag{A.21}$$

So far, only the jitter buffer impact on the overall loss rate was discussed. With regard to the loss distributions in real network traces before and after jitter-buffering, Jiang and Schulzrinne (2000) report a possible increase in loss burstiness[3] due to the merging of network bursts with the loss introduced by the jitter buffer (depending on its implementation). For example, an individual loss of packet j in the network and a discard of packet $j + 1$ by the jitter buffer results in two consecutively lost packets.

The situation is different, when sender-based error correction mechanisms like FEC are applied, both for overall loss rate and loss burstiness. With FEC, a particular packet j is played out if it is: (a) received before its playout time or (b) recovered before its playout time. In this case, the choice of D_i is much more dependent on the network loss p_n, as the reception of the restoring redundant data may possibly occur later than the expected arrival of the packet. Buffer algorithms taking the effect of FEC into consideration were proposed by Rosenberg *et al.* (2000), and their performance was investigated by Jiang and Schulzrinne (2000). Here, it was shown that (5,3) Reed-Solomon FEC reduces network loss, but may increase loss due to jitter buffering, if the two are controlled independently (as with buffer algorithms like in Ramjee *et al.*, 1994). In spite of almost perfect loss recovery by the FEC, the total packet loss was found to be still bursty, however, due to bursty discard by the jitter buffer. For the combined FEC and jitter buffering approach suggested by Rosenberg *et al.* (2000), the network loss can be reduced as well as discards by the buffer due to late packets be minimized, without introducing bursty packet discard.

A.3   Impairment under GSM Bit Errors

Another type of time-varying distortion closely related to that of packet loss is that of bit errors typical of GSM networks. As it is discussed in Section 3.3.6, bit errors on the GSM channel encoded speech data stream are translated into frame erasures and residual bit errors during channel and speech decoding: Only when bit errors hit the most important parts of the encoded data, frames are erased (see Section 3.3.6). The corresponding perceptual degradation mainly pertains to the quality dimension 'smoothness', which corresponds to one of the underlying dimensions of speech transmission quality under packet loss (Bernex and Barriac, 2002; Mattila, 2002a). In GSM, bit errors are caused by signal interference

---

[3]Bursty and burstiness here relate to the short-term behavior of packet loss, that is, a Burst Ratio greater than 1, see Equation (3.14).

on the radio link, which are expressed by the so-called carrier-to-interference ratio C/I [dB]. As part of the listening-only test conducted at IKA to assess the speech transmission quality under random interruptions (see Appendix G.1), a number of conditions with GSM-codecs under random bit errors were presented to the subjects. A parametric approach for predicting these test results by means of the E-model was proposed in Möller and Raake (2002):

$$Ie_{BER} = Ie + (95 - Ie) \cdot \left(\frac{BER^3}{500} - \frac{BER^2}{65} + \frac{BER}{x} \right), \qquad (A.22)$$

where Ie_{BER} is a new impairment factor to be considered in the basic function of the E-model, Equation (2.5). Ie is the codec-specific equipment impairment factor under error-free conditions, BER the bit error rate (%), and x a codec-specific parameter. In Möller and Raake (2002), a value of $x = 10$ was determined for the GSM-EFR, and of $x = 11.5$ for the GSM-FR.

B

Simulation of Quality Elements

In Chapter 3, packet loss, wideband (WB) transmission and user interfaces were identified as the quality elements, which we wanted to investigate in more detail in this book: both individually, and in combination with other types of quality elements, for which a relationship to perceived quality has already been established (such as echo or noise).

A telephone line simulation tool has been used throughout this work to implement all quality elements relevant to Voice over Internet Protocol (VoIP) quality (Fig. B.1). In the tool, the quality elements are implemented in a parametric way, reflecting the approach followed in this book, namely, to relate parametric descriptions of quality elements to perceived quality. The real-time tool enables simulating most quality elements of public switched telephone network (PSTN)/integrated services digital network (ISDN) as well as VoIP and mobile networks. With this tool, it was possible to conduct a large number of listening-only as well as conversation tests under controlled laboratory conditions.

It has the following general properties and abilities:

- Realistic reproduction of the acoustic signals resulting from the telephone connection, including all the relevant transmission paths (including sidetone and echo).

- Ability to adjust all the parameters of the transmission within the relevant limits (see Figure 2.8). The system is fast and relatively easy to use, so that connections with different transmission characteristics can be tested within one auditory test session.

- Real-time transmission, to be able to carry out conversation tests (except mobile network simulation).

- The same system can be used for both listening and conversation tests.

- Different user interfaces can be connected to the system.

Depending on whether a conversation or a listening test is to be conducted with the simulation system, different system configurations are typically employed. The configuration used for conversation tests is illustrated in Figure B.2(a), and that used in

Speech Quality of VoIP: Assessment and Prediction   Alexander Raake
© 2006 John Wiley & Sons, Ltd

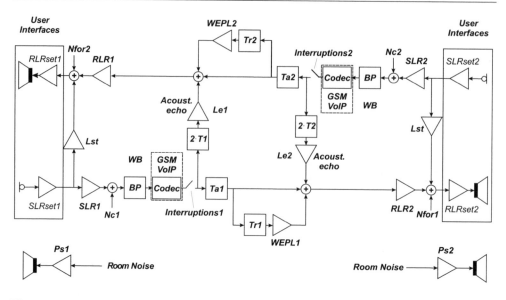

Figure B.1: Integral on line telephone line simulation tool. △: Programmable filters; □: Delay lines (T, Ta, Tr), channel bandpass (BP), external codecs, external GSM-simulation (off-line), and external VoIP simulation (on line) (see text for details; Möller, 2000; Möller and Raake, 2002; Raake, 2004; Rehmann *et al.*, 2002).

listening-only tests in Figure B.2(b). Amplifiers and programmable attenuators at the electric interfaces of the PSTN/ISDN core simulation serve for global level adjustments.

B.1 PSTN/ISDN

The core system of the simulation tool implements all the analog or digital PSTN quality elements taken into account by the E-model in its version from 2000 (Möller, 2000, see Figure 2.8, page 44; ITU–T Rec. G.107, 2000).

The system consists of a software-programmable digital signal processor (DSP) unit from Tucker Davis Technologies (TDT), a personal computer (PC) to program the DSPs, and DSP boards with commercial software to implement standard ITU-T codecs (aspi ELF-boards). The software carries out a direct coding-decoding process, so that no access to the encoded bitstreams is available. The codec-DSPs can be concatenated to realize codec-tandems. Also implemented on the codec-DSPs is a modulated noise reference unit (MNRU) according to ITU–T Rec. P.810 (1996). The narrowband (NB) channel bandpass (BP) filter defined in ITU–T Rec. G.712 (1992) was implemented. The corresponding frequency response and group delay distortion are shown in Section C.1, Figures C.1 and C.2.

The parameters that can be controlled using the simulating model for both conversation and listening-only tests are summarized in Figure B.1 (they are in accordance with the E-model parameters depicted in Figure 2.8, p. 44):

- Send and receive loudness ratings (*SLR*, *RLR*), including different frequency responses (flat; Intermediate Reference System (IRS) acc. to ITU–T Rec. P.48, 1989; IRSmod acc. to ITU–T Rec. P.830, 1996).

- Talker echo loss, delay and frequency responses of the echo path (EL, T).

- Listener echo loss and delay ($WEPL$, Tr).

- Overall propagation time (Ta).

- Sidetone path attenuation and frequency response (Lst).

- Stationary (Gaussian) white noise (Nc, $Nfor$).

- Codecs (e.g. G.711, G.726 (40, 32, and 24 kbit/s), G.728, G.729, IS-54, G.722)[1].

- Room noise at send and at receive side ($Ps1$, $Ps2$).

From this list and Figure B.1, it can be seen that all relevant speech paths are considered. Only the implementation of talker echo differs from real networks: To prevent a closed loop and thus instability, the echo signal undergoes only one coding-decoding cycle. With the programmable filters (represented by the triangles), different transmission characteristics and attenuation-settings can be realized. Noise is created over the band 0–16 kHz (Nc and $Nfor$), and is psophometrically weighted to determine the noise level (in dBm0p). With regard to the speech level on the line, the simulation system rests on the assumption that the overload point of digital networks is of 6 dBr; if it is further assumed that the speech level typically found in real-life networks can be approximated as $-11 - SLR$ [dB] (ETSI Technical Report ETR 250, 1996, p. 67), with $SLR = 8$ dB for standard connections, the average speech level lies 25 dB below the overload point of the digital system.

An actual call signaling is not implemented in the simulation tool. In conversation tests carried out with the system, the physical call setup and termination are taken care of by the experimenter: A subject places a call by pressing a button on the telephone, causing the

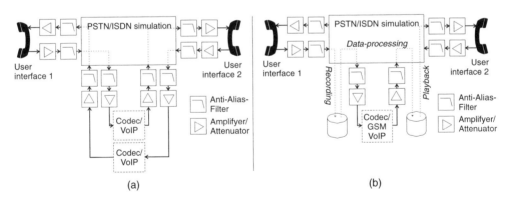

(a)   (b)

Figure B.2: Schematic illustration of the overall system configurations in case of a conversation test (a), and in case of the data processing for a listening test (b) (modified from Möller and Jekosch, 1998).

---

[1]EIA/TIA/IS–54 (1990); ITU–T Rec. G.711 (1988); ITU–T Rec. G.722 (1988); ITU–T Rec. G.726 (1990); ITU–T Rec. G.728 (1992); ITU–T Rec. G.729 (1996).

other subject's phone to ring; the experimenter follows the conversation test via headphones and can establish and stop the physical connection accordingly in a timely manner.

B.1.1   Instrumental Verification

The PSTN/ISDN-simulation system was instrumentally validated by two approaches:

(a) In an industry-funded project, the simulation system was used to verify the measurements and quality predictions made by a commercial intrusive monitoring tool for global system for mobile communications (GSM) networks (Raake and Möller, 1999). The measurements made by this tool delivered implicit information on the accuracy of the simulation settings.

(b) All monitoring measurements made under (a) were counterchecked by independent instrumental measurements.

In the verification, all parameter values were found to be in agreement with the desired settings.

B.2   Mobile Networks

The simulation tool implements degradations typical of GSM networks, such as time-varying degradations due to bit errors on the (source and channel-) coded speech data. These characteristics could not be implemented on the core PSTN/ISDN system or the external DSP unit otherwise used for coding/decoding. As an efficient implementation for auditory tests, a twofold approach was chosen:

(a) An off-line implementation of the GSM-codecs to process speech samples for listening-only tests (GSM-FR, GSM-EFR, and GSM-HR; ETSI GSM 06.10, 1988; ETSI GSM 06.20, 1996; ETSI GSM 06.60, 1996). The off-line GSM-codecs had been made available to us by Tektronix Padova SpA. The implementations include full-rate channel coding and discontinued transmission (DTX), and an additional program for introducing random bit errors on the encoded data stream. The codecs can be used instead of the external codecs of the core PSTN/ISDN-simulation (see Figure B.1).

(b) For conversation and listening tests, the on-line insertion of interruptions was chosen as an efficient reference type of time-varying distortions. The interruptions are performed using a cosine-switch, which is triggered by a wave signal created beforehand using standard signal processing software. The use of the cosine switch allows the selection among different switching options, such as $\cos^{2n}$, linear ramps and abrupt switching, where the rise time (from $0.1 \cdot max$ to $0.9 \cdot max$) can be chosen between 0.5 and 99 ms, according to the required interruption shape. The wave signal can be designed freely, and this way, the distribution, occurrence rate and length of the interruptions individually chosen. The interruptions can be inserted at different points in the simulation system. To prevent interruption of the noise floor $Nfor$, and to reflect the realistic situation that (acoustic) talker echo is affected by possible interruptions too, the point behind the codec was chosen for the interruption insertion (see Figure B.1).

B.3 VoIP

The VoIP system is integrated into the core PSTN/ISDN-simulation described in Section B.1.

Different options exist for the combination of the VoIP and the PSTN/ISDN system. Depending on the envisaged network configuration (e.g. PSTN–VoIP–PSTN or VoIP–PSTN–VoIP, see Section 3.1, Figure 3.2), the VoIP system can be placed at one of the outer edges of the PSTN/ISDN system, or embedded between PSTN/ISDN trunks. In this work, the latter option has been exclusively adopted. This choice was made in order to reflect the frequent situation that users are connected to the telephone network via a PSTN link, while subsequent parts of the network are realized based on VoIP – without the user being aware of it. In this case, the VoIP simulation is integrated into the PSTN/ISDN-simulation (Figure B.1) by replacing the external 'Codec' modules of the tool with the VoIP simulation module, as outlined in Figure B.3. In real life, this situation may perceptually differ from others such as VoIP–PSTN–VoIP only in the interface(s) available to the user.

The VoIP system consists of three types of components (Rehmann, 2001):

(a) The interfaces carrying out the analog-to-ISDN signal conversion.

(b) The VoIP gateways, which provide the speech compression, packetization and routing.

(c) The Internet protocol (IP) network simulation, introducing IP-typical degradations like jitter or packet loss.

B.3.1 Analog to ISDN Conversion

Transformation of the four-wire[2] analog speech signal into ISDN is performed by two A2P modules provided by T-Systems, Berlin, Germany. The A2P modules are implemented as

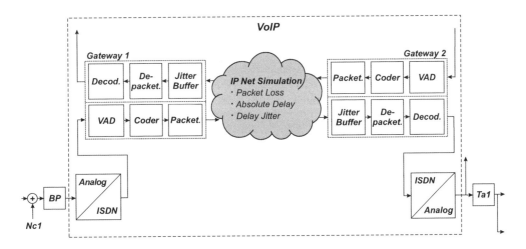

Figure B.3: VoIP simulation tool (Raake, 2004; Rehmann *et al.*, 2002).

---

[2]Two for each direction to avoid line echoes.

two PCI cards mounted in a personal computer. The cards convert the analog speech signals into logarithmic Pulse Code Modulation (PCM) coded data (A-law) according to ITU–T Rec. G.711 (1988). Initially, the A2P cards provided the microphone interfaces. To yield the impedance matching necessary for loss-less interconnection with the PSTN/ISDN-simulation, the microphone interfaces were transformed into line-interfaces. The A2P cards are equipped with a Siemens DSP ARCOFI PSB 2163. Apart from signal conversion, the A2P modules also perform the ISDN call signaling for communication with the VoIP gateways, which is employed during call buildup and release. The A2P cards are connected to the VoIP gateways by two ISDN basic rate interfaces (BRI). To setup a call, a multiple subscriber number (MSN) is dialed using one of the software dialers controlling the A2P cards. Via the two gateways, the call is routed to the second A2P card, so that a full-duplex VoIP connection is established.

B.3.2   VoIP Gateways

The gateway components of the VoIP simulation are responsible for packetizing and routing the ISDN speech signal, and for initially building up a VoIP call according to the H.323 protocol (ITU–T Rec. H.323, 1998). They are implemented using two Cisco 2611 routers, each equipped with an ISDN BRI voice port and an Ethernet port. They are connected via 10BaseT Ethernet to a PC running the IP network simulation. A VoIP connection can be established by dialing a certain number at the dialer software of the A2P cards. The gateways were programmed in such a way that different multiple subscriber numbers can be chosen, which are each associated with a particular connection configuration. Depending on the called line, both the calling and the called router are used with corresponding settings for the codec, the packet size, voice activity detection enabled/disabled, and so on. (Krebber, 2002). This way, different conditions can easily be selected in the course of one test session.

The codecs supported by the VoIP gateways are summarized in Table B.1. A proprietary implementation of voice activity detection is generally made available by the gateways for

Table B.1: Codecs supported by the Cisco 2611 gateways. Information is provided on the bit-rate, the frame size or minimal packet size (some codecs like the G.711 operate sample-wise; here, the minimal packet size is specified), the packet loss concealment algorithm (PLC), the availability of voice activity detection (VAD) implemented in the encoder, and the transmission delay the VoIP simulation introduces for the minimum packet size.

Codec	Bit-rate [kbit/s]	Frame Size/ Minimum Packet Size [ms]	PLC	VAD	Minimum Delay [ms]
G.711	64	10	Proprietary	–	56
G.723.1	5.3; 6.3	30	Native	Annex A	107–134
G.726	16; 24; 32	10	Proprietary	–	59
G.728	16	10	Native	–	56
G.729A	8	10	Native	Annex B	50
GSM-EFR	12.2	20	Native	Internal	52–82
GSM-FR	13	20	Native	–	73

all codecs. Since the algorithm was not disclosed and was operational only in one direction, it was not used in the tests carried out for this book.

B.3.3   IP Network Simulation

The main focus of this book is on conversational speech quality. Consequently, the network simulation used had to operate in real time (note that since the employed software tools do not in all aspects mimic an actual network, the term 'network simulation' is used here rather than 'network emulation'). Since time-varying distortions were one of the two main classes of quality elements aimed at, another requirement was the ability to cover a broad range of different loss or jitter distributions. Because of the fact that in real networks jitter buffers translate jitter into additional absolute delay and packet discard (i.e. packet loss, as discussed in more depth in Section 3.3.3), this requirement could be limited to the aspect of packet loss. Finally, this book aims to establish a link between parametric descriptions of quality elements and quality or quality features. Hence, a third and last requirement for the network simulation was the free control of the implemented loss or jitter models based on scalable parameters.

The above requirement was met by two of different publicly available software network simulations that were thus found suitable for this book.

B.3.3.1   NIST Net

An easily accessible and cost-efficient network simulation is the public domain software NIST Net (NIST Net, 2002). This simulation is often used in the framework of IP-component evaluation and testing, and has already been employed in VoIP speech quality tests (e.g. ETSI Tiphon 11 Temporary Document 64, 1999). It operates as a kernel module for the Linux operating system and is installed on a PC acting as IP router, which is equipped with two Ethernet cards. NIST Net allows controlled insertion of packet loss, packet doubling, delay and jitter into one or more of the IP routes configured on the router. Hence, in addition to the Microsoft (MS) Windows-based PC employed to steer the PSTN/ISDN-simulation, a second, Linux-based PC is used for the integrated simulation. It serves for both running NIST Net, and for configuring the Cisco VoIP gateways (via serial interface cables).

The 'functionality' of the simulators most important for this book was packet loss insertion. Two options for packet loss distributions are provided by NIST Net: Random loss (i.e. independent, uniform loss) or correlated loss (where an additional parameter defines the correlation between losses).

NIST Net can be controlled both in a command line mode and with a graphical user interface (GUI) it is distributed with. Since the GUI provided by NIST Net was found to be too unstable on the employed PC to be used for example in conversation tests, a Perl-based GUI was written, to permit quick and intuitive parameter changes between test conditions of a conversation test (Rehmann, 2001).

B.3.3.2   NetDisturb

With only random loss and correlated loss, NIST Net has some limitations in the choice of loss distributions. Hence, an alternative network simulation software was sought extending the available range of packet loss distributions. From all other tested network simulators, the commercial tool NetDisturb from ZTI, Lannion, France, was found to be the most flexible and scalable solution.

Apart from several already implemented packet loss and jitter distributions, it offers the opportunity to drop or delay packets on the basis of packet lists provided to the software as text-files. Such lists contain one entry for each packet, which specifies whether the packet is to be dropped or to be transported further (loss-files). This approach can be seen as a reversal of the collection of packet traces from real networks: Instead of observing whether a certain packet was lost, the text-files define whether the packet is to be lost. A typical representation is a list of zeros and ones (e.g. ...0001101101001...), where ones indicate loss, and zeros indicate the receipt of the respective packets.

Such text-files can be easily created offline, and thus a sheer unlimited range of packet-loss distributions can be realized, consisting both of artificially created traces as well as real packet traces collected in operating networks. For this book, random packet loss was inserted with NIST Net, and for more complex loss distributions, NetDisturb was applied. Using matrix algebra and 2-state and 4-state Markov models, different packet lists were created offline with MATLAB, using the parametric descriptions given in Section 3.3.5.1.

For an n-state Markov-model, the applied algorithm uses the $n \times n$ transition probability matrix $\mathcal{P}$ and the total number N of packets in the desired packet list as input parameters (see Appendix A). As output, it delivers a line vector T of length N, consisting of zeros and ones, which identify the lost and found packets. Additional selection criteria for the resulting packet trace can be specified, such as the error-limits in the overall loss rate defined by the trace. For example, if an overall loss rate of 5% is to be achieved with a given loss list, an error-limit can assure that for that trace the overall loss rate does not differ by more than 3% (relative to the 5%) from the envisaged rate.

In the VoIP simulation described here, NetDisturb runs on a MS Windows-based PC equipped with two network cards, which it renders transparent to crossing IP-packets. Hence, the corresponding network configuration differs from the one implemented with NIST Net: In the latter case, packet routing between the VoIP gateways is provided by the NIST Net Linux PC; since the PC running NetDisturb is transparent to IP-packets, it cannot provide routing.

Consequently, the integral PSTN/VoIP simulation system is realized with three PCs:

(1) The MS Windows-based control-PC of the core PSTN-simulation system, which also hosts the A2P cards and software used for ISDN call-setup and analog-to-ISDN conversion.

(2) The Linux PC running NIST Net, which serves as router between the two VoIP gateways.

(3) The 'transparent' MS Windows-based PC equipped with NetDisturb (merging the two MS Windows-PCs was prevented by the fact that different MS Windows versions were required by the external codecs of the PSTN-simulation on one hand, and by NetDisturb on the other hand).

B.3.4 Instrumental Verification

The VoIP simulation was subject to an extensive instrumental characterization and verification. Therefore, the system was connected to the core PSTN/ISDN-simulation as depicted in Figure B.1, to include level effects due to the interconnection. The main results are (Krebber, 2002):

- The delay introduced by the VoIP simulation in case of different codec–packet size combinations were determined based on the maximization of the cross-covariance function between the iteratively delayed original and the transmitted signal. The delays introduced by different codecs in conjunction with the entire VoIP simulation are listed in Table B.1 for the minimum packet sizes. For some codecs, like the G.723.1, a switching of the delay between different levels was observed, which may be due to router timing-drift.

- The attenuation due to the VoIP system as a function of the codec was determined based on three different methods:

    (1) Active speech level measurement according to ITU–T Rec. P.56 (1993).

    (2) Measurement of the amplitude spectrum of the transmission channel using a babble-noise signal, and calculation of the loudness rating according to ITU–T Rec. P.79 (1999). Note that this approach is a very limited one, when strongly nonlinear codecs are addressed.

    (3) In an auditory adjustment test, eight normal-hearing subjects recruited from the IKA staff were asked to adjust the attenuation of the channel under test to that of a known reference channel (based on perceived loudness, using the 'method of adjustment': in both directions, i.e. once with the reference considerably louder and once considerably lower than the system under test; see e.g. Blauert, 1997, pp. 14–22). The loudness loss due to the channel was derived from the relative attenuator settings, averaged over subjects and test direction.

The overall attenuation/amplification introduced by the VoIP system was adjusted to 0 dB using the amplifiers and programmable attenuators available at the interfaces of the core PSTN-simulation (see Figure B.2). When the levels obtained from the three methods (1)–(3) were in good agreement (i.e. did not differ by more than 1.5 dB), their average was taken as the appropriate setting. When the three levels differed considerably (i.e. by more than 1.5 dB, as in some cases of low bit-rate codecs), more weight was given to the level determined in the listening test, as auditory tests were the main focus of this work.

In the process of studying the suitability of different network simulation software for this work, the file-based loss insertion by NetDisturb has been verified instrumentally. Using the public domain packet-sniffer Ethereal, the packet trace of a certain connection was compared with the loss file imposed by NetDisturb. All the 10 artificial input packet lists (consisting of 50000 packets each) were identical to the corresponding list collected with the sniffer software.

B.4   Wideband Transmission

Operating of the system in a WB transmission mode (50–7000 Hz) currently is limited by the codec and VoIP-trunk implementations. For example, the interface between the VoIP-trunk and the PSTN-trunk was realized with an ISDN-card employing NB logarithmic PCM, see Section B.3.1. Consequently, we considered a WB-capable PSTN-simulation to be a good starting point for studying WB speech quality (Figure B.1).

With a cutoff frequency of 13.3 kHz, the hardware anti-alias filters of the system principally enable a larger transmission bandwidth than the 50–7000 kHz that was implemented for this book. The WB transmission channel BP filter was designed to meet the tolerance scheme defined in ITU–T Rec. G.722 (1988). The transfer characteristics and group delay distortion introduced by the filter are depicted in Figures C.3 and C.4.

When changing between the WB and the NB mode of the system, the signal and noise levels have to be adjusted accordingly. The adaptation of the signal levels is described in Section B.5. In the WB case, the noise levels were adjusted based on level information provided in ITU–T Rec. P.830 (1996), Annex A. All levels were verified instrumentally with an Ono-Sokki CF-6400 FFT-Analyzer and a psophometer (B&K Measurement amplifier Type 2610 and Wandel& Goltermann Universal-Network UN-1).

B.5   User Interfaces

The resulting system was adapted to different WB user interfaces. In case of WB telephone speech, the choice of user interface is important for two reasons (Section 3.3.14):

(i) Only user interfaces of sufficiently good electro-acoustic properties provide the full advantage of WB speech.

(ii) The user interface has an influence on the user's expectation. (Section 5.4).

The electro-acoustic transmission characteristics of all user interfaces in send and receive direction were determined instrumentally by HEAD acoustics, Herzogenrath, Germany. The measurements were made with a Head and Torso Simulation (i.e. dummy head) of the type HMS II.3 with ear simulator according to ITU–T Rec. P.51 (1989), Pinna Type 3.4, and the units MFE II and CAS 2.76, in an anechoic room, according to the procedure described in ITU–T Rec. P.64 (1999). From the characteristics obtained this way, loudness ratings were derived according to ITU–T Rec. P.79 (1999), employing the wideband coefficients defined in ITU–T Rec. P.79, Annex G (2001). The measured frequency responses can be adjusted to a desired shape using the programmable filters *SLR*1/*SLR*2 and *RLR*1/*RLR*2 shown in Figure B.1.

Table B.2 provides a list of the user interfaces that can be connected to the WB PSTN/ISDN-simulation system, indicating what type of tests they can be used in. Moreover, the table gives information on the shape of the amplitude spectrum the user interfaces have been adapted to. The loudness ratings can be selected parametrically with the control software of the PSTN/ISDN-simulation. The individual user interfaces will be described in the following subsections.

Table B.2: Wideband user interfaces implemented with the simulation system.

User Interface	Test Type	Spectral Shape	
		Send	Receive
Handset '7'	CT/LOT	IRSmod	IRSrec
Headset (diotic)	"	Flat	'Stax'
Headset (monotic)	"	"	"
HFT (ideal)	"	Native	Native
WB handset '7'	"	"	"
'Hi-fi phone'	LOT	–	"

B.5.1 Handset Telephones

The default user interfaces of the simulation tool are two conventional handset telephones ('Type 7', a model used in Germany for years). Both telephones are operated with the piezoelectric microphone capsules and loudspeakers they are typically equipped with. As can be seen from Figure B.1, the microphone and loudspeaker of each phone are connected to the simulation system in a four-wire mode, so that undesired line echo can be avoided, and the sidetone path be controlled. Using the programmable filters denoted by SLR and RLR, the frequency responses of the handsets $SLRset$ and $RLRset$ are typically adjusted to that of the intermediate reference system (IRS acc. to ITU–T Rec. P.48, 1989; IRSmod acc. to ITU–T Rec. P.830, 1996). The transmission characteristics of each phone in send and receive direction ($SLRset$ and $RLRset$) had therefore been determined instrumentally (according to ITU–T Rec. P.64, 1999).

B.5.2 Headsets

The headsets that were adapted to the system were selected based on the criteria of little weight and a broad transmission bandwidth. Monotic headsets of the type m@b25 and diotic headsets of type m@b30 from Sennheiser, Germany, were chosen. Both types are equipped with open headphones and microphones of spherical directivity, and have (according to the technical specifications) frequency ranges of 20–22000 Hz (headphone) and 40–12500 Hz (microphone), respectively. Different correction filters were implemented. For the correction filters in send direction, flat responses were aimed at. A number of superimposed BP filters were designed, so that conditions of frequency band limitations can be presented in auditory tests. The corresponding frequency characteristics are depicted in Section C.2. At receive side, the frequency response of a high-quality headphone (Stax) can optionally be set using appropriate correction filters. The amplitude spectra of one monotic and one diotic headset – before and after correction – are depicted in Section C.2.

B.5.3 Wideband Handset and Hi-fi Phone

With commercial handset telephones, which are designed for NB telephony, the coupling especially of low frequency components is problematic (Section 3.3.14). However, handset telephones are the user interfaces most readily associated with telephony. Hence, we equipped a standard handset with a high-quality receiver. Therefore, the ear cap of a 'type 7'-handset was replaced with an AKG-headphone cap (Type K 240 DF, open construction, diffuse-field equalized). In the mouthpiece, a standard piezo-cap is used. The transfer characteristics of the WB phone (here referred to as phone 'B5') in receive direction are shown in Section C.2 (Figure C.9). An adaptation of the frequency response in receive direction to a reference flat spectrum can optionally be used. However, as the positioning of the handset, and to some extent the shape of the pinna play a role for the receive frequency response, such a reference spectrum may not be a realistic choice. Therefore, in the tests conducted for this book, no correction of the transfer characteristics of the WB handset were employed.

Since the WB phone 'B5' did not provide the desired quality, a better acoustic coupling was aimed at. To enable WB listening tests with a user interface that subjects can easily relate to telephony, we constructed a second, more dedicated WB handset. It is referred to as the 'Hi-fi phone' in this book. For its construction, the listening half of a 'type 7' handset was cut off and replaced with one earphone of an electrostatic, open headphone ('Stax Lambda Pro'; see Figure B.4). The angular position $\alpha = 35°$ was chosen in accordance with the angle between the two surfaces of the listening and speaking parts of the 'type 7' handset. The additional angular rotation by $\beta = 22°$ in the other plane was used to assure that the users' (left) ear is well covered by the earphone (Figure B.5). Here, the usage of the left ear was implied as the most natural one for our tests, since most right-handed subjects

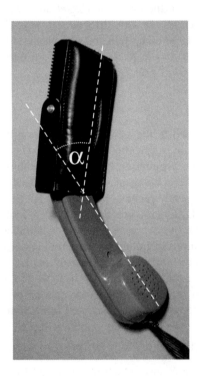

Figure B.4: 'Hi-fiphone' realized as a combination of a 'type 7' handset telephone with a Stax earphone (Raake, 2000).

Figure B.5: Angular position of listening part and mouthpiece for the 'Hi-fi phone' (Raake, 2000, see text for details).

use the left ear to hold the handset, while the right one is used for controlling the listening test program with the mouse.

The value for β was determined empirically: In an informal test, seven members of IKA were asked to use a normal 'type 7' handset telephone and hold it to their left ear in a comfortable listening position. We took pictures of the individual handset positions, and $\beta = 22°$ was derived as the average position.

As the underlying Stax phones are of very high quality, and their receive characteristics were used as the reference system for the adjustment of other WB user interfaces, we did not equalize the 'Hi-fi phone' to a reference spectrum. However, it has to be noted that some sensitivity to the handset position could be observed also for this phone (see Appendix C, Section C.2.2, Figure C.10).

B.5.4   Bandpass Filters

For usage, for example, with the 'Hi-fi phone', different bandpass-filters were implemented in receive direction (using the programmable filters *SLR*1/*SLR*2, Figure B.1). A list of currently available filters is provided in Table B.3. Most filters were designed to show a flat spectral shape in the passband; additional, superimposed *IRSmod* filtering can be chosen for the indicated filters (ITU–T Rec. P.48, 1989; ITU–T Rec. P.830, 1996, Annex D). All filters were designed with a 50 dB stop-band attenuation and the cutoff frequencies specified in the table. The different BP filters were loudness adjusted according to ITU–T Rec. P.79 (1999) for the BP filters falling into the NB range (300–3400 Hz), and according to ITU–T Rec. P.79, Annex G (2001) for those lying in the WB range (50–7000 Hz). The lower and upper cutoff frequencies and stop-band attenuations of the filters can be refined further using the channel BP filters (again 300–3400 Hz for NB, and 50–7000 Hz for WB, see above).

Table B.3: Bandpass filters implemented for the speech path at send side. All filters have a flat frequency response in the passband. Where indicated, an additionally *IRSmod*-filtered version was implemented, according to Annex D of ITU–T Rec. P.830 (1996).

f_{up} [Hz]	f_{low} [Hz]					
	50	100	200	300	400	600
2000	X			X		X
2400				X	X/IRS	
3400	X	X	X	X/IRS	X	X
4000		X	X			
5000	X	X		X		
7000	X	X	X	X	X	X

B.5.5 Hands-free Terminals (HFTs)

The acoustic properties of hands-free terminals (HFT) impact both the quality in send and in receive direction:

- In send direction, the main aspects are the combined characteristics of the room and the microphone, and acoustic echoes resulting from a coupling between loudspeaker and microphone.

- In receive direction, quality mainly depends on the combined acoustic properties of the employed loudspeaker system and the room.

The assessment of the quality related to either of the two directions in a conversational situation is associated with technical difficulties:

- With real HFTs, the signal in send direction is affected by both the room and acoustic echo. This difficulty is typically referred to as 'HFT dilemma'. To avoid acoustic echoes, echo cancellers (EC) may be employed. However, these may introduce new, unwanted effects such as, level switching. If the impact of echo cancellation on quality is outside the scope of an assessment task, this solution is inappropriate. Instead, the presentation of the acoustic signal at send side can be carried out with headphones so that no echo signal is coupled to the microphone. With this solution, however, valid quality ratings can only be obtained from the subject seated at receive side, since the subject at send-side wears headphones.

- If a general assessment of speech quality in receive direction is desired, an acoustic echo will result for the interlocutor at the other end. If an EC is employed to avoid the echo, additional effects may be introduced, as stated earlier. Consequently, no controlled conditions can be realized for the interlocutor at the sending end.

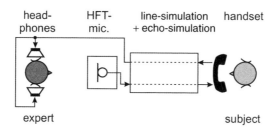

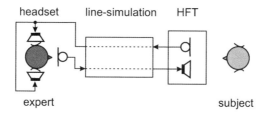

Figure B.6: Set-up for the assessment of an HFT in send direction by means of a conversation test.

Figure B.7: Set-up for the assessment of an HFT in receive direction by means of a conversation test.

In the framework of this book, two scenarios were implemented that are depicted in Figures B.6 and B.7.

As can be seen from these figures, asymmetric network conditions were chosen in both cases. With the implemented configurations, valid quality judgments can only be obtained from one of the two interlocutors, who is identified as 'subject' in Figures B.6 and B.7. The conversation partner at the other end is referred to as 'expert' in these figures.

For the quality assessment of HFTs in send direction, the expert interlocutor seated at the HFT wears headphones to avoid uncontrolled acoustic echo. This way, echo can be introduced in a controlled manner using the line simulation (Figure B.1).

For the assessment of speech quality in receive direction, the user interface employed by the expert interlocutor at send side is of minor importance. In the setup implemented for this book (see Figure B.7), the usage of a headset is foreseen. Due to the fact that no echo cancellation is employed, the expert listener may be exposed to considerable amounts of talker echo. In the following text, more details on the two implementations are provided.

B.5.5.1 Hand-Free Terminal in Receive Direction

The applied HFT was realized with a high-quality loudspeaker (Type JBL Control 1C) and a condenser microphone (Type AKG C414 B-TL). The 'ideal' HFT was positioned as depicted in Figure B.8, according to ITU–T Rec. P.340 (2000). The microphone is placed aside the loudspeaker, for example on an office table, so that the subjects may orient themselves toward it during the test conversations. It has to be noted, though, that an exact positioning during conversation tests is not possible, due to the movements of the conversation partners. This configuration enables the quality assessment of WB speech.

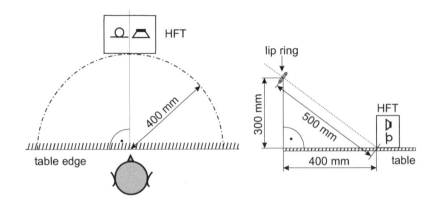

Figure B.8: Positioning of single-unit HFT according to ITU–T Rec. P.340 (2000).

B.5.5.2 Hands-Free Terminal in Send Direction

This simulation scenario was implemented with a high quality HFT-microphone to focus on the quality impacts due to the room and due to acoustic talker echo. Therefore, an HFT mouse-microphone type AKG Q 400 was implemented with the system. It is typically positioned as depicted in Figure B.8. For sound presentation to the expert listener, a headphone of type AKG K 240 DF is employed. In the current implementation, the subject at receive side uses a normal wireline handset telephone.

The controlled introduction of acoustic echoes was implemented using an HFT-simulation realized by HEAD acoustics, Herzogenrath, Germany. A schematic illustration of the system is depicted in Figure B.9. It allows long echo impulse responses to be employed, so that the effect of room echoes can be simulated. The echo impulse responses provided by HEAD acoustics were measured in a real office room of the company. Moreover, the HFT-simulation permits to present the room acoustic conditions at receive side for the expert listener: By convolving the single-channel speech signal with two different, fixed impulse responses, the effect of dichotic listening in a room can be simulated. Note, that the HFT simulation from HEAD acoustics also enables the simulation of level switching, which was not addressed in this work.

However, wearing headphones introduces attenuation that impacts the perception of one's own voice (Pörschmann, 2001). This may result in an undesirable, adapted speaking style. The damping introduced by the headphone can be compensated for by introducing appropriately filtered sidetone. Sidetone implemented with the help of the PSTN/ISDN-simulation, in turn, cannot be presented in this setup, as a closed echo loop would result. Since the expert seated at the HFT could not deliver meaningful quality ratings anyhow, it was considered more important to maintain an appropriate speaking style, to avoid an impact of the quality ratings obtained from the subject. Consequently, the correct acoustic representation of the HFT in the room at receive side is abandoned by the implemented setup, and the speech signal is presented only to one ear of the expert interlocutor. This way, the other ear could remain free so that an appropriate own-voice-perception could be achieved.

The HFT-simulation system can be coupled with the PSTN/ISDN-simulation system realized in this book, so that the combined effect of an HFT in send direction and additional

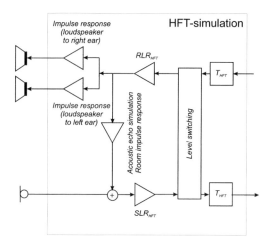

Figure B.9: Schematic illustration of the HFT-simulation provided by HEAD acoustics.

quality elements like noise or delay can be investigated. Since the PSTN/ISDN-simulation itself enables the controlled introduction of talker echo of a short echo impulse response, the effect of different types of echo on speech quality can also be compared with this system.

B.6   Test Rooms

Except where explicitly stated, all listening and conversation tests were carried out in a lab environment set up for auditory quality tests by Wiegelmann (1997) and Möller (2000). The test rooms are two office rooms of the institute, which are characterized by a low background noise level (approx. 35 dB(A)). In acoustic measurements, the requirements of NC25 were well met, and the requirements according to NC20 just missed (Beranek, 1971). The reverberation times in the relevant frequency range are in the range from 0.37 to 0.50 ms. Thus, the test rooms fulfills all prerequisites imposed by ITU–T Rec. P.800 (1996) for auditory test environments. The two adjacent office rooms are acoustically shielded against each other, so that each can be occupied by one test subject, in a conversation test as well as in a listening test. The experimenter is seated in a third room, also acoustically shielded, where the simulation system is located and can communicate with the test subjects over an intercom system.

B.7   Simulation of Quality Elements: Summary

An online simulation system has been implemented that allows all relevant quality elements of PSTN/ISDN as well as VoIP networks to be simulated in a controlled way. With this tool, conversation tests can be carried out under realistic network and user-interface conditions. The tool can also be employed for the processing of speech data to be used in listening tests. Therefore, an additional offline simulation of degradations typical of GSM-networks is available. In total, the following quality elements are covered by the simulation system:

Analog and Digital PSTN/ISDN

- Overall attenuation/frequency distortion of the speech path including the user interface.

- Stationary circuit noise and noise induced by subscriber lines.

- Talker echo attenuation and delay.

- Listener echo attenuation and delay.

- Talker sidetone.

- Overall transmission delay.

- Room noise at send and receive side, including the effects of listener sidetone.

- Handset characteristics, especially sensitivities to direct and diffuse sound.

- Nonlinear distortion due to low-bitrate codecs.

Mobile Networks

- GSM-coding and random bit errors (off-line).

- Introduction of interruptions of scalable duration, distribution and overall rate (on line).

VoIP

- Different VoIP codecs.

- Different options for packet size.

- Packet loss of arbitrary distribution and overall rate.

- Jitter of arbitrary distribution (not used in this work).

Wideband Transmission

- Implementation of WB channel BP (50–7000 Hz).

User Interfaces

- NB and WB handsets.

- 'Hi-fi' telephone (for listening tests).

- Headsets.

- Hands-free terminals (send- and receive-direction).

 – Simulation of acoustic echoes.

 – Realization of the electro-acoustic properties of HFTs in send-direction (room influence).

Due to the parametric layout and the modularity of the integral system, a broad range of circuit conditions and network types can be simulated. In the meantime, the system has been used in different projects and research contexts, such as the following:

• Assessment of speech quality in human-to-human communication (e.g. Möller, 2000; Möller and Raake, 2002; Raake, 2004).

• Verification of a network monitoring system (Raake and Möller, 1999).

• Assessment of speech quality for the interaction with a spoken dialogue system over the phone (speech recognition performance (e.g. Möller *et al.*, 2004), speech synthesis performance (e.g. Möller, 2004), and overall spoken dialogue system performance (e.g. Möller, 2005a)).

The simulation has been employed for all of the listening and conversation tests described in Chapters 4 and 5.

C

Frequency Responses

In the following appendix, the frequency responses of the employed filters are provided. Note that the programmable DSP-farm used for the PSTN line simulation only allows FIR-filters to be implemented.

C.1 Transmission Bandpass

C.1.1 Narrowband

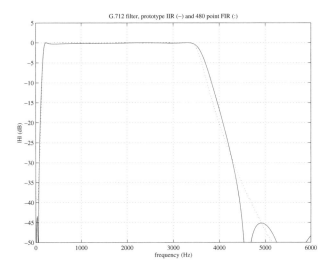

Figure C.1: Narrowband channel bandpass-filter designed according to ITU–T Rec. G.712 (1992).

Speech Quality of VoIP: Assessment and Prediction Alexander Raake
© 2006 John Wiley & Sons, Ltd

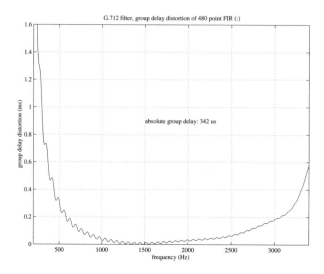

Figure C.2: Group delay distortion introduced by the narrowband channel bandpass.

C.1.2   Wideband

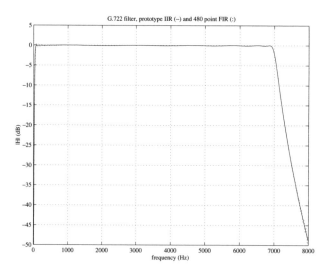

Figure C.3: Wideband channel bandpass-filter designed according to ITU–T Rec. G.722 (1988).

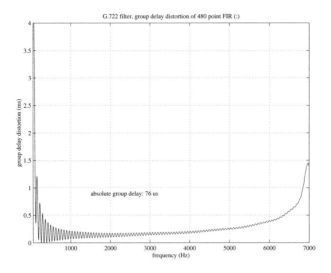

Figure C.4: Group delay distortion introduced by the wideband channel bandpass.

C.2   User Interfaces

C.2.1   Headsets

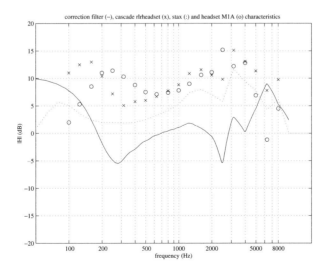

Figure C.5: Frequency response of monotic headset 'MA1' in receive direction.

C.2.2   Wideband Handsets

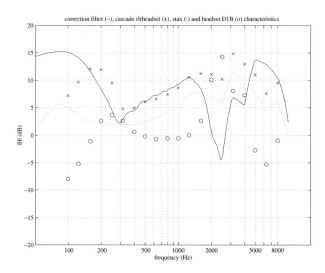

Figure C.6: Frequency response of diotic headset 'DB1' in receive direction.

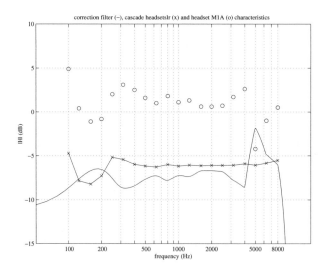

Figure C.7: Frequency response of monotic headset 'MA1' in send direction.

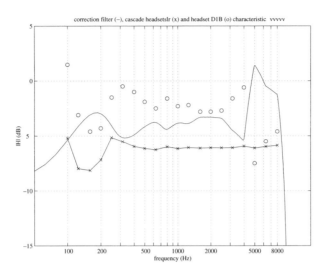

Figure C.8: Frequency response of diotic headset 'DB1' in send direction.

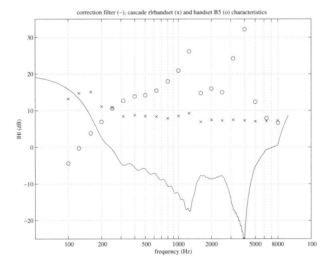

Figure C.9: Frequency response of wideband handset 'B5' in receive direction.

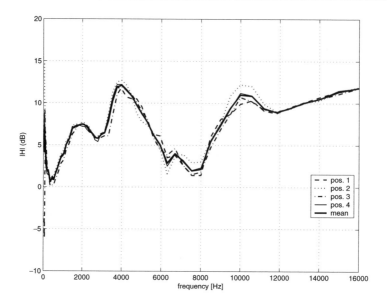

Figure C.10: Frequency response of wideband handset 'Hifi-phone' in receive direction for different variations of holding the phone to the Head and Torso Simulator (HATS), and mean response.

D

Test Data Normalization and Transformation

In this book, model predictions are compared with results of the author's own auditory tests and tests compiled from the literature, which were predominantly collected employing the mean opinion score (MOS) scale (ITU–T Rec. P.800, 1996). To allow the results obtained by different labs to be compared, a test data normalization is commonly carried out. Therefore, particular reference conditions are presented in the test. For these references, quality or the way quality depends on them is assumed to be known. Due to the usage of particular reference conditions, normalization is closely linked to the role of reference and expectation for speech quality perception (see Section 3.7; Möller, 2000, pp. 116–129): Perceived speech quality depends on several factors such as the situation, and the user's experience from the connections typically faced with, but also on network conditions experienced more recently.

The reference conditions used for normalization may have a twofold impact on the test data analysis:

- The reference conditions may impact the rating behavior of subjects during the test, since they may direct the subjects' attention to particular features, emphasizing their importance for quality ratings.

- The transformation or normalization procedure itself makes strong assumptions on the quality relation of the reference conditions, which may however depend on the specific test situation (e.g. whether quality of mobile networks or of business, wireline connections are to be assessed).

Three normalization procedures are considered in this book:

D.1 Equivalent-Q Method

In the past, the 'equivalent-Q method' was used to normalize auditory test data obtained on the MOS scale (ETSI Technical Report ETR 250, 1996, Annex G, but with updated

Speech Quality of VoIP: Assessment and Prediction Alexander Raake
© 2006 John Wiley & Sons, Ltd

versions of equations G.7–G.9 as in ITU–T Rec. G.107 (2005)). According to this procedure, different conditions with a modulated noise reference unit (MNRU, see ITU–T Rec.
P.810, 1996) of different signal-to-correlated-noise ratios are presented as reference degradations in the test. The method assumes a fixed relation between the signal-to-correlated
noise ratio Q of the MNRU and the resulting MOS-ratings:

$$MOS = 1 + A + B \cdot \tanh\left(\frac{Q - Q_m}{C}\right) \tag{D.1}$$

In the normalization process, the constants A, B, C and Q_m are derived by curve-fitting of
the MOS-ratings obtained in the test for the MNRU-conditions. The constants can then be
exploited to estimate an *equivalent-Q* value for the (non-MNRU) test conditions:

$$Q_{eq} = Q_m + \frac{C}{2} \cdot \ln\left(\frac{B - A - 1 + MOS}{B + A + 1 - MOS}\right) \tag{D.2}$$

Finally, transformed MOS-values can be calculated employing the corresponding E-model
equations (Equation E.7, Equations E.12–E.15, and Equation 2.5).

D.2   Method According to ITU-T Rec. P.833

Since the degradation due to MNRU is perceptually different from that due to low-bitrate
coding or packet loss (see Section 3.4), a new procedure for deriving equipment impairment
factors Ie is now recommended by the by the International Telecommunication Union (ITU-
T Rec. P.833, 2001). It is based on a linear transformation of listening test results on the
R-scale. Instead of a modulated noise source, it employs perceptually more similar reference
conditions of known impairment, namely, different codecs in single and tandem operation
(for error free codecs), or additional packet- loss conditions (if, for example, the packet-
loss robustness factor of a certain codec under random packet loss is to be determined;
ITU–T Rec. P.833, 2001). P.833 recommends the usage of 14 codec and codec-tandeming
conditions (plus a number of conditions for checking the impairment factor estimate for
tandems involving the codec under test), and 10 additional reference conditions including
transmission errors. Since these reference conditions are perceptually more similar to the
codecs under test, they are preferred to the usage of the equivalent-Q method involving
MNRU degradations.

   The procedure can be outlined as follows: (i) The ratings obtained for both the test
and the reference conditions are transformed to the R-scale (e.g. in case of MOS-ratings
using Equation E.33). (ii) The impairments due to the test and reference conditions are
determined as the decrease in R relative to a clean reference. (iii) A linear interpolation
curve between the impairment values of the reference conditions obtained from the test and
their known impairment values is derived by least-square fitting. (iv) Using the offset and
slope of the linear interpolation, the transformed impairment values for the test conditions
are calculated from their initial test values.

D.3   Linear Transformation

Another method employs direct, linear test data transformation on the applied rating scale.
For example, in case a clean reference condition of known ('expected') quality MOS_r is

presented in the test, test results obtained on the MOS scale (MOS_t) can be transformed to values MOS_l to cover the whole upper scale range:

$$MOS_l = \frac{MOS_t - 1}{MOS_t(C_r) - 1} \cdot (MOS_r - 1) + 1. \tag{D.3}$$

Here, $MOS_t(C_r)$ is the quality rating obtained in the test for the clean reference condition C_r. In this book, the results of some tests were linearly transformed using Equation (D.3) so that $MOS_r = 4.41$ was met in case of a clean channel with G.711, as it is predicted by the E-model.

A similar linear transformation can be employed so that the transformed test data MOS_l cover the whole 'expected' scale-range:

$$MOS_l = \frac{MOS_t - MOS_t(C_{r,min})}{MOS_t(C_{r,max}) - MOS_t(C_{r,min})} \cdot (MOS_{r,max} - MOS_{r,min}) + 1. \tag{D.4}$$

In this case, $C_{r,max}$ is the clean reference condition expected to yield highest quality, and $C_{r,min}$ the reference condition expected to be associated with lowest quality. $MOS_{r,max}$ and $MOS_{r,min}$ are the respective 'expected' MOS-values, and the corresponding MOS-values obtained in the test are referred to as $MOS_t(C_{r,min})$ and $MOS_t(C_{r,max})$.

Such a simple, *linear* transformation may allow a better comparison of different test results to be made than without employing any transformation. However, the accuracy of this method is reduced as compared with the more complex linear transformation involved in the P.833 methodology (Section D.2). In turn, the simple linear normalization procedure, too, avoids the usage of the equivalent-Q method, which may lead to erroneous transformations, especially when the features related to the system under test differ considerably from those related with an MNRU-type signal-correlated noise. The simplified transformation also reduces the very large number of reference conditions required by the P.833 methodology, which is barely manageable, if a large number of actual test conditions is to be assessed, or when the P.833 listening test methodology is applied to conversation tests.

D.4   Note on MOS-terminology

In the framework of standardization bodies like the ITU-T, speech quality is often expressed in terms of MOS (Mean Opinion Score, see Section 2.1.1.2). To better distinguish MOS-values obtained by different means such as auditory tests or instrumental models, a dedicated MOS-terminology was introduced in ITU–T Rec. P.800.1 (2003), as summarized in Table D.1. Here, 'LQ' refers to listening quality, and 'CQ' to conversational quality. The

Table D.1: MOS-terminology according to ITU–T Rec. P.800.1 (2003).

	Listening-only	Conversational
Subjective	MOS-LQS	MOS-CQS
Objective	MOS-LQO	MOS-CQO
Estimated	MOS-LQE	MOS-CQE

third identifier 'S', 'O' or 'E' specifies, whether the MOS was derived by auditory testing ('S' for 'subjective'), by a signal-based measure such as PESQ (ITU–T Rec. P.862, 2001), where 'O' stands for 'objective'. Predictions obtained by network planning models like the E-model (ITU–T Rec. G.107, 2005) are here referred to by the letter 'E', which stands for 'estimate'.

E

E-model Algorithm

In the following appendix, we provide a description of the E-model algorithm. A summary of an earlier version of the algorithm is given in (Möller, 2000, Appendix G), which has been adjusted to the current model version (ITU–T Rec. G.107, 2005).

The Transmission Rating Factor R is defined by Equation 2.5. The additive terms are derived based on the following formulae:

The basic signal–to–noise ratio Ro is calculated as

$$Ro = 15 - 1.5\,(SLR + No) \tag{E.1}$$

with No [dBm0p] determined by addition of the different noise sources (in terms of power):

$$No = 10 \cdot \log_{10}[10^{Nc/10} + 10^{Nos/10} + 10^{Nor/10} + 10^{Nfo/10}] \tag{E.2}$$

Nc [dBm0p] expresses the sum of the circuit noise powers, referred to the 0 dBr point. The equivalent circuit noise at the 0 dBr point is referred to by the parameter Nos [dBm0p], and reflects the noise on the line generated by room noise at the send side of level Ps [dB(A)]:

$$Nos = Ps - SLR - Ds - 100 + 0.004\,(Ps - OLR - Ds - 14)^2 \tag{E.3}$$

Accordingly, the equivalent circuit noise for the room noise Pr [dB(A)] at receive side is derived as:

$$Nor = RLR - 121 + Pre + 0.008\,(Pre - 35)^2 \tag{E.4}$$

Here, Pre [dBm0p] is the level Pr including the listener sidetone:

$$Pre = Pr + 10\,\log_{10}[1 + 10^{(10-LSTR)/10}] \tag{E.5}$$

Correspondingly, the noise floor $Nfor$, which is set to $Nfor = -64dBmp$, is denoted by Nfo when referred to the 0 dBr point:

$$Nfo = Nfor + RLR \tag{E.6}$$

Speech Quality of VoIP: Assessment and Prediction Alexander Raake
© 2006 John Wiley & Sons, Ltd

Is, the simultaneous impairment factor, is composed of three contributions:

$$Is = Iolr + Ist + Iq \qquad (E.7)$$

Here, *Iolr* is the impairment caused by an overly loud connection, and is defined as

$$Iolr = 20\,[\{1 + (X/8)^8\}^{1/8} - X/8], \qquad (E.8)$$

where

$$X = OLR + 0.2\,(64 + No - RLR). \qquad (E.9)$$

The impairment for non-optimum sidetone, *Ist*, is defined as:

$$Ist = 12\left[1 + \left(\frac{STMRo - 13}{6}\right)^8\right]^{\frac{1}{8}} \qquad (E.10)$$

$$-28\left[1 + \left(\frac{STMRo + 1}{19.4}\right)^{35}\right]^{\frac{1}{35}}$$

$$-13\left[1 + \left(\frac{STMRo - 3}{33}\right)^{13}\right]^{\frac{1}{13}} + 29,$$

with

$$STMRo = -10\,\log_{10}\,[10^{-STMR/10} + e^{-T/4}\,10^{-TELR/10}]. \qquad (E.11)$$

The one-way delay *T* [ms] and the talker echo loudness rating (*TELR*) [dB] characterize the echo the talker receives of his own voice.

Iq is a measure of the impairment caused by quantizing distortion

$$Iq = 15\,\log_{10}\left[1 + 10^Y + 10^Z\right]. \qquad (E.12)$$

Here,

$$Y = \frac{Ro - 100}{15} + \frac{46}{8.4} - \frac{G}{9}, \qquad (E.13)$$

$$Z = \frac{46}{30} - \frac{G}{40}. \qquad (E.14)$$

The intermediate parameter *G* represents the equivalent continuous circuit noise. According to Richards (1973), it can be calculated as:

$$G = 1.07 + 0.258\,Q + 0.0602\,Q^2. \qquad (E.15)$$

Here, *Q* is the signal–to–quantizing–noise ratio in terms of 'quantizing distortion units', *qdu*.

$$Q = 37 - 15\,\log_{10}\,(qdu) \qquad (E.16)$$

Note that 1 *qdu* quantifies the quantizing distortion introduced by a single PCM codec. The above relation has been derived empirically (Coleman *et al.*, 1988; South and Usai, 1992).

Similar to the simultaneous impairment factor, the delayed impairment factor *Id* is composed of three factors:

$$Id = Idte + Idle + Idd. \tag{E.17}$$

Now, *Idte* is the impairment caused by talker echo (characterized by the one–way delay *T* [ms] and the *TELR* [dB], see preceding text), and includes the masking by noise:

$$Idte = [(Roe - Re)/2 + \sqrt{(Roe - Re)^2/4 + 100} - 1] (1 - e^{-T}). \tag{E.18}$$

Here,

$$Roe = -1.5 \, (No - RLR) \tag{E.19}$$

$$Re = 80 + 2.5 \, (TERV - 14) \tag{E.20}$$

$$TERV = TELR - 40 \, \log_{10} \frac{1 + T/10}{1 + T/150} + 6 \, e^{-0.3T^2} \tag{E.21}$$

As discussed earlier in this book, talker echo can be interpreted as sidetone if $T < 1$ ms; then, $Idte = 0$. In case of a too loud sidetone (i.e. $STMR < 9$ dB), the talker echo may partly be masked by the sidetone, and the parameter *TERV* is exchanged by *TERVs*:

$$TERVs = TERV + Ist/2. \tag{E.22}$$

In turn, in case the sidetone is very low (i.e. $STMR > 15$ dB), the talker echo may become more audible, and the term *Idtes* is used instead of *Idte*:

$$Idtes = \sqrt{Idte^2 + Ist^2} \tag{E.23}$$

The impairment due to listener echo is accounted for by the factor *Idle*:

$$Idle = (Ro - Rle)/2 + \sqrt{(Ro - Rle)^2/4 + 169}, \tag{E.24}$$

where

$$Rle = 10.5 \, (WEPL + 7)(Tr + 1)^{-0.25}. \tag{E.25}$$

In its current version, the E-model expresses the impairment caused by absolute delay higher than 100 ms as

$$Idd = \begin{cases} 0 & \text{for } Ta \le 100 \text{ ms} \\ 25 \, \{(1 + X^6)^{1/6} - 3 \, (1 + [X/3]^6)^{1/6} + 2\} & \text{for } Ta > 100 \text{ ms} \end{cases}, \tag{E.26}$$

with

$$X = \frac{\log_{10}(Ta/100)}{\log_{10} 2}. \tag{E.27}$$

The effective equipment impairment factor *Ie,eff* is used to capture the quality impact of packet loss (see Equation (4.14), Chapter 4).

$$Ie,eff = Ie + (95 - Ie)\frac{Ppl}{\dfrac{Ppl}{BurstR} + Bpl}, \tag{E.28}$$

with

$$BurstR = \frac{\text{average No. consecutively lost packets}}{\text{average No. consecutively lost packets for random loss}}. \tag{E.29}$$

The corresponding input values for the equipment impairment factor *Ie* and the packet-loss robustness factor *Bpl* are tabulated for different codecs in ITU–T Rec. G.113 Appendix I (2002). The Packet Loss Percentage *Ppl* and the Burst Ratio *BurstR* (Equation (E.29)) are determined based on the packet-loss distribution (see Chapter 4).

Additional estimates of user opinion (%GoB, %PoW, and MOS) can be directly derived from the transmission rating *R* with the Gaussian error function

$$E(X) = \frac{1}{\sqrt{2\pi}} \int_{-\infty}^{x} e^{-\frac{t^2}{2}} dt \tag{E.30}$$

and the relations:

$$\%GoB = 100 \cdot E\left(\frac{R - 60}{16}\right)[\%], \tag{E.31}$$

$$\%PoW = 100 \cdot E\left(\frac{45 - R}{16}\right)[\%], \tag{E.32}$$

and

$$MOS = \begin{cases} 1 & \text{for } R < 0 \\ 1 + 0.035\,R + R(R - 60)(100 - R) \cdot 7 \cdot 10^{-6} & \text{for } 0 \leq R \leq 100 \\ 4.5 & \text{for } R > 100 \end{cases}. \tag{E.33}$$

An inversion of equation (E.33) is permissible only in the range $6.5 \leq R \leq 100$, and can be performed based on equations (E.34)–(E.36) (proposed by Hoene *et al.*, 2003; ITU–T Rec. G.107, 2005, Appendix I):

$$R = \frac{20}{3}\left(8 - \sqrt{226}\cos\left(h + \frac{\pi}{3}\right)\right), \tag{E.34}$$

with

$$\begin{aligned} h = \quad &\tfrac{1}{3}\arctan 2\Big(18566 - 6750MOS, \\ &15 \cdot (-903522 + 1113960MOS \\ &-202500MOS^2)^{\frac{1}{2}}\Big) \end{aligned} \tag{E.35}$$

and

$$\arctan 2\,(x, y) = \begin{cases} \arctan\left(\frac{y}{x}\right), & x \geq 0; \\ \pi - \arctan\left(\frac{y}{-x}\right), & x < 0. \end{cases} \tag{E.36}$$

Table E.1: Default Settings of the E-model characterizing an average, clear telephone channel with PCM-coding, yielding a Transmission Rating Factor of $R = 93.2$ (Modified from Table 2 of ITU–T Rec. G.107 2005; for more details on the parameters see also ETSI Technical Report ETR 250, 1996).

Parameter	Abbreviation	Unit	Default Value
Sending Loudness Rating	*SLR*	dB	8
Receiving Loudness Rating	*RLR*	dB	2
Sidetone Masking Rating	*STMR*	dB	15
Listener Sidetone Rating	*LSTR*	dB	18
D-Factor Handset, Send Side,	*Ds*	–	3
D-Factor Handset, Receive Side,	*Dr*	–	3
Talker Echo Loudness Rating	*TELR*	dB	65
Weighted Echo Path Loss	*WEPL*	dB	110
Round Trip Delay in a 4-Wire Loop	*Tr*	ms	0
Absolute Delay	*Ta*	ms	0
Mean One-Way Delay	*T*	ms	0
Number of *Quantizing Distortion Units*	*qdu*	–	1
Equipment Impairment Factor	*Ie*	–	0
Circuit Noise relative to 0 dBr-point	*Nc*	dBm0p	−70
Noise Floor at Receive Side	*Nfor*	dBmp	−64
Room Noise at Send Side	*Ps*	dB(A)	35
Room Noise at Receive Side	*Pr*	dB(A)	35
Packet Loss Percentage	*Pp*	%	0
Packet Loss Robustness Factor	*Bpl*	%	1
Burst Ratio	*BurstR*	%	1
Advantage Factor	*A*	–	0

The default settings of the E-model, corresponding to a clean connection with log-PCM coding according to ITU–T Rec. G.711 (1988), are shown in Table E.1.

F

Interactive Short Conversation Test Scenarios (iSCTs)

Speech Quality of VoIP: Assessment and Prediction   Alexander Raake
© 2006 John Wiley & Sons, Ltd

F.1   Example

 Network Central Company X

 Exchange of identification numbers and email-addresses of new company members

Name	Personal-Nr	Email
Fachmann	536-952487	ks.fachmann
Bauer		fp.bauer
Dreierlein		ps.dreierlein
Gerhards		jf.gerhards
Kobalt		kh.kobalt
Tuchmeyer		ag.tuchmeyer

Figure F.1: Example of interactive Short Conversation Test scenario (iSCT); speaker A.

Example   263

Human Resource Department Company X

Exchange of identification numbers and email-addresses of new company members

Name	Personal-Nr.	Email
Fachmann	536-952487	ks.fachmann
Bauer	258-761926	
Dreierlein	536-879177	
Gerhards	258-327431	
Hamberg	668-215623	
Tuchmeyer	536-412142	

Figure F.2: Example of interactive Short Conversation Test scenario (iSCT); speaker B.

G

Auditory Test Settings and Results

G.1 Global System for Mobile (GSM): Short Conversation and Listening Only Test

The short conversation test 'SCT-GSM' and the listening test 'LOT-GSM' were conducted with the online-system described in Appendix B. A number of $n = 16$ naïve subjects participated in the conversation test, and $n = 13$ naïve subjects in the listening test. In both cases, the ratings were collected on the 5-point absolute category rating (ACR) scale shown in Figure 2.1. The source speech material used in the listening test was recorded from four speakers (2 female, 2 male). In the listening test, an additional group of $n = 13$ subjects made ratings on the CR-10 scale (Borg, 1982; ITU–T Rec. P.833, 2001, Appendix I). Owing to the similarity to the ACR ratings, only these are presented in this book. The test conditions and results can be found in Tables G.1, G.2 and G.3.

Speech Quality of VoIP: Assessment and Prediction   Alexander Raake
© 2006 John Wiley & Sons, Ltd

Table G.1: Short conversation test 'SCT-GSM' on speech quality under interruptions (and others): Test conditions and results (Raake and Möller, 1999).

No.	Codec / MNRU	T [ms]	TELR [dB]	Interr. Length [ms]	Interr. Rate [%]	MOS	Std.
1	MNRU, Q = 5 dB	–	–	–	–	1.44	0.51
2	MNRU, Q = 20 dB	–	–	–	–	3.44	1.08
3	G.711	–	–	–	–	4.19	0.66
4	IS-54	–	–	–	–	3.44	0.96
5	G.711	100	25	–	–	2.13	0.81
6	G.711	100	35	–	–	3.00	1.03
7	G.711	100	45	–	–	3.31	1.25
8	G.711	–	–	40	10	2.63	0.89
9	G.711	–	–	40	20	1.69	0.48
10	G.711	–	–	160	5	2.69	0.95
11	G.711	–	–	160	10	2.13	0.72
12	G.711	–	–	160	20	1.75	0.86
13	G.711	–	–	160	30	1.19	0.40
14	G.711	–	–	2000	20	1.50	0.89

All other parameter values were adjusted to the E-model default values depicted in Table E.1.

Table G.2: Listening-only test 'LOT-GSM' on speech transmission quality under interruptions and bit errors on GSM-encoded speech: Test conditions and results (Raake and Möller, 1999).

No.	Codec/ MNRU	BER [%]	Interr. Length [ms]	Interr. Rate [%]	MOS	MOS From CR-10	CR-10	MOS Std	MOS (CR10) Std
1	MNRU, Q = 5 dB	–	–	–	1.6	1.5	7.8	0.63	0.54
2	EFR	4	–	–	3.2	3.7	2.5	0.86	0.70
3	EFR	6	–	–	2.2	2.5	5.2	0.70	0.83
4	G.711	–	–	–	4.6	4.3	0.9	0.72	0.37
5	IS-54	–	–	–	3.5	4.0	1.8	0.90	0.54
6	FR	–	–	–	3.6	3.9	2.0	0.89	0.68
7	HR	–	–	–	3.5	3.9	2.1	0.90	0.62
8	EFR	–	–	–	4.3	4.3	0.9	0.92	0.35
9	FR	4	–	–	3.0	3.5	3.1	0.90	0.89
10	FR	6	–	–	2.0	1.9	6.7	0.82	0.80
11	FR	8	–	–	1.4	1.2	8.7	0.69	0.36
12	FR	10	–	–	1.1	1.0	14.2	0.49	0.06
13	HR	8	–	–	1.2	1.2	9.7	0.43	0.29
14	EFR	8	–	–	1.1	1.1	12.4	0.41	0.18
15	G.711	–	40	10	2.5	2.9	4.5	0.67	0.89
16	G.711	–	40	20	1.9	1.9	6.6	0.56	0.70
17	G.711	–	160	5	2.9	3.2	3.7	0.86	0.78
18	G.711	–	160	10	2.4	3.0	4.1	0.69	0.77
19	G.711	–	160	20	1.5	1.7	7.4	0.58	0.60
20	G.711	–	160	30	1.4	1.4	8.7	0.80	0.50

All other parameter values were adjusted to the E-model default values depicted in Table E.1.

Table G.3: ANOVA of listening test results obtained in "LOT-GSM" for interruptions, using the interruption percentage, the interruption rate and the speaker as fixed factors (derived with SPSS).

Tests of Between-Subjects Effects

Dependent Variable: MOS

Source	Type III Sum of Squares	df	Mean Square	F	Sig.
Corrected Model	112.074[a]	23	4.873	10.719	0.000
Intercept	1247.713	1	1247.713	2744.675	0.000
SPEAK	2.308	3	0.769	1.692	0.169
INT_LENG	2.543	1	2.543	5.595	0.019
INT_PERC	90.428	3	30.143	66.307	0.000
SPEAK * INT_LENG	4.014	3	1.338	2.944	0.033
SPEAK * INT_PERC	8.168	9	0.908	1.996	0.040
INT_LENG * INT_PERC	1.389	1	1.389	3.056	0.081
SPEAK * INT_LENG * INT_PERC	0.245	3	0.082	0.180	0.910
Error	130.923	228	0.455		
Total	1593.000	312			
Corrected Total	242.997	311			

[a]R Squared = 0.461 (Adjusted R Squared = 0.418)

G.2 2-state Markov Loss: Listening Only Test

Table G.4 provides the results of the analysis of variance (ANOVA) carried out on the listening test results described in more detail in Section 4.1.2.

Table G.4: ANOVA of listening test results obtained in LOT on 2-state Markov loss using the loss percentage *Ppl*[%], the conditional loss rate *pc* and the speaker as fixed factors

Tests of Between-Subjects Effects

Dependent Variable: MOS

Source	Type III Sum of Squares	df	Mean Square	F	Sig.
Corrected Model	1197.354[a]	59	20.294	33.784	0.000
Intercept	11861.804	1	11861.804	19746.40	0.000
PPL	1008.009	4	252.002	419.510	0.000
PC	6.547	2	3.274	5.450	0.004
SPEAKER	77.141	3	25.714	42.805	0.000
PPL * PC	14.221	8	1.778	2.959	0.003
PPL * SPEAKER	44.548	12	3.712	6.180	0.000
PC * SPEAKER	10.610	6	1.768	2.944	0.007
PPL * PC * SPEAKER	31.565	24	1.315	2.189	0.001
Error	999.577	1664	0.601		
Total	14058.000	1724			
Corrected Total	2196.930	1723			

[a]R Squared = 0.545 (Adjusted R Squared = 0.529)

G.3   Random Loss: Conversation Test

G.3.1   Test Setup

In the conversation tests, the online simulation tool outlined in Appendix B was used. For the tests, the subsystems of the simulation were combined to reflect the VoIP-typical (Voice over Internet Protocol) usage scenario PSTN–VoIP–PSTN, with telephone handsets as user interfaces. For the introduction of random packet loss, the NIST Net network simulation was used (NIST Net, 2002).

The standard *SLR* setting used in the E-model is 8 dB, and this value has been adopted both in SCT11 and SCT3. Due to the nonperfect terminal coupling loss (TCLw) of the applied handsets, an *SLR* of 13 dB was used in SCT2 (packet loss and delay), in order to minimize additional effects of talker echo. Therefore, SCT1 – that was supposed to serve for comparison to both SCT2 and SCT3 – was carried out in two different versions, one with *SLR* set to 13 dB (SCT11, higher attenuation), and one with *SLR* = 8 dB (SCT12, lower attenuation). As a side effect, a comparison of the two series allows conclusions to be drawn on the effect of speech level in case of packet loss and noise.

As codec, the VoIP-typical G.729A was chosen, and as an anchor-point a 'clean' connection with logarithmic pulse code modulation (PCM) (ITU-T Rec. G.711, 1988). For each SCT, 14 different conditions were used, with 12 actual test conditions (four rates of random packet loss: 0%, 3%, 5% and 15%; three levels of the additional test parameter). Owing to the limited number of network conditions that can be tested in a conversation test without subject fatigue, no additional reference conditions were presented. Instead, the aim was to cover the whole range of the applied rating scales by choosing appropriate parameter settings relying on E-model predictions. The G.729A codec was used with its built-in packet loss concealment (PLC). A packet size of $Tp = 20$ ms was chosen (corresponding to two frames per packet). No voice activity detection was applied, to avoid additional artifacts (such as clipping). The general simulation settings for the experiments are listed in Table G.5. All other more test specific settings are listed in Table G.6.

Table G.5: General simulation settings used in the different short conversation tests (SCT11–SCT3).

| | SCT11   SCT12 | | | All SCTs |
	SCT2 SCT3			
Codecs	G.711 (20 ms)		*Ta*	66 ms (G.711)
(packet size)	G.729A (20 ms)			60 ms (G.729A)
Ie; *Bpl*	0; 4.3 (G.711)		*T*	33 ms (G.711)
	11; 19.0 (G.729A)		*Ta*	30 ms (G.729A)
VAD	disabled		*Tr*	0 ms
SLR	13 dB	8 dB		*WEPL*	110 dB
RLR	2 dB		*TELR*	65 dB
STMR	15 dB		*Nc*	−70 dBm0p
LSTR	16 dB		*N for*	−64 dBmp
Ds = Dr	1		*Ps = Pr*	35 dB(A)

Table G.6: Test parameter settings for the different short conversation tests (SCT11–SCT3). The nonzero delays T and Ta are due to the processing time of the VoIP system.

No.	All SCTs		SCT11/SCT12 N for [dBmp]	SCT2 Ta [ms]	SCT3	
	Ppl	Codec [%]			T [ms]	$TELR$ [dB]
1	0	G.711	−64	66	33	65
2	0	G.711 (SCT1x) G.729 (SCT2/3)	−40	60	30	65
3	0	G.729	−64	200	100	50
4	3	G.729	−64	200	100	50
5	5	G.729	−64	200	100	50
6	15	G.729	−64	200	100	50
7	0	G.729	−50	400	100	35
8	3	G.729	−50	400	100	35
9	5	G.729	−50	400	100	35
10	15	G.729	−50	400	100	35
11	0	G.729	−40	600	100	20
12	3	G.729	−40	600	100	20
13	5	G.729	−40	600	100	20
14	15	G.729	−40	600	100	20

G.3.2 Test Procedure

Two conversation partners participated in each test run. They were placed in separate, acoustically shielded office rooms, each of which was equipped with a traditional handset-telephone. The subjects were instructed to imagine a normal wireline telephone connection. They were asked to carry out a number of dialogue tasks according to the short conversation test scenarios described in Section 2.1 (Möller, 2000). Test instructions were handed out in a written form. Two different scales were used: The 5-point ACR scale [mean opinion score (MOS) scale, see Figure 2.1]) for rating the integral quality ('overall impression'), and – for direct rating of the impairment – the CR-10 degradation scale (Borg, 1982; ITU–T Rec. P.833, 2001, Appendix I). In principle, the CR-10 ratings can be transformed into the R-scale by using a simple linear transformation. Hence, they can be considered as impairment ratings (Möller, 2000, pp. 147-155). For the tests described here, the ratings on the MOS- and on the CR-10-scales were highly correlated (Pearson's correlation coefficient between −0.8 and −0.9 for the different SCTs). For clarity, the focus of the following analysis thus will be on the MOS-ratings.

The scenarios as well as the rating scales were presented in paper form. Each test pair of subjects was asked to carry out 15 conversations over 14 different connections – the clean G.711 connection was presented twice, once at the beginning of each test run to familiarize the subjects with the test scenarios, and to serve as reference connection. Prior to the actual test, the subjects were asked to read a text dialogue with their conversation partner across four selected connections, so that they would become acquainted with the quality ranges to be experienced in the actual test. Note that the subjects were not instructed to pay any attention to a specific type of impairment, to reflect a realistic telephone situation. After each conversation, the subjects were asked to judge the overall impression (MOS) and the degradation (CR-10) of the line they had just been using. In the test, every pair was presented a different randomized order of test conditions.

G.3.3 Subjects

Each test series was carried out with 22 test subjects (11 pairs of interlocutors), who were naïve with respect to the test task and conditions. They were by their own account of normal hearing. The subjects were paid for their participation in the tests. Most subjects recruited were university students, which resulted in a low average age of 25.8 years (age range 18 – 63 years). Of the 88 subjects, 41 were female and 47 male. After the tests, a screening of the subjects with respect to the consistency of their ratings was carried out. It was found that two subjects in each SCT11 and SCT2, one subject in SCT3 and three subjects in SCT12 showed rating behaviors inconsistent in themselves (i.e. in the ratings for the G.711 reference condition presented twice in each test run, and/or in the ratings for different packet loss percentages). Hence, they were excluded from the more detailed analysis of the results.

G.3.4 Details on Selected Results

In Table G.7, the results of an ANOVA for the test 'SCT12' are shown, with the packet loss percentage Ppl (%) and and the noise level $Nfor$ (dBmp) as fixed factors. In the same way, Table G.8 shows the results of an ANOVA for the test 'SCT3', here with Ppl (%) and $TELR$ (dB) as fixed factors.

Table G.7: ANOVA of conversation test results obtained in 'SCT12' on random packet loss and noise, using the loss percentage Ppl [%] and the noise level $Nfor$ [dBmp] as fixed factors.

Tests of Between-Subjects Effects

Dependent Variable: MOS

Source	Type III Sum of Squares	df	Mean Square	F	Sig.
Corrected Model	147.579[a]	11	13.416	25.894	0.000
Intercept	1253.049	1	1253.049	2418.406	0.000
N FOR	71.894	2	35.947	69.378	0.000
PL_PERC	66.474	3	22.158	42.765	0.000
N FOR * PL_PERC	11.610	6	1.935	3.734	0.001
Error	116.061	224	0.518		
Total	1513.000	236			
Corrected Total	236.640	235			

[a] R Squared = 0.560 (Adjusted R Squared = 0.538)

Table G.8: ANOVA of conversation test results obtained in 'SCT3' on random packet loss and talker echo, using the loss percentage Ppl [%] and the talker echo loudness rating $TELR$ [dB] as fixed factors.

Tests of Between-Subjects Effects

Dependent Variable: MOS

Source	Type III Sum of Squares	df	Mean Square	F	Sig.
Corrected Model	131.919[a]	12	10.993	15.938	0.000
Intercept	1483.533	1	1483.533	2150.848	0.000
PL_PERC	33.714	3	11.238	16.293	0.000
TELR	76.282	3	25.427	36.865	0.000
PL_PERC * TELR	5.500	6	0.917	1.329	0.244
Error	179.333	260	0.690		
Total	1936.000	273			
Corrected Total	311.253	272			

[a]R Squared = 0.424 (Adjusted R Squared = 0.397)

G.4   3-state Markov Loss: Conversation Test

G.4.1   Test Setup

For the conversation tests, the line simulation tool described in Appendix B was used. It was connected according to the VoIP-typical configuration PSTN-VoIP-PSTN, with traditional handset-telephones as user interfaces. Packet loss was introduced with the Network Simulation NetDisturb by ZTI, Lannion, France (see Section B.3.3). Therefore, different loss traces of 100 s duration were created offline as text files using MATLAB. Since the actual duration of a test conversation could not be foreseen, each trace was looped after the 100 s had passed. In order to avoid considerably higher or lower overall loss rates than those defined by the Ppl-settings, the traces were selected so that either in the beginning or in the end of the 100 s a period of 10 s without loss was assured. In case of a repetition of the 100 s trace, a burst at the end of the trace and a burst in the beginning of the trace would combine to a longer burst. This was prevented with the loss-free passage of 10 s duration at either the beginning or the end of the trace. In addition, only those traces were selected, which showed at least one transition from 'bad' to 'good', and one from 'good' to 'bad'. Moreover, only those traces were used, which did not deviate from the desired overall loss percentage by more than 3% (relative to the desired loss percentage).

For each connection of one test series, a different trace was used, reflecting the model-oriented approach: Instead of investigating the quality impact of a particular profile, the aim was to assess speech quality under packet loss introduced by a particular loss model. Moreover, with the creation of a large number of traces, an effect of recency could be avoided (on average), as was the intention of this test.

As the codec, the G.729A was employed, with a packet size of 20 ms. An error-free, 'clean' condition with logarithmic PCM coding was presented as anchor point (G.711, A-law). The built-in PLC of the G.729A codec was used, with voice activity detection (VAD) disabled. Consequently, packet loss was not only affecting packets containing actual voice

Table G.9: Test conditions for conversation test series on macroscopic loss (3-state Markov model) under combinations with other types of degradations.

No.	CT-macro1			CT-macro2			CT-macro3			
	Cod.	Ppl [%]	N for [dBmp]	Cod.	Ppl [%]	Ta [ms]	Cod.	Ppl [%]	T [ms]	TELR [dB]
1	G.711	0	−64	G.711	0	66	G.711	0	33	65
2	G.711	0	−40	G.729	0	1000	G.729	0	100	65
3	G.729	0	−64	G.729	0	60	G.729	3	100	65
4	G.729	3	−64	G.729	3	60	G.729	5	100	65
5	G.729	5	−64	G.729	5	60	G.729	15	100	65
6	G.729	15	−64	G.729	15	60	G.729	0	100	45
7	G.729	0	−50	G.729	0	400	G.729	5	100	45
8	G.729	3	−50	G.729	3	400	G.729	15	100	45
9	G.729	5	−50	G.729	5	400	G.729	0	100	34
10	G.729	15	−50	G.729	15	400	G.729	3	100	34
11	G.729	0	−40	G.729	0	600	G.729	5	100	34
12	G.729	3	−40	G.729	3	600	G.729	15	100	34
13	G.729	5	−40	G.729	5	600	G.729	0	100	20
14	G.729	15	−40	G.729	15	600	G.729	5	100	20

data, but also packets containing silence[1]. All other additional signal processing options provided by the VoIP gateways were disabled (e.g. echo cancellation).

The conditions of the conversation tests (CT-macro1, CT-macro2, and CT-macro3) are summarized in Tables G.9 and G.10. Table G.9 also indicates the additional degradations presented to the test subjects (see Section 4.4.3).

G.4.2   Test Procedure and Test Subjects

Two conversation partners participated in each test run. The subjects were instructed to imagine a normal telephone connection. In two of the tests (CT-macro1 and CT-macro3), they were asked to carry out a dialogue task for each condition according to the SCT-scenarios described in Section 2.1.1.4 (Möller, 2000, pp. 75-81). In the conversation test on combinations of macroscopic packet loss and delay (CT-macro2), specifically designed interactive short conversation test (iSCT) scenarios were employed, which are described in more detail in Section 4.3.

In all tests, the subjects were asked to provide quality ratings on the 5-point ACR scale (MOS; Figure 2.1, p. 27), and degradation ratings on the CR-10 scale (Borg, 1982; ITU–T Rec. P.833, 2001, Appendix I). The ratings were compiled on paper.

Each test series was carried out with 22 test subjects (11 test-pairs) who were naïve with respect to the test task and conditions. They were by their own account of normal hearing. The subjects were paid for their participation in the tests. Most subjects recruited were university students, which resulted in a low average age of 26.1 years (age range 16–58 years). Of the 66 subjects, 35 were female, and 31 male. After the tests, a screening of the subjects, with respect to the consistency of their ratings, was carried out. It was found

---

[1]This was accepted, since it was considered important to maintain a trace duration of 100 s, which is impossible when VAD is used: In that case, it depends on the interlocutors, at which point in time the trace will be fully processed. Moreover, burst and gap lengths were chosen long enough to assure that each interlocutor was able to perceive the occurrence of burst and gap sequences (see text).

Table G.10: General simulation settings for conversation tests (CT-macro1 – CT-macro3) on macroscopic loss combined with additional degradations. Owing to the usage of the VoIP-simulation, the baseline transmission delay could not fall below 60 ms in case of the G.729A, and 66 ms for the G.711.

CT-macro1 – CT-macro3		CT-macro1 – CT-macro3	
Codecs	G.711 (20 ms)	Ta	66 ms (G.711)
	G.729A (20 ms)		60 ms (G.729A)
Ie; Bpl	0; 4.3 (G.711)	T	33 ms (G.711)
	11; 19.0 (G.729A)		30 ms (G.729A)
VAD	disabled	Tr	0 ms
SLR	8 dB (CT-macro1/3)	$WEPL$	110 dB
	13 dB (CT-macro2)	$TELR$	65 dB
RLR	2 dB	Nc	−70 dBm0p
$STMR$	15 dB	$N\,for$	−64 dBmp
$LSTR$	16 dB	$Ps = Pr$	35 dB(A)
$Ds = Dr$	1		

that two subjects in CT-macro1, and one subject in CT-macro3 showed intraindividually inconsistent rating behaviors (i.e. in the ratings for the G.711 reference condition presented twice in each test run, and/or in the ratings for different levels of degradation). Hence, they were excluded from the more detailed analysis of the results.

G.4.3 Details on Selected Results

To analyze the significance of the two degradation types presented in the different tests, and whether they interact in a statistically significant manner, an ANOVA was conducted for all tests 'CT-macro1' – 'CT-macro3'. In this section, only the ANOVA results for 'CT-macro1' and 'CT-macro3' are presented (Tables G.11 and G.12), since the delay presented in 'CT-macro2' did not show a significant effect on quality.

Table G.11: ANOVA of conversation test results obtained in 'CT-macro1' on 3-state Markov model type packet loss and additional noise, using the loss percentage $Ppl[\%]$ and the noise level $N\,for$ [dBmp] as fixed factors.

Tests of Between-Subjects Effects

Dependent Variable: MOS

Source	Type III Sum of Squares	df	Mean Square	F	Sig.
Corrected Model	144.146[a]	11	13.104	24.560	0.000
Incercept	1617.204	1	1617.204	3031.012	0.000
NFOR	21.233	2	10.617	19.898	0.000
PL_PERC	109.079	3	36.360	68.146	0.000
NFOR * PL_PERC	13.833	6	2.306	4.321	0.000
Error	121.650	228	0.534		
Total	1883.000	240			
Corrected Total	265.796	239			

[a]R Squared = 0.542 (Adjusted R Squared = 0.520)

Table G.12: ANOVA of conversation test results obtained in 'CT-macro3' on 3-state Markov model type packet loss and additional talker echo, using the loss percentage $Ppl[\%]$ and the talker echo loudness rating $TELR$ [dB] as fixed factors.

Tests of Between-Subjects Effects

Dependent Variable: MOS

Source	Type III Sum of Squares	df	Mean Square	F	Sig.
Corrected Model	159.718[a]	12	13.310	21.945	0.000
Intercept	1450.066	1	1450.066	2390.888	0.000
TELR	40.252	3	13.417	22.123	0.000
PL_PERC	125.054	3	41.685	68.730	0.000
TELR * PL_PERC	10.480	6	1.747	2.880	0.010
Error	155.263	256	0.606		
Total	2039.000	269			
Corrected Total	314.981	268			

[a]R Squared = 0.507 (Adjusted R Squared = 0.484)

G.5 Speech Sound Quality and Content

To analyze whether quality in the tests described in Section 5.4.2 was affected by the genre of the text presented to the subjects, and whether interactions such as between the genre and the speaker can be observed, an ANOVA was conducted both for the test run with the French subjects, and the tests run with the German subjects (Tables G.13 and G.14).

Table G.13: ANOVA of listening test results obtained from the French subjects in the test on the speech sound quality-impact of the speech content under different bandwidth limitations. The speaker, the text genre and the bandwidth were used as fixed factors.

Tests of Between-Subjects Effects

Dependent Variable: SOUND

Source	Type III Sum of Squares	df	Mean Square	F	Sig.
Corrected Model	4436.899[a]	71	62.492	16.001	0.000
Intercept	32988.369	1	32988.369	8446.774	0.000
SPEAKER	36.130	5	7.226	1.850	0.100
GENRE	106.313	3	35.438	9.074	0.000
F_BAND	3934.018	2	1967.009	503.659	0.000
SPEAKER * GENRE	98.398	15	6.560	1.680	0.049
SPEAKER * F_BAND	76.160	10	7.616	1.950	0.035
GENRE * F_BAND	9.709	6	1.618	0.414	0.870
SPEAKER * GENRE * F_BAND	176.170	30	5.872	1.504	0.040
Error	4499.067	1152	3.905		
Total	41924.335	1224			
Corrected Total	8935.966	1223			

[a]R Squared = 0.497 (Adjusted R Squared = 0.465)

Table G.14: ANOVA of listening test results obtained from the German subjects in the test on the speech sound quality-impact of the speech content under different bandwidth limitations. The speaker, the text genre and the bandwidth were used as fixed factors.

Tests of Between-Subjects Effects

Dependent Variable: SOUND

Source	Type III Sum of Squares	df	Mean Square	F	Sig.
Corrected Model	2890.171[a]	71	40.707	21.359	0.000
Intercept	20369.298	1	20369.298	10687.65	0.000
SPEAKER	53.681	5	10.736	5.633	0.000
GENRE	3.866	3	1.289	0.676	0.567
BANDPASS	2640.838	2	1320.419	692.816	0.000
SPEAKER * GENRE	72.065	15	4.804	2.521	0.001
SPEAKER * BANDPASS	25.137	10	2.514	1.319	0.216
GENRE * BANDPASS	15.786	6	2.631	1.380	0.220
SPEAKER * GENRE * BANDPASS	78.798	30	2.627	1.378	0.087
Error	1509.452	792	1.906		
Total	24768.921	864			
Corrected Total	4399.622	863			

[a]R Squared = 0.657 (Adjusted R Squared = 0.626)

H

Modeling Details

H.1   Time-Varying Distortions

H.1.1   Macroscopic Loss Behavior

H.1.1.1   Prediction: Time Averaging versus Average Loss

For all loss profiles studied in Section 4.2.1 (for references to the corresponding documents, see Table H.1), Table H.2 shows a comparison between the integral ratings (column #3: 'integral MOS', 'iMOS'), and the Mean Opinion Score predicted by time-averaging over random loss segments (column #4: '$\overline{MOS}(R)$'). Column #6 provides the differences between the integral test ratings and the quality averaging estimates ('$A = iMOS - \overline{MOS}(R)$'). As can be seen from this column, two outliers do not fulfil the requirement set by item point (c), namely profiles #19 and #25. For these profiles, the absolute difference between prediction and test result ($|A|$) is larger than 0.35 MOS (column #8: $|A| > 0.35$). For several profiles similar to these two, good agreement between integral test MOS and estimated time-averaged MOS was achieved. Hence, the mismatch may be ascribed to the individual test results rather than to the method for estimating integral quality by segment-wise averaging.

Table H.1: Macroscopic loss; study on the relation between segments of microscopic (here random) loss behavior and integral quality: List of test profiles employed from Gros and Chateau's work (see Section 4.2.1).

Number in this Book	Number in Original Document	Original Document
#1–8	#3–10	(ITU–T Contribution COM 12–94, 1999)
#9–14	#4–6; #13–15	(ITU–T Delayed Contribution D.139, 2000)
#15–26	#2, 5, 8, 10, 12, 15, 18, 21, 24, 26, 28, 31	(Gros, 2001, Experiment 3, pp. 98–104)
#27–30	P3, D1E2, D2E2, D3E2	(Gros, 2001, Experiment 4, pp. 114–120)
#31–35	P8, P15, P16, P17, P18	(ITU–T Contribution COM 12–22, 2000)

Speech Quality of VoIP: Assessment and Prediction   Alexander Raake
© 2006 John Wiley & Sons, Ltd

Table H.2: Integral quality obtained from the listening tests by Gros and Chateau using different methods.

No.	$\overline{Ppl}$ [%]	iMOS (test)	$\overline{MOS}$ (R)	MOS ($\overline{Ppl}$)	A	B	$|A| >$ 0.35	$|B| >$ 0.35
1	6.3	3.2	3.0	2.9	0.2	0.3		
2	12.5	2.1	2.5	2.2	−0.3	−0.1		
3	8.5	2.5	2.7	2.6	−0.2	−0.1		
4	5.5	3.0	3.1	3.0	−0.1	0.0		
5	8.8	2.7	2.8	2.6	−0.1	0.1		
6	13.5	2.0	2.3	2.1	−0.3	−0.1		
7	20.5	2.1	1.9	1.7	0.2	0.3		
8	22.3	1.9	1.8	1.7	0.1	0.2		
9	2.4	3.9	3.7	3.5	0.3	0.4		X
10	4.7	3.5	3.4	3.1	0.1	0.4		X
11	9.5	3.3	3.0	2.5	0.3	0.8		X
12	27.6	1.8	1.6	1.5	0.1	0.3		
13	25.3	1.9	1.8	1.6	0.1	0.3		
14	20.5	2.4	2.1	1.7	0.2	0.6		X
15	5.0	3.8	3.5	3.1	0.3	0.7		X
16	10.0	3.3	3.1	2.4	0.2	0.9		X
17	15.0	2.8	2.7	2.0	0.1	0.8		X
18	15.0	2.6	2.8	2.0	−0.2	0.5		X
19	20.0	2.0	2.4	1.8	−0.5	0.2	X	
20	25.0	2.1	2.1	1.6	0.0	0.5		X
21	5.0	3.5	3.4	3.1	0.1	0.4		X
22	10.0	3.1	2.9	2.4	0.1	0.6		X
23	15.0	2.8	2.5	2.0	0.2	0.7		X
24	15.0	2.7	2.6	2.0	0.1	0.7		X
25	20.0	1.8	2.2	1.8	−0.4	0.0	X	
26	25.0	1.8	1.9	1.6	−0.1	0.2		
27	10.0	3.3	3.1	2.4	0.2	0.8		X
28	10.0	2.7	2.9	2.4	−0.2	0.3		
29	10.0	3.3	3.0	2.4	0.3	0.9		X
30	6.3	2.9	3.0	2.9	−0.2	0.0		
31	10.0	2.9	2.9	2.4	−0.1	0.4		X
32	10.0	2.8	2.8	2.4	0.0	0.4		X
33	10.0	2.7	2.8	2.4	−0.1	0.3		
34	6.7	3.0	3.1	2.8	0.0	0.2		
35	10.0	3.0	2.7	2.4	0.3	0.6		X

2nd column, '$\overline{Ppl}$ [%]': Average packet loss percentage of the profile; 3rd column, 'iMOS(test)': Integral MOS-scores obtained in the listening tests by Gros and Chateau; 4th column, '$\overline{MOS}$(R)': Segment-wise time-average (including exponential behavior) over R estimated with E-model for random loss, and subsequent transformation to MOS scale; 5th column, 'MOS($\overline{Ppl}$)': MOS estimated with the random loss E-model based on the mean loss rate, which has been averaged over the individual profile; 6th column, '$A = iMOS - \overline{MOS}(R)$': Difference between integral quality ratings (intMOS) and segment-wise time-averaging model ($\overline{MOS}(R)$); 7th column, '$B = iMOS - MOS(\overline{Ppl})$': Difference between integral quality ratings (iMOS) and MOS estimated based on the mean loss rate ($MOS(\overline{Ppl})$).

In column #5 of Table H.2, MOS estimates are listed that were calculated from the average loss percentage of each profile ($MOS(\overline{Ppl})$), by using the random loss formula of the E-model (Equation (4.5)). The discrepancies between the integral test ratings and the estimates from the average loss percentage are listed in column #7 ($B = intMOS - MOS(\overline{Ppl})$). Column #9 indicates those profiles, for which the prediction based on average packet loss deviates by more than 0.35 MOS from the integral quality test results.

H.1.1.2   Prediction from Average Loss Rate: Limitations

A theoretical investigation was undertaken in order to study the limits for accurately estimating integral quality associated with macroscopic loss patterns from the average loss percentage $\overline{Ppl}$ (see Section 4.2.1.2). To this aim, a set of virtual loss profiles was created, with the following motivation:

- The time-averaging over passages of random loss behavior was found to yield acceptable predictions of integral quality, when appropriate decay times are chosen (Section 4.2.1.1). In this case, the discrepancy between integral quality judgments and estimates by time-averaging are below 0.35 MOS.

- Estimates of integral quality based on the average loss rate deviate most from integral quality ratings, when the loss profiles show an expressed *granularity*, that is when sparsely, but strongly varying loss levels are considered.

A two-step procedure was employed:

(a) Different loss profiles are generated, which are defined by two interchanging levels of packet loss percentage ($Ppl1$ and $Ppl2$), and a given periodicity T. These three values define the granularity of the macroscopic loss pattern. An illustration of the profiles is provided in Figure H.1.

(b) The profiles were created iteratively, using different values for $Ppl1$ and $Ppl2$. For each loss setting, the periodicity T is iterated as $T = 2^i, i \in [0, 8]$. At each step, a quality estimate is calculated using the segment-wise averaging of time-varying quality as outlined in Section 4.2.1.1. Also calculated is a quality estimate based on the average

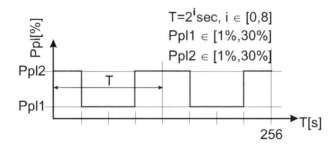

Figure H.1: Profiles used to estimate the maximal granularity of loss profiles that allows integral quality to be predicted with reasonable accuracy, based on the average loss rate (ITU–T Delayed Contribution D.179, 2003).

loss percentage of the profile. The iteration was stopped, when one or both of two criteria were met:

- The quality estimated from the average loss percentage is smaller than the MOS obtained by time-averaging by more than 0.34 MOS.

- The quality estimated from the average loss percentage is larger than the MOS obtained by time-averaging by more than 0.05 MOS.

Since predictions by time averaging are assumed to reflect integral quality judgements $intMOS$ with $\pm 0.35\ MOS$ accuracy (Section 4.2.1.1), the above criteria assure that the MOS estimated from the average loss percentage lies in the range $(intMOS - 0.7) < MOS(\overline{Ppl}) < (intMOS + 0.4)$. According to the observations in Section 4.2.1.2, however, overestimation by the predicted $MOS(\overline{Ppl})$ cannot be expected.

For each combination of packet loss levels ($Ppl1$ and $Ppl2$), the periodicity T at which the iteration stops provides a measure of the maximum spread of loss periods, that still allows reasonable quality predictions to be made from the average loss percentage.

As can be seen from Figure H.2, almost all studied packet loss combinations [$Ppl1$, $Ppl2$] yield differences between average MOS ($\overline{MOS}(R)$) and $MOS(\overline{Ppl})$, that is MOS of average packet loss, which meet the criteria described above. Only if passages of very low packet losses (e.g. $Ppl1 < 3\%$) are combined with very high ones (e.g. $Ppl2 > 23\%$), the maximal period T is reduced below 256 s.

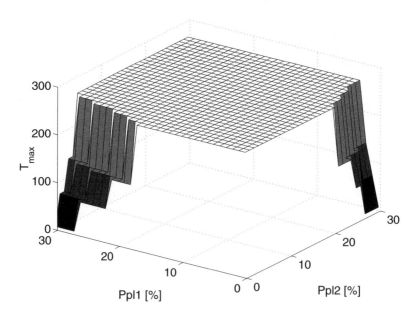

Figure H.2: Maximum loss-period T_{max} that still allows acceptable quality estimates to be made based on the average loss rate (ITU–T Delayed Contribution D.179, 2003).

I

Glossary

Quality	Result of judgement of the perceived composition of an entity with respect to its desired composition (Jekosch, 2000, 2005b, pp. 15).
Perceived Composition (Perceived Nature)	Totality of features of an entity (Jekosch, 2005b, pp. 16).
Feature	Recognizable and nameable characteristic of an entity (Jekosch, 2005b, p. 14).
Desired Composition (Desired Nature)	Totality of features of individual expectations and/or relevant demands and/or social requirements (Jekosch, 2005b, p. 16).
Assessment	Measurement of system performance with respect to one or more criteria. Typically used to compare like with like, whether two alternative implementations of a technology, or successive generations of the same implementation (Jekosch, 2000, 2005b, p. 109).
Subject-oriented (Quality) Test	Test with the (perceiving) test subject as the measurement object.
Object-oriented (Quality) Test	Test with the perceived object (e.g. speech sound or transmission system) as the measurement object.
Utilitarian Quality Test	Test aiming at a descriptive quality rating typically obtained on a uni-dimensional quality or impairment scale (Hecker and Guttman, 1967; Quackenbush *et al.*, 1988, pp. 15-16).

Analytical Quality Test	Test aiming at a description of all or of certain features of the perceived entity, without necessarily collecting quality judgements.
Quality Element	Contribution to the quality 1) of an immaterial or a material product as the result of an action/activity or a process in one of the planning, execution or usage phases or 2) of an action or of a process as the result of an element in the course of this action or process (according to Jekosch, 2005b, p. 22, modified from DIN 55350, Part 11).
Quality Feature	Recognized and designated characteristic of an entity that is relevant to the entity's quality (Jekosch, 2000, 2004, 2005b, p. 17).
Speech Quality	Quality perceived in a conversational situation.
Speech Transmission Quality	Quality perceived of a transmitted speech signal in a listening-only situation.
Integral (Speech) Quality	Quality due to the totality of quality features, related to an entire conversation or speech passage, taking the history and evolution of the conversation or passage into consideration; corresponds to the final quality judgement obtained at the end of the conversation.
Instantaneous Quality	Quality instantaneously judged by subjects (Gros and Chateau, 2001).
Average Instantaneous Quality	Mathematically obtained time average of instantaneous quality.
Quality of Service (QoS)	The collective effect of objective service performance which ultimately determines the degree of satisfaction of a user of the service (ITU–T Delayed Contribution D.197, 2004; ITU–T Rec. E.800, 1994).
Quality of Experience (QoE)	Measure of the overall acceptability of an application or service, as perceived subjectively by the end-user (cf. ITU–T Delayed Contribution D.197, 2004).
Intelligibility	Identifiability of the content of an utterance based on the form (i.e. the carrier); strongly depends on lexical, syntactic and semantic context.
Communicability	Capacity of a speech message to serve to communicate and to be fully understood by a recipient, ideally as it was intended by the sender; related to functional aspects of speech and the entire communication process.

Coloration	Directed change in the original timbre of a sound source (see e.g. Brüggen, 2001, p. 7).
Speech Sound Quality	Speech quality associated with a particular coloration.
Talkspurt	Time period judged by a listener to contain a sequence of speech sounds unbroken by silence (Brady, 1965).
Packet Trace	Collection of packet records containing a particular amount of packet header information from subsequently collected packets (e.g. timestamp, packet number, etc.).
Microscopic Loss Behavior	Packet loss behavior clearly perceived as time-varying on a perceptual feature level, without leading to time-varying quality, i.e. considerable level-changes in instantaneous quality judgements.
Macroscopic Loss Behavior	Packet loss behavior leading to time-varying instantaneous quality, i.e. associated with clearly recognizable changes in instantaneous speech quality.

Bibliography

3GPP TR 26.976 2002 *Performance Characterization of the Adaptive Multi-Rate Wideband (AMR-WB) Speech Codec*. 3rd Generation Partnership Project (3GPP), F–Sophia Antipolis.

Allnatt JW 1975 Subjective rating and apparent magnitude. *Int. J. Man Mach. Stud.* **7**, 801–816.

Andersen SV, Kleijn W, Hagen R, Linden J, Murthi M and Skoglund J 2002 iLBC – a linear predictive coder with robustness to packet losses. In: *Proc. IEEE Speech Coding Workshop 2002*, October 6–9, JP-Japan, pp. 23–25.

Appel R and Beerends JG 2002 On the quality of hearing one's own voice. *J. Audio Eng. Soc. (AES)* **50**(4), 237–246.

Arai T and Greenberg S 1997 The temporal properties of spoken Japanese are similar to those of English. In: *Proc. 5th European Conference Speech Communication and Technology. (Eurospeech 97)*, GR-Rhodes, pp. 1011–1014.

Baddeley A 1997 *Human Memory – Theory and Practice*. Taylor & Francis: Psychology Press, UK–East Sussex.

Bappert V and Blauert J 1994 Auditory quality evaluation of speech-coding systems. *Acta Acustica* **2**, 49–58.

Barber CB, Dobkin DP and Huhdanpaa HT 1996 The Quickhull algorithm for convex hulls. *ACM Trans. Math. Softw.* **22**(4), 469–483.

Barriac V, Le Saout JY and Lockwood C 2004 Discussion on unified objective methodologies for the comparison of voice quality of narrowband and wideband scenarios. In: *Proc. Workshop on Wideband Speech Quality in Terminals and Networks: Assessment and Prediction*, D-Mainz.

Bashford JA and Warren JRM 1987 Multiple phonemic restoration follows the rules of auditory induction. *Percept. Psychophys.* **42**, 114–121.

Bashford JA Jr, Meyers MD, Brubaker BS and Warren RM 1988 Illusory continuity of interrupted speech: Speech rate determines durational limits. *J. Acoust. Soc. Am.* **84**(5), 1635–1638.

Beerends J, Hekstra A, Rix A and Hollier M 2002 Perceptual evaluation of speech quality (PESQ), the new itu standard for end-to-end speech quality assessment, part II – psychoacoustic model. *J. Audio Eng. Soc. (AES)* **50**(10), 765–778.

Behrens E 2000 Introduction to Markov Chains. *Advanced Lectures in Mathematics*. Vieweg, D–Braunschweig.

Benoît C 1990 An intelligibility test using semantically unpredictable sentences: Towards the quantification of linguistic complexity. *Speech Commun.* **9**(4), 293–304.

Benoît C, Grice M and Hazan V 1996 The SUS test: A method for the assessment of text-to-speech synthesis intelligibility using semantically unpredictable sentences. *Speech Commun.* **18**(4), 381–392.

Beranek LL 1971 *Noise and Vibration Control*. McGraw–Hill, USA–New York.

Speech Quality of VoIP: Assessment and Prediction   Alexander Raake
© 2006 John Wiley & Sons, Ltd

Berger J 1998 *Instrumentelle Verfahren zur Sprachqualitätsschätzung – Modelle auditiver Tests*. Doctoral dissertation, Christian–Albrechts–Universität Kiel (Arbeiten über Digitale Signalverarbeitung No. 13, U. Heute, ed.). Shaker Verlag, D–Aachen.

Bernex E and Barriac V 2002 Architecture of non-intrusive perceived voice-quality assessment. In: *Proc. MESAQIN 2002* (J. Holub and R. Smid, eds.). Czech Technical University, CZ–Prague, pp. 13–16,

Blauert J 1997 *Spatial Hearing: The Psychophysics of Human Sound Localization*. MIT Press, USA-Cambridge, MA.

Blauert J and Jekosch U 2003 Concepts behind sound quality: Some basic considerations. In: *Proc. of Internoise 2003*, Vol. 1, KR-Jeju, pp. 72–79.

Blauert J and Lindemann W 1986 Auditory spaciousness: Some further psychoacoustic analyses. *J. Acoust. Soc. Am.* **80**(2), 533–542.

Bodden M and Jekosch U 1996 *Entwicklung und Durchführung von Tests mit Versuchspersonen zur Verifizierung von Modellen zur Berechnung der Sprachübertragungsqualität*. Final report to a project funded by Deutsche Telekom AG (unpublished). Institut für Kommunikationsakustik, Ruhr–Universität, D–Bochum.

Bolot JC 1993 Characterizing end-to-end delay and loss in the Internet. *J. High Speed Netw.* **2**(3), 305–323.

Bolot JC and Vega-García A 1996 Control mechanisms for packet audio in the Internet. In: *Proc. 15th Annual Joint Conference of the IEEE Computer and Communications Societies (INFOCOM '96)*, April, Vol. 1, USA–San Francisco, CA, pp. 232–239.

Bolot JC, Fosse-Parisis S and Towsley D 1999 Adaptive FEC-based error control for Internet telephony. In: *Proc. 18th Annual Joint Conference of the IEEE Computer and Communications Societies (INFOCOM '99)*, March, USA–New York, Vol. 3, pp. 1453–1460.

Borg G 1982 A category rating scale with ratio properties for intermodal and interindividual comparisons. In: *Psychophysical Judgement and the Process of Perception* (H.-G. Geissler and P. Petzold, eds.). VEB Deutscher Verlag der Wissenschaften, D-Berlin. pp. 25–34.

Brady P 1965 A technique for investigating on-off patterns of speech. *Bell Syst. Tech. J.* **44**(1), 1–22.

Brady P 1968 A statistical analysis of on-off patterns in 16 conversations. *Bell Syst. Tech. J.* **47**(1), 73–91.

Bregman AS 1990 *Auditory Scene Analysis*. MIT Press, USA-Cambridge, MA.

Bronkhorst A 2000 The Cocktail Party phenomenon: A review of research on speech intelligibility in multi-talker conditions. *Acta Acustica* **86**(1), 117–128.

Bronkhorst AW, Bosman AJ and Smoorenburg GF 1993 A model for context in speech recognition. *J. Acoust. Soc. Am.* **93**(1), 499–509.

Bronkhorst AW, Brand T and Wagener K 2002 Evaluation of context effects in sentence recognition. *J. Acoust. Soc. Am.* **111**(6), 2874–2886.

Brüggen M 2001 *Klangverfärbung durch Rückwürfe und ihre auditive und instrumentelle Kompensation*. dissertation.de, www.dissertation.de, D–Berlin.

Bücklein R 1962 Hörbarkeit von Unregelmäßigkeiten in Frequenzgängen bei akustischer Übertragung. *Frequenz* **16**, 103–108.

Carlyon RP, Deeks J, Norris D and Butterfield S 2002 The continuity illusion and vowel identification. *Acta Acustica* **88**, 408–415.

Carroll JD 1972 Individual preferences and multidimensional scaling. In: Multidimensional Scaling: Theory and Applications in the Behavioral Sciences (R. N. Shepard, A. K. Romney and S.B. Nerlove, eds.), Vol. I, pp. 105–55.

Carroll JD and Chang JJ 1970 Analysis of individual differences in multidimensional scaling via an n-way generalization of Eckart-Young decomposition. *Psychometrika* **35**, 283–319.

Clark A 2001 Modeling the effects of burst packet loss and recency on subjective voice quality. In: *Internet Telephony Workshop (IPtel 2001)*, April 2–3, Columbia University, USA–New York.

Cole R and Rosenbluth J 2001 Voice over IP performance monitoring. *Comput. Commun. Rev.* **31**(2), 9–24.

Coleman AE, Gleiss N and Usai P 1988 A subjective testing methodology for evaluating medium rate codecs for digital mobile radio applications. *Speech Commun.* **7**, 151–166.

Cowan N 1984 On short and long auditory stores. *Psychol. Bull.* **96**(2), 341–70.

Crovella ME and Bestavros A 1996 Self-similarity in world wide web traffic: Evidence and possible causes. In: *Proceedings 1996 IEEE Conference on Network Protocols (ICNP'96)*, USA–Los Alamitos, CA, pp. 171–180.

Crovella ME and Bestavros A 1997 Self-similarity in world wide web traffic: Evidence and possible causes. *IEEE/ACM Trans. Netw.* **5**(6), 835–846.

Crowder RG 1993 Auditory memory. In: *Thinking in Sound – The Cognitive Psychology of Human Audition* (S. McAdams, E. Bigand, eds.). Oxford University Press, UK–Oxford. pp. 113–145.

Crowder RG and Morton J 1969 Precategorical acoustic storage (PAS). *Percept. Psychophys.* **5**, 365–73.

DIN 55350, Part 11 1995 *Begriffe zu Qualitätsmanagement und Statistik*. Beuth, D–Berlin.

Ding L and Goubran RA 2003 Speech quality prediction in VoIP using the extended E-model. In: *Proc. IEEE GLOBECOM 2003*, December 1–5 2003, Vol. 7, San Francisco, USA, pp. 3974–3978.

Duncanson JP 1969 The average telephone call is better than the average telephone call. *Public. Opin. Q.* **33**(1), 112–116.

EIA/TIA/IS–54 1990 *Cellular System Dual-Mode Mobile Station - Base Station Compatibility Standard*. Telecommunications Industry Association, USA-Arlington, VA.

EIA/TIA/IS-54-B 1990 *Cellular System Dual-Mode Mobile Station - Base Station Compatibility Standard*. Telecommunications Industry Association (TIA), USA-Washington, WA.

Eiken HK 2004 *Influence of Delay on Conversational Quality in IP Telephony*. Masters Thesis (unpublished), Norwegian University of Science and Technology (NTNU), NO–Trondheim.

Elliott E 1963 Estimates of error rates for codes on burst noise channels. *Bell Syst. Tech. J.* **42**(2), 1977–1997.

ETSI GSM 06.10 1988 *GSM Full Rate Speech Transcoding*. European Telecommunications Standards Institute, F–Sophia Antipolis.

ETSI GSM 06.20 1996 *GSM Half Rate Speech Transcoding*. European Telecommunications Standards Institute, F–Sophia Antipolis.

ETSI GSM 06.60 1996 *Digital Cellular Telecommunications System; Enhanced Full Rate Speech Transcoding*. European Telecommunications Standards Institute, F–Sophia Antipolis.

ETSI Guide EG 201 013 1997 *Human Factors (HF); Definitions, Abbreviations and Symbols*. European Telecommunications Standards Institute, F–Sophia Antipolis.

ETSI Technical Report ETR 250 1996 *Transmission and Multiplexing (TM); Speech Communication Quality from Mouth to Ear for 3,1 kHz Handset Telephony across Networks*. European Telecommunications Standards Institute, F–Sophia Antipolis.

ETSI Tiphon 11 Temporary Document 64 1999 *Results of VoIP Simulation*. European Telecommunications Standards Institute, F–Sophia Antipolis.

ETSI TS 126 071 2002 *Universal Mobile Telecommunications System (UMTS); AMR Speech Codec; General Description*. European Telecommunications Standards Institute, F–Sophia Antipolis.

Fletcher H and Munson W 1937 Relation between loudness and masking. *J. Acoust. Soc. Am.* **9**(1), 1–10.

Flohrer W 1968 Die Beeinträchtigung der Natürlichkeit von Sprache durch Löcher im Frequenzband. *Frequenz* **22**, 175–178.

French NR and Steinberg JC 1947 Factors governing the intelligibility of speech sounds. *J. Acoust. Soc. Am.* **19**(1), 90–119.

Frossard P 2001 FEC performance in multimedia streaming. *IEEE Commun. Lett.* **5**(3), 122–124.

Gabrielsson A and Sjogren H 1979 Perceived sound quality of sound-reproduction systems. *J. Acoust. Soc. Am.* **65**(4), 1019–1033.

Gabrielsson A, Schenkman BN and Hagerman B 1985 *The Effects of Different Frequency Responses on Sound Quality Judgments and Speech Intelligibility*, Report Technical Audiology No. 112, Karolinska Institute, Department of Technical Audiology, KTH, S–Stockholm.

Gibbon D 1992 *EUROM.1 German Speech Database*, ESPRIT Project 2589 Report (SAM, Multi–Lingual Speech Input/Output Assessment, Methodology and Standardization), Universität Bielefeld, D–Bielefeld.

Gibbon D, Moore R and Winski R 1997 *Handbook on Standards and Resources for Spoken Language Systems*. Mouton de Gruyter, D–Berlin.

Gierlich HW 1996 The auditory perceived quality of hands-free telephones: Auditory judgements, instrumental measurements and their relationship. *Speech Commun.* **20**, 241–254.

Gilbert EN 1960 Capacity of a burst-noise channel. *Bell Syst. Tech. J.* **34**(5), 1253–1265.

Glanzer M and Cunitz AR 1966 Two storage mechanisms in free recall. *J. Verbal Learn. Verbal Behav.* **5**, 351–360.

Gleiss N 1970 The effect of bandwidth restriction on speech transmission quality in telephony. *Proceedings 4th International Symposium on Human Factors in Telephony (D–Bad Wiessee, 1968)*. VDE–Verlag, D-Berlin. pp. 1–6.

Gleiss N 1989 *Desirable Sending Frequency Response of Telephone Sets*, Vol. 1/89, TELE (English edition). Swedish Telecommunications Administration, S-Stockholm, pp. 18–23.

Greenberg S 1999 Speaking in shorthand – a syllable-centric perspective for understanding pronunciation variation. *Speech Commun.* **29**, 159–176.

Greenberg S 2004 Temporal properties of spoken language. In: *Proc. International Congress on Acoustics*, Vol. 1, J-Kyoto, pp. 441–444.

Gros L 2001 *Evaluation Subjective de la Qualité Vocale Fluctuante*. Doctoral dissertation, Université de la Méditerranée Aix-Marseille II, Equipe d'acceuil, France Telecom R&D, F-Lannion.

Gros L and Chateau N 2001 Instantaneous and overall judgements for time-varying speech quality: Assessments and relationships. *Acta Acustica* **87**(3), 367–377.

Gros L and Chateau N 2002 The impact of listening and conversational situations on speech perceived quality for time-varying impairments. In: *Proc. MESAQIN 2002* (J. Holub and R. Smid, eds.). Czech Technical University, CZ-Prague, pp. 17–19.

Gruber JG and Strawczynski L 1985 Subjective effects of variable delay and speech clipping in dynamically managed voice systems. *IEEE Trans. Commun.* **COM–33**(8), 801–8.

Guilford JP 1954 *Psychometric Methods*. 2nd edition. McGraw-Hill, USA–New York.

Halka U 1993 *Objektive Qualitätsbeurteilung von Sprachkodiersystemen unter Anwendung von Sprachmodellprozessen*. Doctoral dissertation, Ruhr–Universität Bochum (Arbeiten über Digitale Signalverarbeitung No. 3, U. Heute, ed.). Shaker Verlag, D–Aachen.

Hall JL 2001 Application of multidimensional scaling to subjective evaluation of coded speech. *J. Acoust. Soc. Am.* **110**(4), 2167–2182.

Hammer F, Reichl P and Raake A 2004a Elements of interactivity in telephone conversations. In: *Proc. International Conference Spoken Language Processing (ICSLP 2004)*, October 4–8, 2004, KR–Jeju Island.

Hammer F, Reichl P and Ziegler T 2004b Where packet traces meet speech samples: An instrumental approach to perceptual QoS evaluation of VoIP. In: *Proc. International Workshop on Quality of Service (IWQoS)*, June, CDN-Montreal, pp. 273–280.

Hammer F, Reichl P, Nordström T and Kubin G 2004c Corrupted speech data considered useful: Improving perceived speech quality of VoIP over error-prone channels. *Acta Acustica* **90**(6), 1052–1060.

Hammer F, Reichl P, Nordström T and Kubin G 2003 Corrupted speech data considered useful. In: *Proc. 1st ISCA Tutorial and Research Workshop on Auditory Quality of Systems*, April 23–25, Akademie Mont-Cenis, D–Herne, pp. 51–4.

Hands D, Gale C and Bruine AD 2001 *User Experiments and Trials* unpublished paper, M3I Project www.m3I.org.

Hansen M and Kollmeier B 1999 Continuous assessment of time-varying speech quality. *J. Acoust. Soc. Am.* **106**(5), 2888–2899.

Hansen M and Kollmeier B 2000 Objective modeling of speech quality with a psychoacoustically validated auditory model. *J. Audio Eng. Soc. (AES)* **48**(5), 395–409.

Hardman V, Sasse MA, Handley M and Watson A 1995 Reliable audio for use over the Internet. In: *Proc. International Networking Conference, Internet Socity (INET'95)* Vancouver, CDN–Raston.

Hauenstein M 1997 *Psychoakustisch motivierte Maße zur instrumentellen Sprachgütebeurteilung.* Doctoral dissertation, Christian–Albrechts–Universität Kiel (Arbeiten über Digitale Signalverarbeitung No. 10, U. Heute, ed.). Shaker Verlag, D–Aachen.

Hecker MHL and Guttman N 1967 Survey of methods for measuring speech quality. *J. Audio Eng. Soc. (AES)* **15**(4), 400–403.

Heute U, Moeller S, Raake A, Waeltermann M and Scholz K 2005 Integral and diagnostic speech-quality measurement: State of the art, problems, and new approaches. In: *Proc. EAA/Forum Acusticum 2005*, H–Budapest.

Hoene C, Rathke B and Wolisz A 2003 On the importance of a VoIP packet. In: *Proc. 1st ISCA Tutorial and Research Workshop on Auditory Quality of Systems*, April 23–25, Akademie Mont-Cenis, D-Herne. pp. 55–62.

Holmes P, Aarhus L and Maus E 2001 Tolerance of highly degraded network conditiond for an H.323-based VoIP service. In: *Proc. Interactive Distributed Multimedia Systems. 8th International Workshop (IDMS 2001)* (D. Shepard, J. Finney, L. Mathy and N. Race (eds.). Springer-Verlag, D–Berlin, pp. 74–85.

House AS, Williams CE, Hecker MHL and Kryter KD 1965 Articulation testing methods: Consonantal differentiation with a closed response set. *J. Acoust. Soc. Am.* **37**(1), 158–166.

Houtgast T and Steeneken HJM 1985 A review of the MTF concept in room acoustics and its use for estimating speech intelligibility in auditoria. *J. Acoust. Soc. Am.* **77**(3), 1069–1077.

Huggins A and Nickerson R 1985 Speech quality evaluation using "phoneme-specific" sentences. *J. Acoust. Soc. Am.* **77**(5), 1896–1906.

IEC Publ. 268-13 date unknown *Listening Test on Loudspeakers*. International Electrotechnical Commission, CH–Geneva.

IEEE Std 802.11 2005 *Wireless LAN Medium Access Control (MAC) and Physical Layer (PHY) Specifications*. IEEE, USA–New York.

IETF RFC 3550 2003 *RTP: A Transport Protocol for Real-Time Applications* (Schulzrinne H, Casner S, Frederick R, Jacobson, V). Internet Engineering Task Force (IETF), http://www.ietf.org.

IETF RFC 2198 1997 *RTP Payload for Redundant Audio Data* (Perkins C, Kouvelas I, Hodson O, Hardman V, Handley M, Bolot J, Vega-Garcia A, Fosse-Parisis S). Internet Engineering Task Force (IETF), http://www.ietf.org.

IETF RFC 2205 1997 *Resource ReSerVation Protocol (RSVP)* (Authors: Braden R, Zhang L, Berson S, Herzog S, Jamin S). Internet Engineering Task Force (IETF), http://www.ietf.org.

IETF RFC 2474 1998 *Definition of the Differentiated Services Field (DS Field) in the IPv4 and IPv6 Headers* (Nichols K, Blake S, Baker F, Black D). Internet Engineering Task Force (IETF), http://www.ietf.org.

IETF RFC 2475 1998 *An Architecture for Differentiated Services* (Blake S, Black D, Carlson M, Davies E, Wang Z, Weiss W). Internet Engineering Task Force (IETF), http://www.ietf.org.

IETF RFC 2597 1999 *Assured Forwarding PHB Group* (Heinanen J, Baker F, Weiss W, Wroclawski J). Internet Engineering Task Force (IETF), http://www.ietf.org.

IETF RFC 2733 1999 *An RTP Payload Format for Generic Forward Error Correction* (Rosenberg J, and Schulzrinne H). Internet Engineering Task Force (IETF), http://www.ietf.org.

IETF RFC 3031 2001 *Multiprotocol Label Switching Architecture* (Rosen E, Viswanathan A, Callon R). Internet Engineering Task Force (IETF), http://www.ietf.org.

IETF RFC 3095 2001 *RObust Header Compression (ROHC): Framework and Four Profiles: RTP, UDP, ESP, and Uncompressed* (Bormann C, Burmeister C, Degermark M, Fukushima H, Hannu H, Jonsson L.-E, Hakenberg R, Koren T, Le K, Liu Z, Martensson A, Miyazaki A, Svanbro K, Wiebke T, Yoshimura T, Zheng H). Internet Engineering Task Force (IETF), http://www.ietf.org.

IETF RFC 3261 2002 *SIP: Session Initiation Protocol* (Authors: Rosenberg J, Schulzrinne H, Camarillo G, Johnston A, Peterson J, Sparks R, Handley M, Schooler E). Internet Engineering Task Force (IETF), http://www.ietf.org.

IETF RFC 3357 2002 *One-way Loss Pattern Sample Metrics* (Koodli R. and Ravikanth R.) Internet Engineering Task Force (IETF), http://www.ietf.org.

IETF RFC 3611 2003 *RTP Control Protocol Extended Reports (RTCP XR)* (Friedman T, Caceres R, Clark A). Internet Engineering Task Force (IETF), http://www.ietf.org.

IETF RFC 768 1980 *User Datagram Protocol* (Postel J). Internet Engineering Task Force (IETF), http://www.ietf.org.

Institut de la Communication Parlée (ICP) 1994 *Eurom 1 Multilingual Database; 40 blocks of 5 thematically linked sentences.* Institut de la Communication Parlée (ICP), http://www.icp.grenet.fr/ F–Grenoble.

ITU–R BT-500-8 1998 *Methodology for the Subjective Assessment of the Quality of Television.* International Telecommunication Union, CH–Geneva.

ITU–R Rec. M.1457-5 2006 *Detailed specifications of the radio interfaces of International Mobile Telecommunications-2000 (IMT-2000).* International Telecommunication Union, CH–Geneva.

ITU–T Contribution COM 12–04, 2004 *Modeling the Impact of Linear Frequency Response Distortions using a P.862 PESQ based Approach* Source: TNO, The Netherlands (J.G. Beerends). International Telecommunication Union, CH–Geneva.

ITU–T Contribution COM 12–11 1993 *Difference in Loudness and Speech Quality for 3.1 kHz and 7 kHz Handset Telephony* Source: Swedish Telecom, Sweden (N. Gleiss). International Telecommunication Union, CH–Geneva.

ITU–T Contribution COM 12–20 2005 *Speech Degradation Decomposition using a P.862 PESQ based Approach* Source: TNO, The Netherlands (J.G. Beerends, J. van Vugt). International Telecommunication Union, CH–Geneva.

ITU–T Contribution COM 12–21 1997 *An Experimental Investigation of the Accumulation of Perceived Erros is Time-Varying Speech Distortions* Source: British Telecom, UK (M. Hollier). International Telecommunication Union, CH–Geneva.

ITU–T Contribution COM 12–22 2000 *Assessment of Time-Varying Speech Quality: Comparison between Listening Situation and Conversational Situation* Source: France Télécom R&D, France (L. Gros). International Telecommunication Union, CH–Geneva.

ITU–T Contribution COM 12–23 2000 *Non-Intrusive Assessment of Perceived Voice Quality Impairments* Source: France Télécom R&D, France (E. Bernex). International Telecommunication Union, CH–Geneva.

ITU–T Contribution COM 12–28 2001 *The Effect of Packet Losses on Speech Quality* Source: Telia AB, Sweden (C. Karlsson). International Telecommunication Union, CH–Geneva.

ITU–T Contribution COM 12–37 1997 *The E–Model: An Analysis of the Sources and Comparison with Published and New Test Results* Source: Germany (S. Möller). International Telecommunication Union, CH–Geneva.

ITU–T Contribution COM 12–42 1997 *Listening Only Test Results for Hands-Free Telephones and their Dependence upon Room Surroundings* Source: Germany (E. Diedrich). International Telecommunication Union, CH–Geneva.

ITU–T Contribution COM 12–47 2002 *Rapporteur Report - Workshop on Headsets/Sending Part* Source: Rapporteur Q.5/12 (L. Madec). International Telecommunication Union, CH–Geneva.

ITU–T Contribution COM 12–54 2003 *Conclusions of the Round Robin Test on Headsets : Receiving Side* Source: Rapporteur Q.5/12 (L. Madec). International Telecommunication Union, CH–Geneva.

ITU–T Contribution COM 12–62 1990 *Subjective Evaluation of Pure Delay on Echo–Free Telephone Connections* Source: Bellcore, USA. International Telecommunication Union, CH–Geneva.

ITU–T Contribution COM 12–63 1990 *Two Subjective Tests of Transmission Quality of 3- and 7-kHz Bandlimited Speech and Music Signals* Source: Bellcore, USA. International Telecommunication Union, CH–Geneva.

ITU–T Contribution COM 12–69 1998 *E–Model Predictions and the Impairment Factor Principle for Low–Bitrate Codecs and Quantizing Distortion: Analysis of Test Results* Source: Germany (S. Möller). International Telecommunication Union, CH–Geneva.

ITU–T Contribution COM 12–94 1999 *Continuous Assessment of Time-Varying Subjective Vocal Quality and its Relationship with Overall Subjective Quality* Source: France Télécom R&D, France (N. Chateau). International Telecommunication Union, CH–Geneva.

ITU–T Contribution COM 12–102 1999 *Proposal For The Definition of Different Types of Hands-Free Telephones Based on Double Talk Performance* Source: Germany (H.W. Gierlich). International Telecommunication Union, CH–Geneva.

ITU–T Delayed Contribution D.009 1997 *User Perception of Total Quality for Wired vs. Wireless Telephones* Source: Nortel Networks, Canada (L. Thorpe). International Telecommunication Union, CH–Geneva.

ITU–T Delayed Contribution D.012 1997 *Subjective Evaluation of Echo Canceller Performance: Sidetone and Ambient Noise; Impairment Factors* Source: Rapp. Q14/12 (M. Perkins). International Telecommunication Union, CH–Geneva.

ITU–T Delayed Contribution, D.020 2001 *The Burst Ratio: A Measure of Bursty Packet Loss* Source: Lucent Technologies, USA (J. McGowan). International Telecommunication Union, CH–Geneva.

ITU–T Delayed Contribution D.027 2005 *General Prediction of the Impairment due to Dependent (Non-Random) Packet Loss for Inclusion in the E-Model* Source: Germany (A. Raake). International Telecommunication Union, CH–Geneva.

ITU–T Delayed Contribution D.029 2005 *Preliminary Equipment Impairment Factors for Wideband Speech Codecs* Source: Germany (S. Möller, A. Raake). International Telecommunication Union, CH–Geneva. Jan. 2005.

ITU–T Delayed Contribution D.044 2001 *Modeling Impairment due to Packet Loss for application in the E-model* Source: Deutsche Telekom, Germany (A. Raake). International Telecommunication Union, SEN–Dakar.

ITU–T Delayed Contribution D.061 2005 *Shortcomings of P.862 and its Proposed Wideband Extension, Evaluation of Speech Mixed by Environmental Noise* Source: Nokia Corporation, Finland (J. Selmela, N. Zacharov). International Telecommunication Union, CH–Geneva.

ITU–T Delayed Contribution D.064 1998 *Testing the Quality of Connections Having Time Varying Impairments* Source: AT&T, USA (J. H. Rosenbluth). International Telecommunication Union, CH–Geneva.

ITU–T Delayed Contribution D.064 2005 *Examples of Ie and Bpl Values for Wideband Codecs* Source: NTT, Japan (A. Takahashi). International Telecommunication Union, CH–Geneva.

ITU–T Delayed Contribution D.067 2005 *Performance Evaluation of Wideband Extension of P.862* Source: NTT, Japan (A. Takahashi). International Telecommunication Union, CH–Geneva.

ITU–T Delayed Contribution D.068 2002 *Effect of Consecutive Packet Losses on Voice Quality* Source: Nortel Networks, Canada (K. Bharrathsingh). International Telecommunication Union, CH–Geneva.

ITU–T Delayed Contribution D.071 1995 *Impairment Factors for Speech Clipping* Source: Ellemtel, Sweden (N.O. Johannesson). International Telecommunication Union, CH–Geneva.

ITU–T Delayed Contribution D.071 2005 *Perceptual Correlates of the E-model's Impairment Factors* Source: Germany (M. Wältermann, S. Möller). International Telecommunication Union, CH–Geneva.

ITU–T Delayed Contribution D.105 2003 *Description of VQmon Algorithm* Source: Telchemy, USA (A. Clark). International Telecommunication Union, CH–Geneva.

ITU–T Delayed Contribution D.110 1999 *Subjective Results on Impairment Effects of IP Packet Loss* Source: Nortel Networks, Canada (A. Patrick). International Telecommunication Union, CH–Geneva.

ITU–T Delayed Contribution D.113 1999 *E-Model Equipment Impairment Factors for G.711 with Frame Erasure Concealment* Source: AT&T, USA (M. Perkins). International Telecommunication Union, CH–Geneva.

ITU–T Delayed Contribution D.139 2000 *Study of the Relationship between Instantaneous and Overall Subjective Speech Quality for Time-Varying Quality Speech Sequences: Influence of a Recency Effect* Source: France Télécom R&D, France (N. Chateau). International Telecommunication Union, CH–Geneva.

ITU–T Delayed Contribution D.179 2003 *E-Model: Average Quality vs. Quality as a Function of Average Packet Loss* Source: Germany (A. Raake). International Telecommunication Union, CH–Geneva.

ITU–T Delayed Contribution D.197 2004 *Definition of Quality of Experience* Source: Nortel Networks, Canada (P. Coverdale). International Telecommunication Union, CH–Geneva.

ITU–T Delayed Contribution D.214 2004 *Echo-Free Delay, VoIP Speech Quality and the E-model* Source: AT&T, USA (C. Dvorak). International Telecommunication Union, CH–Geneva.

ITU–T Delayed Contribution D.221 2004 *E-Model: Additivity of Burst Packet Loss Impairment with other Impairment Types* Source: Germany (A. Raake). International Telecommunication Union, CH–Geneva.

ITU–T Delayed Contribution D.222 2004 *Speech Transmission Quality under 2-state Markov Loss and Implications for the E-model* Source: Germany (A. Raake). International Telecommunication Union, CH–Geneva.

ITU–T Rec. E.800 1994 *Terms and Definitions Related to Quality of Service and Network Performance Including Dependability.* International Telecommunication Union, CH–Geneva.

ITU–T Rec. G.107 2000 *The E–Model, a Computational Model for Use in Transmission Planning.* International Telecommunication Union, CH–Geneva. Withdrawn.

ITU–T Rec. G.107 2005 *The E–Model, a Computational Model for Use in Transmission Planning.* International Telecommunication Union, CH–Geneva.

II ITRGA 2006 *Provisional Impairment Factor Framework for Wideband Speech Transmission.* International Telecommunication, Union CH–Geneva.

ITU–T Rec. G.113 1996 *Transmission Impairments*. International Telecommunication Union, (withdrawn) CH–Geneva.

ITU–T Rec. G.113 Appendix I 2001 *Provisional Planning Values for the Equipment Impairment Factor Ie*. International Telecommunication Union, CH–Geneva.

ITU–T Rec. G.113 Appendix I 2002 *Provisional Planning Values for the Equipment Impairment Factor Ie and Packet-Loss Robustness Factor Bpl*. International Telecommunication Union, CH–Geneva.

ITU–T Rec. G.114 2000 *One–Way Transmission Time*. International Telecommunication Union, CH–Geneva.

ITU–T Rec. G.122 1993 *Influence of National Systems on Stability and Talker Echo in International Connections*. International Telecommunication Union, CH–Geneva.

ITU–T Rec. G.126 1993 *Listener Echo in Telephone Networks*. International Telecommunication Union, CH–Geneva.

ITU–T Rec. G.165 1993 *Echo Cancellers*. International Telecommunication Union, CH–Geneva.

ITU–T Rec. G.167 1993 *Acoustic Echo Controllers*. International Telecommunication Union, CH–Geneva.

ITU–T Rec. G.168 2002 *Digital Network Echo Cancellers*. International Telecommunication Union, CH–Geneva.

ITU–T Rec. G.711 1988 *Pulse Code Modulation (PCM) of Voice Frequencies*. International Telecommunication Union, CH–Geneva.

ITU–T Rec. G.712 1992 *Transmission Performance Characteristics of Pulse Code Modulation*. International Telecommunication Union, CH–Geneva.

ITU–T Rec. G.722 1988 *7 kHz Audio-Coding Within 64 kbit/s*. International Telecommunication Union, CH–Geneva.

ITU–T Rec. G.722.2 2002 *Wideband Coding of Speech at Around 16 kbit/s Using Adaptive Multi-Rate Wideband (AMR-WB)*. International Telecommunication Union, CH–Geneva.

ITU–T Rec. G.723.1 1996 *Dual Rate Speech Coder for Multimedia Communications Transmitting at 5.3 and 6.3 kbit/s*. International Telecommunication Union, CH–Geneva.

ITU–T Rec. G.726 1990 *40, 32, 24, 16 kbit/s Adaptive Differential Pulse Code Modulation (ADPCM)*. International Telecommunication Union, CH–Geneva.

ITU–T Rec. G.728 1992 *Coding of Speech at 16 kbit/s Using Low-delay Code Excited Linear Prediction (LD-CELP)*. International Telecommunication Union, CH–Geneva.

ITU–T Rec. G.729 1996 *Coding of Speech at 8 kbit/s Using Conjugate-structure Algebraic-Code-Excited Linear-prediction (CS-ACELP)*. International Telecommunication Union, CH–Geneva.

ITU–T Rec. G.729.1 2006 *An 8-32 kbit/s Scalable Wideband Coder Bitstream Interoperable with G.729*. International Telecommunication Union, CH–Geneva.

ITU–T Rec. H.323 1998 *Packet–Based Multimedia Communication Systems*. International Telecommunication Union, CH–Geneva.

ITU–T Rec. O.41 1994 *Psophometer for Use on Telephone–Type Circuits*. International Telecommunication Union, CH–Geneva.

ITU–T Rec. P.340 2000 *Transmission Characteristics of Hands–Free Telephones*. International Telecommunication Union, CH–Geneva.

ITU–T Rec. P.342 2000 *Transmission Characteristics for Telephone Band (300–3400 Hz) Digital Loudspeaking and Hands-free Telephony Terminals*. International Telecommunication Union, CH–Geneva.

ITU–T Rec. P.48 1989 *Specifications for an Intermediate Reference System (Blue Book edition)*. International Telecommunication Union, CH–Geneva.

ITU–T Rec. P.51 1989 *Artificial Mouth and Artificial Ear*. International Telecommunication Union, CH–Geneva.

ITU–T Rec. P.56 1993 *Objective Measurement of Active Speech Level*. International Telecommunication Union, CH–Geneva.

ITU–T Rec. P.562 2000 *Analysis and Interpretation of INMD Voice-Services Measurements*. International Telecommunication Union, CH–Geneva.

ITU–T Rec. P.563 2004 *Single-Ended Method for Objective Speech Quality Assessment in Narrow-Band Telephony Applications*. International Telecommunication Union, CH–Geneva.

ITU–T Rec. P.59 1993 *Artificial Conversational Speech*. International Telecommunication Union, CH–Geneva.

ITU–T Rec. P.64 1999 *Determination of Sensitivity/Frequency Characteristics of Local Telephone Systems*. International Telecommunication Union, CH–Geneva.

ITU–T Rec. P.79 1999 *Calculation of Loudness Ratings for Telephone Sets*. International Telecommunication Union, CH–Geneva.

ITU–T Rec. P.79, Annex G 2001 *Wideband Loudness Rating Algorithm*. International Telecommunication Union, CH–Geneva.

ITU–T Rec. P.800 1996 *Methods for Subjective Determination of Transmission Quality*. International Telecommunication Union, CH–Geneva.

ITU–T Rec. P.800.1 2003 *Mean Opinion Score (MOS) Terminology*. International Telecommunication Union, CH–Geneva.

ITU–T Rec. P.810 1996 *Modulated Noise Reference Unit (MNRU)*. International Telecommunication Union, CH–Geneva.

ITU–T Rec. P.830 1996 *Subjective Performance Assessment of Telephone-Band and Wideband Digital Codecs*. International Telecommunication Union, CH–Geneva.

ITU–T Rec. P.831 1998 *Subjective Performance Evaluation of Network Echo Cancellers*. International Telecommunication Union, CH–Geneva.

ITU–T Rec. P.832 2000 *Subjective Performance Evaluation of Hands-Free Terminals*. International Telecommunication Union, CH–Geneva.

ITU–T Rec. P.833 2001 *Methodology for Derivation of Equipment Impairment Factors from Subjective Listening-only Tests*. International Telecommunication Union, CH–Geneva.

ITU–T Rec. P.834 2002 *Methodology for the Derivation of Equipment Impairment Factors from Instrumental Models*. International Telecommunication Union, CH–Geneva.

ITU–T Rec. P.835 2003 *Recommendation on the Subjective Evaluation of Noise Suppression Algorithms*. International Telecommunication Union, CH–Geneva.

ITU–T Rec. P.861 1996 *Objective Quality Measurement of Telephone–Band (300–3400 Hz) Speech Codecs*. International Telecommunication Union, CH–Geneva.

ITU–T Rec. P.862 2001 *Perceptual Evaluation of Speech Quality (PESQ), an Objective Method for End-to-end Speech Quality Assessment of Narrowband Telephone Networks and Speech Codecs*. International Telecommunication Union, CH–Geneva.

ITU–T Rec. P.862.1 2003 *Mapping Function for Transforming P.862 Raw Result Scores to MOS-LQO*. International Telecommunication Union, CH–Geneva.

ITU–T Rec. P.862.2 2005 *Wideband Extension to Recommendation P.862 for the Assessment of Wideband Telephone Networks and Speech Codecs*. International Telecommunication Union, CH–Geneva.

ITU–T Rec. P.862.3 2005 *Application Guide for Objective Quality Measurement Based on Recommendations P.862, P.862.1 and P.862.2*. International Telecommunication Union, CH–Geneva.

ITU–T Rec. P.880 2004 *Continuous Evaluation of Time-Varying Speech Quality*. International Telecommunication Union, CH–Geneva.

ITU–T Rec. X.200 1994 *Information Technology – Open Systems Interconnection – Basic Reference Model: The Basic Model*. International Telecommunication Union, CH–Geneva.

ITU–T Rec. Y.1541 2006 *Network Performance Objectives for IP-Based Services*. International Telecommunication Union, CH–Geneva.

ITU–T Suppl. 3 to P–Series Rec. 1993 *Models for Predicting Transmission Quality from Objective Measurements*. International Telecommunication Union, CH–Geneva.

IV ITRGA 2006 *Provisional Planning Values for the Wideband Equipment Impairment Factor Ie,wb*. International Telecommunication Union, June, CH–Geneva.

Janssen J, Vleeschauwer DD, Büchli M and Petit GH 2002 Assessing voice quality in packet-based telephony. *IEEE Internet Comput. Mag.* **6**(3), 48–57.

Jekosch U 1992 The Cluster-Identification test. In: *Proc. 2th International Conference on Spoken Language Processing (ICSLP'92)*, Vol. 1, CDN-Banff, pp. 205–209.

Jekosch U 2000 *Sprache hören und beurteilen: Ein Ansatz zur Grundlegung der Sprachqualitätsbeurteilung*, Habilitation thesis, Universität/Gesamthochschule, D–Essen.

Jekosch U 2004 Basic concepts and terms of 'quality', reconsidered in the context of product sound quality. *Acta Acustica* **90**(6), 999–1006.

Jekosch U 2005a Assigning meaning to sounds: Semiotics in the context of product-sound design. In: *Communication Acoustics* (J. Blauert, ed.). Springer, D–Heidelberg.

Jekosch U 2005b *Voice and Speech Quality Perception – Assessment and Evaluation Signals and Communication Technology*. Springer, D–Berlin.

Jiang W and Schulzrinne H 2000 Modeling of packet loss and delay and their effect on real-time multimedia service quality. *Proc. International Workshop on Network and Operating System Support for Digital Audio and Video (NOSSDAV)*, USA–Chapel Hill.

Jiang W and Schulzrinne H 2002 Comparison and optimization of packet loss repair methods on VoIP perceived quality under bursty packet loss. *Proc. 12th International Workshop on Network and Operating Systems Support for Digital Audio and Video (NOSSDAV'02)*, USA–Miami Beach.

Johannesson NO 1996 *Echo Canceller Performance Characterized by Impairment Factors*. Document IP–16 to ITU–T Speech Quality Experts Group (unpublished), International Telecommunication Union, (presented at the UK–Ipswich meeting, Sept. 23–27), CH-Geneva.

Johannesson NO 1997 The ETSI computation model: A tool for transmission planning of telephone networks. *IEEE Commun. Mag.* **35**(1) 70–79.

Jones BL and McManus PR 1986 Graphic scaling of qualitative terms. *J. Soc. Motion Pict. Telev. Eng.* **95**, 1166–1171.

Josephson D 1999 A brief tutorial on proximity effect. In: *Proc. 107th Audio Engineering Society (AES) Convention*, USA–New York.

Karis D 1991 Evaluating transmission quality in mobile telecommunication systems using conversation tests. *Proc. of the Human Factors Society 35th Annual Meeting*, USA–San Francisco, CA, pp. 217–221.

Katz D 1969 *Gestaltpsychologie*. 4th edition, revised by W. Metzger, M. Stadler and H. Grabus (1st edition 1944). Schwabe, CH–Basel.

Kitawaki N and Itoh K 1990 Delay effect assessment taking into account of human factors in telecommunications. *Proc. 13th International Symposium on Human Factors in Telecommunications*, I-Turin, pp. 555–562.

Kitawaki N and Itoh K 1991 Pure delay effects on speech quality in telecommunications. *IEEE J. Sel. Areas Commun.* **9**(4), 586–93.

Klatt DH 1976 Linguistic uses of segmental duration in English: Acoustic and perceptual evidence. *J. Acoust. Soc. Am.* **59**(5), 1208–21.

Klimo M 1999 Voice over IP: Packet loss and jitter in E-model. In: *QoS Summit '99*, November, F–Paris pp. 17–19.

Köster S 2003 *Modellierung von Sprechweisen für widrige Kommunikationsbedingungen mit Anwendung auf die Sprachsynthese*. Doctoral dissertation, Institut für Kommunikationsakustik, Ruhr-University Bochum, Shaker Verlag, D–Aachen.

Krauss RM and Bricker PD 1966 Effects of transmission delay and access delay on the efficiency of verbal communication. *J. Acoust. Soc. Am.* **41**(2), 286–292.

Krebber W 1995 *Sprachübertragungsqualität von Fernsprech–Handapparaten*, Vol. 357, Series 10, VDI–Verlag GmbH, D–Düsseldorf.

Krebber J 2002 *Simulation der Kombination von Synchronen und Paketbasierten Telefonnetzen für Smart-Home-Anwendungen*. Diploma Thesis (unpublished), Institut für Kommunikationsakustik, Ruhr–Universität, D–Bochum.

Kruskal JB and Wish M 1978 *Multidimensional scaling Vol. 07–011 Quantitative Applications in the Social Sciences* (E. M. Uslaner, ed.). Sage, USA–Newbury Park, CA.

Kwitt R, Fichtel T and Pfeiffenberger T 2006 Measuring perceptual VoIP speech quality over UMTS. In: *Proc. 4th International Workshop on Internet Performance, Simulation, Monitoring and Measurement (IPS-MoMe 2006)*, AT–Salzburg.

Lane HL, Catania AC and Stevens SS 1961 Voice level: Autophonic scale, perceived loudness, and effects of sidetone. *J. Acoust. Soc. Am.* **33**(2), 160–167.

Lane HL, Tranel B and Sisson C 1970 Regulation of voice communication by sensory dynamics. *J. Acoust. Soc. Am.* **47**(2), 618–624.

Lawson GD and Chial MR 1982 Magnitude estimation of degraded speech quality by normal- and impaired-hearing listeners. *J. Acoust. Soc. Am.* **72**(6), 1781–1787.

Lee BS 1950 Effects of delayed speech feedback. *J. Acoust. Soc. Am.* **22**(6), 824–826.

Letowski T 1989 Sound quality assessment: Concepts and criteria. *Preprint 87th Audio Engineering Society (AES) Convention* (Paper D-8, Preprint 2825), USA–New York, October.

Lewis NW and Allnatt JW 1965 Subjective quality of television pictures with multiple impairments. *Electron. Lett.* **1**(7), 187–188.

Liang Y, Färber N and Girod B 2003 Adaptive playout scheduling and loss concealment for voice communication over IP networks. *IEEE Trans. Multimedia* **5**(4), 532–543.

Lu T, Liang L and Wang X 2001 Temporal and rate representations of time-varying signals in the auditory cortex of awake primates. *Nat. Neurosci.* **4**(11), 1131–1138.

Ludwig T 2003 *Messung von Signaleigenschaften zur referenzfreien Qualitätsbewertung von Telefonbandsprache*. Doctoral dissertation, Christian–Albrechts–Universität Kiel (Arbeiten über Digitale Signalverarbeitung No. 23, U. Heute, ed.). Shaker Verlag, D–Aachen.

Markopoulou AP, Tobagi FA and Karam MJ 2002 Assessment of VoIP quality over Internet backbones. In: *Proc. 21st Annual Joint Conference of the IEEE Computer and Communications Societies (INFOCOM '02)*, USA–New York.

Martin R 1995 *Freisprecheinrichtungen mit mehrkanaliger Echokompensation und Störgeräuschreduktion Vol. 3 of Aachener Beiträge zu Digitalen Nachrichtensystemen* (P. Vary, ed.). Verlag Augustinus Buchhandlung, D–Aachen.

Massaro DW 1975 Backward recognition masking. *J. Acoust. Soc. Am.* **58**(5), 1059–1065.

Matta J, Pépin C, Lashkari K and Jain R 2003 A source and channel rate adaptation algorithm for AMR in VoIP using the E-model. *Proc. 13th International Workshop on Network and Operating Systems Support for Digital Audio and Video (NOSSDAV'03)*, USA–Monterey.

Mattila VV 2001 *Perceptual Analysis of Speech Quality in Mobile Communications*, Vol. 340. Doctoral Dissertation, Tampere University of Technology, FIN–Tampere.

Mattila VV 2002a Ideal point modeling of speech quality in mobile communications based on multidimensional scaling. In: *Proc. 112th Audio Engineering Society (AES) Convention*, D–Munich.

Mattila VV 2002b Descriptive analysis and ideal point modeling of speech quality in mobile communications. In: *Proc. 113th Audio Engineering Society (AES) Convention*, USA–Los Angeles.

Mattila VV 2003 Ideal point modeling of the quality of noisy speech in mobile communications based on multidimensional scaling. In: *Proc. 114th Audio Engineering Society (AES) Convention*, NL–Amsterdam.

Mattila VV and Zacharov N 2001 Generalized listener selection (GLS) procedure. In: *Proc. 110th Audio Engineering Society (AES) Convention*, NL–Amsterdam.

McAdams S 1993 Recognition of sound sources and events. In: *Thinking in Sound – The Cognitive Psychology of Human Audition* (S. McAdams, E. Bigand, eds.). Oxford University Press, UK-Oxford, pp. 146–198.

McDermott BJ 1969 Multidimensional analyses of circuit quality judgments. *J. Acoust. Soc. Am.* **45**(3), 774–781.

McGee VE 1964 Semantic components of the quality of processed speech. *J. Speech Hear. Res.* **7**, 310–323.

McGee VE 1965 Determining perceptual spaces for the quality of filtered speech. *J. Speech Hear. Res.* **8**, 23–38.

Mehler J and Segui J 1987 English and French speech processing: Some psycholinguistic investigations. In: *The Psychophysics of Speech Perception NATO ASI Series D: Behavioral and Social Sciences* (M. E. H. Schouten, ed.), Vol. 39, Springer, pp. 405–418.

Mersdorf J 1996 Ein Hörversuch zur perzeptiven Unterscheidbarkeit von Sprechern bei ausschließlich intonatorischer Information. In: *Fortschr. Akust. - DAGA'96*. Deutsche Gesellschaft für Akustik, D–Oldenburg. pp. 482–483.

Miller GA 1956 The magical number seven, plus or minus two: Some limits on our capacity for processing information. *Psychol. Rev.* **63**, 81–97.

Miller GA 1962 Decision units in the perception of speech. *IEEE Trans. Inf. Theory* **8**, 81–83.

Miller GA and Licklider JCR 1950 The intelligibility of interrupted speech. *J. Acoust. Soc. Am.* **22**(2), 167–173.

Modena G, Coleman A, Usai P and Coverdale P 1986 Subjective performance evaluation of the 7 kHz audio coder. In: *Proc. IEEE GLOBECOM 1986*, USA–Houston, pp. 599–604.

Möller S 2000 *Assessment and Prediction of Speech Quality in Telecommunications*. Kluwer Academic Publishers, USA-Boston, MA.

Möller S 2004 Telephone transmission impact on synthesized speech: Quality assessment and prediction. *Acta Acustica* **90**(6), 121–136.

Möller S 2005a *Quality of Telephone-Based Spoken Dialogue Systems*. Springer, USA–New York.

Möller S 2005b Quality of transmitted speech for humans and machines. In: *Communication Acoustics* (J. Blauert, ed.). Springer, D–Heidelberg.

Möller S and Jekosch U 1998 *Verifikation des ETSI–Modells zur Bestimmung der Sprachübertragungsqualität im Zusammenhang unterschiedlicher Systemkonfigurationen*. Final report to a project funded by Deutsche Telekom Berkom GmbH (unpublished), Institut für Kommunikationsakustik, Ruhr–Universität, D–Bochum.

Möller S and Berger J 2002 Describing telephone speech codec quality degradations by means of impairment factors. *J. Audio Eng. Soc. (AES)* **50**(9), 667–680.

Möller S and Raake A 2002 Telephone speech quality prediction: Towards network planning and monitoring models for modern network scenarios. *Speech Commun.* **38**(1–2), 47–75.

Möller S, Krebber J and Raake A 2004 Performance of speech recognition and synthesis in packet-based networks. In: *Proc. International Conference Spoken Language Processing (ICSLP 2004)*, Vol. 2, KR–Jeju Island, pp. 1541–1544.

Möller S, Raake A, Kitawaki N, Takahashi A and Wältermann M (accepted for publication) Impairment factor framework for wideband speech codecs. *IEEE Trans. Audio Speech and Language*.

Moore BCJ and Tan CT 2003 Perceived naturalness of spectrally distorted speech and music. *J. Acoust. Soc. Am.* **114**(1), 408–19.

Munson WA and Karlin JE 1962 Isopreference method for evaluating speech–transmission circuits. *J. Acoust. Soc. Am.* **34**(6), 762–774.

Murray IR and Arnott JL 1993 Toward the simulation of emotion in synthetic speech: A review of the literature on human vocal emotion. *J. Acoust. Soc. Am.* **93**(2), 1097–1108.

Nakatani LH and Dukes KD 1973 A sensitive test of speech communication quality. *J. Acoust. Soc. Am.* **53**(4), 1083–1092.

Narbutt M and Davis M 2005 Assessing the quality of VoIP transmission affected by playout buffer scheme. In: *Proc. MESAQIN 2005* (J. Holub and R. Smid, eds.). Vol 1, Czech Technical University, CZ–Prague.

NIST Net 2002 *NIST Net Network Emulator*. National Institute of Standards and Technology, `http://www.antd.nist.gov/nistnet/` USA-Gaithersburg, MD.

Nöth W 2000 *Handbuch der Semiotik*. Metzler, D–Stuttgart.

Ogden CK and Richards IA 1960 *The Meaning of Meaning*, 10th edition. Routledge & Kegan Paul Limited, UK–London.

Osgood CE, Suci G and Tannenbaum P 1957 *The Measurement of Meaning*. University of Illinois Press, USA-Urbana, IL.

O'Shaughnessy D 2000 *Speech Communication – Human and Machine*, 2nd edition. IEEE Press, USA-Piscataway, NJ.

Parizet E, Hamzaoui N and Ségaud L 2003 Continuous evaluation of noise uncomfort in a bus. *Acta Acustica* **89**, 900–907.

Pascal D 1988 Comparative performances of two subjective methods for improving the fidelity of speech signals. *Preprint 84th Audio Engineering Society (AES) Convention* (Paper L-3, Preprint 2639), F–Paris, March.

Pascal D and Boyer M 1990 Multidimensional perceptive measurement of quality: Comparative performance of two methods. *Proc. 13th International Symposium on Human Factors in Telecommunications*. I-Turin, pp. 519–520.

Payton K, Uchanski R and Braida L 1994 Intelligibility of conversational and clear speech in noise and reverberation for listeners with normal and impaired hearing. *J. Acoust. Soc. Am.* **95**(3), 1581–92.

Peirce CS 1986 *Semiotische Schriften*, Vol. 1, Suhrkamp, D–Frankfurt/Main.

Pennock S 2002 Accuracy of the perceptual evaluation of speech quality (PESQ) algorithm. In: *Proc. MESAQIN 2002* (J. Holub and R. Smid, eds.) Vol. 1, Czech Technical University, CZ–Prague.

Perkins C, Hodson O and Hardman V 1998 A survey of packet loss recovery techniques for streaming audio. *IEEE Netw. Mag.* **12**(5), 40–48.

Pörschmann C 2001 *Eigenwahrnehmung der Stimme in Auditiven Virtuellen Umgebungen*. Vol. 666. Series 10, Dissertation, Institut für Kommunikationsakustik (IKA), Ruhr-University Bochum, VDI–Verlag GmbH, D–Düsseldorf.

Postman L and Philips L 1965 Short-term changes in free recall. *Q. J. Exp. Psychol.* **17**, 132–138.

Powers GL and Wilcox JC 1977 Intelligibility of temporally interrupted speech with and without intervening noise. *J. Acoust. Soc. Am.* **61**(1), 195–199.

Pratt R and Doak P 1976 A subjective rating scale for timbre. *J. Sound Vib.* **45**, 317–328.

Quackenbush SR, Barnwell TP and Clemens MA 1988 *Objective Measures of Speech Quality*. Prentice Hall, USA–Englewood Cliffs, NJ.

Raake A 2000 Perceptual dimensions of speech sound quality in modern transmission systems. In: *Proc. International Conference on Spoken Language Processing (ICSLP 2000)*, Vol. 4, CHN-Beijing, pp. 744–747.

Raake A 2002 Does the content of speech influence its perceived sound quality?. In: *Proc. 3rd International Conference on Language Resources and Evaluation (LREC 2002)*, Vol. 4, ES–Las-Palmas, 1170–1176.

Raake A 2004 Predicting speech quality under random packet loss: Individual impairment and additivity with other network impairments. *Acta Acustica* **90**(6), 1061–1083.

Raake A and Möller S 1999 *Analysis and Verification of the Tektronix M366 GSM Network QoS Analyser's Measurements and Quality Predictions*. Final report of a project funded by Tektronix, Padua, SpA. (unpublished), Institut für Kommunikationsakustik, Ruhr–Universität, D–Bochum.

Rafaeli S 1988 Interactivity: From new media to communication. In: *Sage Annual Review Communication Research: Advancing Communication Science*, Vol. 16, Sage, USA–Beverly Hills, CA. pp. 110–134.

Ramjee R, Kurose J, Towsley D and Schulzrinne H 1994 Adaptive playout mechanisms for packetized audio applications in wide-area networks. In: *Proc. 13th Annual Joint Conference of the IEEE Computer and Communications Societies (INFOCOM '94)*, CA–Toronto, pp. 680–688.

Rehmann S 2001 *Sprachqualität von Voice-over-Internet-Protocol-(VoIP)-Verbindungen bei zeitlich instationären Störungen*. Diploma Thesis (unpublished), Institut für Kommunikationsakustik, Ruhr–Universität, D–Bochum.

Rehmann S, Raake A and Möller S 2002 Parametric simulation of impairments caused by telephone and voice over IP network transmission. In: *Proc. EAA 2002 – Forum Acusticum*, ES–Sevilla.

Richards DL 1973 *Telecommunication by Speech*. Butterworths, UK–London.

Richards DL 1974 Calculation of opinion scores for telephone connections. *Proc. IEE* **121**(5), 313–323.

Rix A and Hollier M 2000 The perceptual analysis measurement system for robust end-to-end speech quality assessment. In: *Proc. IEEE ICASSP'00*, Vol. 3, pp. 1515–1518.

Rix A, Hollier M, Hekstra A and Beerends J 2002 Perceptual evaluation of speech quality (PESQ), the new itu standard for end-to-end speech quality assessment, part I – time-delay compensation. *J. Audio Eng. Soc. (AES)* **50**(10), 755–764.

Rizzo L 1997 Effective erasure codes for reliable computer communication protocols. *ACM Comput. Commun. Rev.* **27**(2), 24–36.

Roberts B and Moore BC 1990 The influence of extraneous sounds on the perceptual estimation of first-formant frequency in vowels. *J. Acoust. Soc. Am.* **88**(6), 2571–2582.

Rosen S 1992 Temporal information in speech: Acoustic, auditory and linguistic aspects. *Philosophical Transactions of the Royal Society of London Series B*, Vol. 336, The Royal Society, pp. 367–373.

Rosenberg JD 2001 *G.729 Error Recovery for Internet Telephony*, Technical Report CUCS-016-01, Department of Computer Science, Columbia University USA–New York.

Rosenberg JD, Qiu L and Schulzrinne H 2000 Integrating FEC into adaptive voice playout buffer algorithms on the Internet. In: *Proc. 19th Annual Joint Conference of the IEEE Computer and Communications Societies (INFOCOM '00)* Vol.3, March 26–30, IL–Tel Aviv, 2000, 1705–14.

Rothauser EH, Urbanek GE and Pachl WP 1968 Isopreference method for speech evaluation. *J. Acoust. Soc. Am.* **44**(2), 408–418.

Samuel AG 1981 Phonemic resoration: Insights from a new methodology. *J. Exp. Psychol. Gen.* **110**(4), 474–494.

Sanneck H and Carle G 2000 A framework model for packet loss metrics based on loss runlengths. In: *Proc. of the SPIE/ACM SIGMM Multimedia Computing and Networking Conference 2000 (MMCN 2000)*, San Jose, CA, January 2000.

Schäfer E 1938 Über die Hörbarkeit von Frequenzbandänderungen bei der Übertragung von Sprache. *Elektrische Nachrichtentechnik* **15**, 237.

Schüssler HW 1987 An objective method for measuring the performance of weakly non-linear and noise systems. *Frequenz* **41**(6), 147–154.

Sebeok TA 1996 Signs, bridges, origins. In: *Origins of Language* (J. Trabant, ed.). Collegium Budapest Workshop Series, Vol. 2, H–Budapest, 89–115.

Shepard RN, Romney AK and Nerlove SB eds 1972 *Multidimensional Scaling: Theory and Applications in the Behavioral Sciences*, Vol. 1, Semniar Press, USA–New York.

Sotscheck J 1982 Ein Reimtest für Verständlichkeitsmessungen mit Deutscher Sprache als ein verbessertes Verfahren zur Bestimmung der Sprachübertragungsgüte. *Der Fernmeldeingenieur* **36**(4/5), 1–84.

South C and Usai P 1992 Subjective performance of CCITT's 16 kbit/s LD–CELP algorithm with voice signals. In: *Communication for Global Users, IEEE Global Telecommunications Conference GLOBECOM '92 (USA–Orlando)*. USA–Piscataway, NJ, pp. 1709–1714.

Stevens SS 1951 *Handbook of Experimental Psychology*. Wiley, USA–New York.

Stevens SS 1957 On the psychophysical law. *Psychol. Rev.* **64**, 153–181.

Stickney GS and Assmann PF 2001 Acoustic and linguistic factors in the perception of bandpass-filtered speech. *J. Acoust. Soc. Am.* **109**(3), 157–165.

Sun L and Ifeachor EC 2002a Perceived speech quality prediction for voice over IP-based networks. In: *Proc. IEEE International Conference on Communication (IEEE ICC'02)*, Vol. 4, USA–New York, pp. 2573–2577.

Sun L and Ifeachor EC 2002b Subjective and objective speech quality evaluation under bursty losses. In: *Proc. MESAQIN 2002* (J. Holub and R. Smid, eds.). Czech Technical University, CZ–Prague.

Sun L and Ifeachor EC 2003 Prediction of perceived conversational speech quality and effect of play-out buffer algorithms. In: *Proc. IEEE International Conference on Communication (IEEE ICC'03)*, Vol. 1(1), USA–Anchorage, pp. 1–6.

Sun L and Ifeachor EC 2004 New models for perceived voice quality prediction and their applications in playout buffer optimization for VoIP networks. In: *Proc. IEEE International Conference on Communication (IEEE ICC'04)*, June 2004, Vol. 1(1), F–Paris, pp. 1478–1483.

Sun L, Wade G, Lines BM and Ifeachor EC 2001 Impact of packet loss location on perceived speech quality. In: *Proc. Internet Telephony Workshop (IPtel 2001)*, USA–New York.

Susini P, McAdams S and Smith BK 2002 Global and continuous loudness estimation of time-varying levels. *Acta Acustica* **88**, 536–548.

T1A1.1/ 2000-014 2000 *Trade-Off Value of VoIP Speech Quality vs. a Messaging Feature* Source: GTE Labs (G.W. Cermak). Committee T1 – Telecommunications, Committee Group T1A1.1 USA-Washington, WA.

Tanenbaum AS 2003 *Computer Networks*. Prentice Hall, USA–Upper Saddle River, NJ.

Thomsen G and Jani Y 2000 Internet telephony: Going like crazy. *IEEE Spectr* **37**(5), 52–58.

TIA/EIA-95-B 1999 *Mobile Station-Base Station Compatibility Standard for Wideband Spread Spectrum Cellular Systems*. Telecommunications Industry Association (TIA), USA-Washington, WA.

Umeda N 1975 Vowel duration in American English. *J. Acoust. Soc. Am.* **58**(2), 434–45.

Umeda N 1977 Consonant duration in American English. *J. Acoust. Soc. Am.* **61**(3), 846–858.

US Patent 6,931,017 2005 *Burst Ratio: A Measure of Bursty Loss on Packet-based Networks* Lucent Technologies, James William McGowan, USA-Murray Hill, .

Vary P and Martin R 2006 *Digital Speech Transmission*. John Wiley and Sons, UK–Chichester.

Vary P, Heute U and Hess W 1998 *Digitale Sprachsignalverarbeitung*. B. G. Teubner, D–Stuttgart.

Västfjäll D 2003 *Affect as a Component of Perceived Sound and Vibration Quality in Aircraft*. Dissertation Thesis, Chalmers University of Technology, S–Göteborg.

Voiers WD 1977 Diagnostic Acceptability Measure for speech communication systems. In: *Proc. IEEE ICASSP '77*, Vol. 2, USA–Washington, 204–207.

Voiers WD, Sharpley A and Hehmsoth C 1975 *Research on Diagnostic Evaluation of Speech Intelligibility*. Research Report AFCRL-72-0694, Air Force Cambridge Research Laboratories, USA-Bedford, MA.

Voran SD 1997 Listener ratings of speech passbands. In: *Proc. IEEE Speech Coding Workshop*, Vol. 1, USA–New York, 81–82.

Voran SD 2003 Perception of temporal discontinuity impairments in coded speech – a proposal for objective estimators and some subjective test results. In: *Proc. Online Workshop Measurement of Speech and Audio Quality in Networks, MESAQIN 2003*. Czech Technical University, CZ–Prague.

Wältermann M 2005 *Bestimmung relevanter Qualitätsdimensionen bei der Sprachübertragung in modernen Telekommunikationsnetzen*. Diploma thesis (unpublished), Institut für Kommunikationsakustik, Ruhr–Universität, D–Bochum.

Wang X, Lu T and Liang L 2003 Cortical processing of temporal modulations. *Speech Commun.* **41**, 107–121.

Warren RM 1970 Perceptual restoration of missing speech sounds. *Science* **167**, 392–3.

Warren RM 1999 *Auditory Perception – A New Analysis and Synthesis*. Cambridge University Press, UK-Cambridge, MA.

Watson A 2001 *Assessing the Quality of Audio and Video Components in Desktop Multimedia Conferencing*. PhD thesis, Department of Computer Science, University College London, UK–London.

Watson A and Sasse A 2000 The good, the bad, and the muffled: The impact of different degradations on Internet speech. In: *Proc. ACM Multimedia 2000*, USA–Los Angeles.

Wiegelmann S 1997 *Einsatzmöglichkeiten und -grenzen der MNRU als Referenzverzerrung bei der Bestimmung von Sprachübertragungsqualität im Telefoniebereich*. Diploma thesis (unpublished), Institut für Kommunikationsakustik, Ruhr–Universität, D–Bochum.

Wijngaarden SJ, Smeele PM and Steeneken HJ 2001 A new method for testing communication efficiency and user acceptability of speech communication channels. In: *Proc. 7th European Conference on Speech Communication and Technology (Eurospeech 2001 Scandinavia)*, Vol. 3, September 3–7, DK–Aalborg, pp. 1675–1678.

Wijngaarden SJ, Tardelli JD, Hassanein H, Ray B and Collura JS 2002 Communicability testing for voice communications. In: *Proc. IEEE Speech Coding Workshop*, J-Ibaraki, pp. 102–104.

Yajnik M, Moon S, Kurose J and Towsley D 1999 Measurement and modeling of the temporal dependence in packet loss. In: *Proc. 18th Annual Joint Conference of the IEEE Computer and Communications Societies (INFOCOM '99)*, March 21–25, Vol. 1, USA–New York, 345–52.

Yamamoto L and Beerends J 1997 Impact of network performance parameters on the end–to–end perceived speech quality. In: *ACTS Project EXPERT (AC094; Platform for Engineering Research and Trials): ATM Traffic Symposium*. GR–Mykonos.

Young ED and Sachs MB 1979 Representation of steady-state vowels in the temporal aspects of the discharge patterns of populations of auditory-nerve fibers. *J. Acoust. Soc. Am.* **66**(5), 1381–1403.

Zwicker E 1961 Subdivision of the audible frequency range into critical bands (Frequenzgruppen). *J. Acoust. Soc. Am.* **33**(2), 248.

Zwicker E and Fastl H 1999 *Psychoacoustics: Facts and Models*. Springer, D–Berlin.

Index

acceptability, 19–22, 35–36, 102
accessability, 20
accommodation, 192
acoustic leakage, coupling, 89–91, 177,
    193–194, 199, 236–237
advantage of access, 46
advantage factor (A), 46, 259
anchor (point), anchoring, 33, 155,
    189–190, 194, 268, 271
articulation, 3–4, 197, 203
  comprehensibility, 9–11, 25
artificial
  ear, 29
  mouth, 29
assessment, 16–18, 23–48, 99, 113, 281
assimilation, 192
asymmetry, 77, 98
asynchronous transfer mode (ATM), 55
attribute scaling, *see* scaling, attribute
    scaling
auditory
  event, 1–3, 86
  perception, 1, 8, 11, 58, 154
  test, 1–3, 11, 16–18, 23–39, 43, 49,
    75–78, 85, 111–127, 130–140,
    147–151, 154–162, 163–168,
    172–173, 176–188, 192–204, 225,
    234, 241–242, 251–254, 265–275

bandpass, 85–91, 93, 175–188, 193–204,
    226–227
bandwidth, 5–7, 57–61, 84–91, 99, 100,
    103–105, 175–197, 200–203,
    208–210, 234–235, 274–275
  critical, 5–6, 87–88, *see also* Bark,
    critical-band rate
  equivalent rectangular bandwidth (ERB),
    187–188, 209–210

Bark, 5–6, 184–188, 208–210
binaural loudness, 194
bit–errors, *see* error, bit–error
Burst Ratio (*BurstR*), 45, 67, 106, 128–129,
    132–133, 139, 205–206
burst(y) (packet) loss, burstiness, 45,
    64–67, 70, 77, 104–108, 128–146,
    162–171, 205–206, 223, 271–273

call
  clarity index (CCI), 43, 47–48
  set–up, 52–53, 100, 227, 230, 233
CDMA 2000, 54
channel coding, 59, 78, 228
clarity, 93
code domain multiple access (CDMA), 54,
    78
coder, codec
  wideband, 59, 61, 189–191, 208–210
coder, codec, coding, 40–42, 46–47, 57,
    58–61, 71–75, 78, 84, 89, 92–96, 100,
    103, 105–106, 116–120, 122–123,
    124–129, 152–158, 170–172,
    189–191, 206–207, 208–210,
    226–231, 233–234, 252, 266, 271
codec impairment, *see* impairment,
    impairment factor, equipment
    impairment factor
color of sound, 60, 93
coloration, 86–89, 93, 184
comfort noise, *see* noise, comfort noise
communication efficiency, 22
composition
  desired composition, *see* nature, desired
    nature
  perceived composition, *see* nature,
    perceived nature
comprehension, 9–11, 13–14,

context, 1, 4, 8, 9–11, 12–13, 15, 20,
   25–27, 30, 39, 51, 87, 91, 99–102,
   177, 184, 197–198
conversation
   effectiveness, 21–22
   task, 30–32
critical bands, 5–6, 83, 87–88, 188
critical-band rate, 5–6, 188
cumulative distribution function (CDF),
   222

delay
   absolute (Ta), transmission delay, delay,
      10, 13, 22, 30, 32, 44, 46, 53,
      55–56, 58, 61–63, 154, 219,
      220–223, 227, 231–234, 242,
      268–273,
   mean one-way (T), 44, 79, 106, 149, 154,
      159, 161, 227
   roundtrip (Tr), 44, 79–80, 226–227, 229,
      268, 273,
   variation, see jitter
device (INMD), see INMD
D–factor (Ds, Dr), 45, 268, 273
diagnostic acceptability measure (DAM),
   35–36
dichotic, 241
digital circuit multiplexing equipment
   (DCME), 48, 114
digital subscriber line (DSL), 78–79
diotic, 193–196, 235–236
distortion
   linear distortion, 85–91, 106–108,
      175–189, 194
   non-linear distortion, 58–61, 81, 91–98,
      106, 189–190, 208–210
   time-varying distortion, 103, 111
distribution, 228, 242–243
   equilibrium, 218
   Pareto, 221
   Weibull, 221, 223
dummy head, see head and torso simulator
duration
   call/conversation duration, 30–32, 52, 61,
      138, 150, 271–272
   memory, 8, 12
   packet loss-related, 72–74, 77, 86, 113,
      115, 124, 132, 137–138, 162, 242,
      271–272
   perceived duration, 15, 72, 86
   phoneme/syllable duration, 4, 74, 79

stimulus duration, 28, 123–124, 145, 182
talkspurt duration, 62

ear reference point (ERP), xviii
ease of communication, 21–22
echo, 30, 46, 61, 83, 86, 98, 235, 256–257
   attenuation, 79, 213, 242,
   canceller (EC), cancellation, xxii, 29–30,
      52, 80–82, 106,
   delay, see delay, mean one-way delay
   echo loss, 44
   listener echo, 44, 48, 80, 106, 225–228,
      256–257
   talker echo, xxv, 22, 43, 44, 56–57,
      79–82, 91, 101, 106, 107, 154–155,
      158–162, 166–170, 225–228,
      238–241, 256–257, 268–271, 274
   talker echo loudness rating (TELR),
      44, 79–80, 106, 154, 157–162,
      166–168, 256–259, 266, 269–274
   weighted echo path loss (WEPL), 44, 48,
      259, 274
effectiveness
   conversation effectiveness, 21–22
efficiency, 149
   communication efficiency, see
      communication efficiency
   service efficiency, 22
E–model, xxiii, xxv, 33, 34, 43–46, 47–48,
   56, 58, 60–61, 79, 82–83, 97, 98,
   additivity property, 45–46, 98, 151–162,
      206–208,
   algorithm, 43–46, 104, 105–108,
      113–123, 128–129, 131–135,
      138–146, 149, 151–162, 163–172,
      175–188, 189–190, 205–210, 224,
      226, 252–254, 255–259, 277–280
emotion, 1, 4, 14, 20, 84, 100
equipment, 58
   terminal equipment, see user interface
equivalent-Q method, 92, 178–179,
   251–253
error, 150, 213, 232, 258
   bit error, 54, 57, 63, 78–79, 94–96,
      223–224, 228, 266
   correction, 59, 61, 70, 223 (see also
      packet, loss, packet loss
      concealment)
   forward error correction (FEC), 57, 63,
      70, 72–73, 219–223